Springer Series in
OPTICAL SCIENCES 94

founded by H.K.V. Lotsch

Springer-Verlag Berlin Heidelberg GmbH

Springer Series in
OPTICAL SCIENCES

The Springer Series in Optical Sciences, under the leadership of Editor-in-Chief *William T. Rhodes*, Georgia Institute of Technology, USA, provides an expanding selection of research monographs in all major areas of optics: lasers and quantum optics, ultrafast phenomena, optical spectroscopy techniques, optoelectronics, quantum information, information optics, applied laser technology, industrial applications, and other topics of contemporary interest.
With this broad coverage of topics, the series is of use to all research scientists and engineers who need up-to-date reference books.

The editors encourage prospective authors to correspond with them in advance of submitting a manuscript. Submission of manuscripts should be made to the Editor-in-Chief or one of the Editors.

K. Inoue K. Ohtaka (Eds.)

Photonic Crystals

Physics, Fabrication and Applications

With 209 Figures

Kuon Inoue, Guest Professor
Chitose Institute of Science and Technology
Bibi, Chitose City, Hokkaido, 066-8655, Japan
E-mail: inoue@phys.spub.chitose.ac.jp

Kazuo Ohtaka, Professor
Chiba University, Center for Frontier Science
1-33 Yayoi, Inage, Chiba City, 263-8522, Japan
E-mail: ohtaka@cfs.chiba-u.ac.jp

ISSN 0342-4111

DOI 10.1007/978-3-540-40032-5

Library of Congress Cataloging-in-Publication Data: Photonic crystals : physics, fabrication and applications / K. Inoue, K. Ohtaka, eds. Includes bibliographical references and index. (acid-free paper) 1. Photons. 2. Crystal optics. I. Inoue, K. (Kuon), 1937– II. Ohtaka, K. (Kazuo), 1941– III. Springer series in optical sciences ; v. 94. QC793.5.P427P78 2004 539.7'217–dc22 2003064916

springeronline.com

Originally published by Springer-Verlag Berlin Heidelberg New York in 2004
MyCopy version of the original edition 2004

Cover concept by eStudio Calamar Steinen using a background picture from The Optics Project. Courtesy of John T. Foley, Professor, Department of Physics and Astronomy, Mississippi State University, USA
Cover production: *design & production* GmbH, Heidelberg

Printed on acid-free paper
www.springer.com/mycopy

Preface

Photonic crystals (PCs) having two (2D)- or three-dimensional (3D), spatial, periodic variations of the dielectric constant on the order of an optical wavelength are very attractive for controlling radiation field and light propagation characteristics. These are dielectrics in a broad sense, and are also attractive in the millimeter or submillimeter wavelength region. This is because electromagnetic (photon) modes in those PCs have special features when compared to a homogeneous material. Making use of those features, a lot of new physical phenomena have already been found on one hand, and novel PC-based devices have also been developed on the other hand. Therefore, PCs have made a great impact on science, in particular, optical science, in the past fifteen years since 1987, when two important papers appeared independently, where E. Yablonovitch and S. John pointed out inhibitation of spontaneous emission and photon localization, respectively.

Being stimulated by the above papers, research works of photonic crystals in Japan also started renewedly around 1991: actually, K. Ohtaka, one of the authors of this book, already did the pioneering works on photonic crystals from 1979 to 1983, prior to the papers mentioned above. According as the importance of photonic crystal was recognized, the number of the researchers in Japan has been growing up until now. In view of the circumstances, the first research project on photonic crystals in Japan started in 1998. The project is called A Grant-in-Aid for Scientific Research in a Priority Area, "Development of Photonic Crystals and Control of the Radiation Field", which was financially supported by the Japanese Ministry of Education, Science, Culture, and Sports. This project was constituted of several research groups that were most active in this field at that time, although it did not necessarily cover all those active ones. The research groups from Hokkaido University, Chiba University, Kyoto University, Yokohama National University, Institute for Physics and Chemistry, Shinshu University, Tokyo Metropolitan University, and The Femtosecond Technology Research Association jointed this project. The project already finished substantially in 2002.

This book is intended to primarily report on the research outcome of this Project, so is not intended for describing all aspects of the research works in this field. However, in order to stress a feature of Japanese researches, we also include some outstanding works performed by other Japanese groups. So, the authors of this book are constituted of several group leaders primarily responsible for this project. By the way, a few books written in English were already published in the field of photonic crystal. Those are prepared rather in a textbook style, and are thereby very useful for a newcomer to understand the basic physics, or for a scientist to study the theoretical background. Considering that those are written by theoretical scientists, we emphasize in this book our experimental works for specialists in this field, but we also include the distinguished theoretical works done in Japan.

We recommend readers to utilize the book in the following way. The book comprises fourteen chapters. Chapters 1 and 2 are prepared so as for readers such as non-specialists or undergraduates to be able to take a survey of photonic crystals. Chapter 3 is also intended for those people to grasp the minimal theoretical background. For this purpose the three chapters are organized in a textbook style: readers who want to understand more advanced theories of photonic crystals are advised to refer to Chap. 4, which covers a variety of theoretical problems and is one of the key chapters in this book. Therefore, the first three chapters may be tedious for specialists, so those people may skip the three. In all other chapters we start, in principle, with the overview of the content described in each chapter. We believe that this should be convenient to readers, because only the abstract is sufficient to know to some of them.

Finally, this book has been published under a support of Grant-in-Aid for Publication of Scientific Research Results from the Japan Society for Promotion of Science (JSPS). We highly appreciate this support.

Hokkaido, November 2003 *Kuon Inoue*

Contents

List of Acronyms

Acronym	Original term
BG	band gap
BZ	Brillouin zone
DFB	distributed feedback
DOS	density of states
DWDM	dense wavelength-division multiplex
FDTD	finite-difference time-domain
LED	light-emitting diode
PB	photonic band
PBG	photonic bang gap
PC	photonic crystal
SEM	scanning electron microscope
SOI	silicon on insulator
SPR	Smith-Purcell radiation
TE	transverse electric
TM	transverse magnetic
VCSEL	vertical-cavity surface-emitting laser

List of Contributors

Kiyoshi Asakawa
The Femtosecond Technology Research Association,
Tokohdai, Tsukuba, Ibaraki,
Postal Code 300-2635, Japan
asakawa@festa.or.jp

Toshihiko Baba
Department of Electrical and Computer Engineering,
Yokohama National University
Hodogaya-ku, Yokohama-city,
Postal Code 240-8501, Japan
baba@dnj.ynu.ac.jp

Kuon Inoue
Chitose Institute of Science and Technology
Bibi, Chitose, Hokkaido,
Postal Code 066-8655, Japan
inoue@phys.spub.chitose.ac.jp*
(Fellow of Toyota Physical and Chemical Research Institute)

Syojiro Kawakami
The New Industry Creation Hatchery Center,
Tohoku University
Aoba-ku, Sendai-city,
Postal Code 980-8579, Japan
kawakami@niche.tohoku.ac.jp

Takayuki Kawashima
The New Industry Creation Hatchery Center,
Tohoku University
Aoba-ku, Sendai-city,
Postal Code 980-8579, Japan
kawashima@niche.tohoku.ac.jp

Susumu Noda
Graduate School of Engineering,
Kyoto University
Saikyou-ku, Kyoto-city,
Postal Code 615-8510, Japan
snoda@kuee.kyoto-u.ac.jp

Kazuo Ohtaka
Center for Frontier Science,
Chiba University
Inage-ku, Chiba-city,
Postal Code 263-8522, Japan
ohtaka@cfs.chiba-u.ac.jp*

Yuzaburo Segawa
Photodynamics Research Center,
Institute of Physical and Chemical Research
Aoba-ku, Sendai-city,
Postal Code 519-1399, Japan
segawa@postman.riken.go.jp

Mitsuo W. Takeda
Faculty of Science,
Shinshu University
Asahi, Matsumoto-city,
Postal Code 390-8621, Japan
wada@azusa.shinshu-u.ac.jp

1 Introduction

K. Inoue

In this chapter, we briefly describe the history of the research on photonic crystals (PCs), and introduce some repesentative PC structures, samples of which have already been fabricated. We also present, on the basis of the concept of eigen-state for light, the concept of the photonic band (PB), which is important to understand for later chapters. In short, the reason that PCs are very attractive in controlling the radiation field and light propagation characteristics relies on the unique and prominent PB that a PC exhibits.

1.1 History of Research on Photonic Crystals

Let us begin with the question about when the research of PCs or this new field started. The answer depends on the definition. It is well known that a quarter-wavelength plate or multiple-layer dielectric mirrors such as those used for a high-Q-value laser cavity have already been widely used in optics for more than 70 years. Those are nothing other than examples of one-dimensional (1D) PCs. However, from the viewpoint of controlling light those are rather exceptional, i.e., 1D PCs are useful only for restricted purposes. In contrast, the fact that one can control light much more freely by using three (3D)- and two (2D)-dimensional PCs was for the first time pointed out and demonstrated theoretically in two independent papers by E. Yablonovitch [1] and S. John [2], and, nowadays, this fact is well known. In this sense it is generally recognized that this field substantially started in 1987 when these papers appeared.

Actually, however, prior to the above two papers, K. Ohtaka, one of the authors of this book, reported in 1979 the calculated result of the band structure (PBS) for a 3D PC [3], as I already mentioned in the Preface. This work was motivated by the fact that a few 3D PC samples were already fabricated around 1968 other than the 1D PCs. One example was a structure with an array of monodispersed polystyrene particles suspended in water [4], making use of the particles that were successfully fabricated and provided in the early 1960s by the Dow Chemical Co. Ltd. The appearance of such a 3D PC sample looks like a jewel in a sense due to Bragg diffraction under white light.

Following this first paper, Ohtaka and his coworkers published a series of related papers from 1980 to 1983 [3, 5–7]. In some of the papers they found by calculation that the density of photon states or modes (DOS) in a PC could be quite large as compared to the homogeneous case [6], and discussed the importance in some physical phenomena such as surface-enhanced Raman scattering [7]. On the other hand, they did not describe anything about the case of the vanishing DOS, i.e., the existence of a full photonic band gap (PBG). Anyway, he or they did not stress that with use of a PC one can control light to a considerable extent freely. Presumably for this reason, those papers unfortunately did not necessarily have an important influence upon physicists at that time. For example, there were not a few experimentalists who were interested in the above 3D PC of arrayed polystylene particles, but even those people including myself were not aware of the importance either. Namely, they paid no attention to the unique band structure, but utilized those samples only for different research purposes.

In contrast, E. Yablonovitch showed by simulation the existence of a PBG where all modes are missing in all directions, and consequently, spontaneous emission corresponding to the gap energy is inhibited inside a 3D PC. On the other hand, S. John pointed out that new phenomena can be observed by using a PC, which includes localization of light. For these reasons, the papers made a great impact on many scientists from physicists to engineers working in the field of applied physics, or optical electronics.

Since 1987 many scientists started their own research work in this field. However, the work done in the first four or five years was, aside from several experimental, mainly theoretical studies, the great concern of which was what kinds of structures were most suited, as well as how large a difference of the relevant refractive indices (n) was needed, for obtaining a full PBG in 3D PCs, or also in 2D ones; in the 2D case, by a full band gap (BG) we mean one in all directions in the 2D plane.

Experimentally, a variety of PBSs were examined by using samples fabricated with the lattice constant a in the millimeter and submillimeter region. Concerning experimental verification of a full PBG, E. Yablonovitch and his coworkers verified in 1991 the existence for a 3D sample also fabricated with a in the micrometer region [8]. According to the scaling law or relation with respect to a versus the involved wavelength λ, it follows from this work that a 3D PC with a full gap in the optical region is also obtained for the same structure and the same dielectric constants with those in the micrometer region, but with the a-value reduced to the order of optical wavelengths. Namely, as will be explained later in more detail, the energy of photon eigenmode can be scaled in units of a/λ [9], indicating that, if the above mentioned requisites are satisfied, the same band structure should also be obtained experimentally in the optical region.

In the near-infrared or optical region, new kinds of 2D and 3D PC samples with a being in such a region, independently of the suspended polystylene-sphere samples, started being successfully fabricated in one way or another in the first half of the 1990s, including those with a 2D PBG [10]. In contrast, a 3D sample with a full band gap, which was one of the main targets, was still technically difficult to fabricate until around 1999, when Noda and his coworkers succeeded, for the first time, in fabricating such a sample by using a special technique [11]; in 1995 a pioneering attempt to develop such a 3D sample, called Yablanovite, was made, but unfortunately the sample quality was not good enough to experimentally observe the gap [12].

The impact of the first works by Yablonovitch and John was so strong that there had been a tendency until the early 1990s, in our opinion, for people to think that samples with a full gap were necessary in order to control light in most cases. However, people gradually became aware that there were many interesting phenomena to be observed with use of a sample without a 2D or 3D PBG. So, the number of research groups or scientists grew at a high rate in the latter half of the 1990s.

In the last seven years a variety of PC samples were fabricated, including 3 D samples with a full gap in the near-infrared and optical regions. Those samples, in particular, of good quality have enabled one to conduct a variety of novel experiments in relation to the unique and distinguished characteristics of the photonic band structures (PBSs). Concomitantly, making good use of those characteristics, a number of new physical phenomena and novel devices, mainly in optoelectronics, have been anticipated or proposed from the theoretical point of view. As a result, a variety of those phenomena have already been observed on one hand, and many attractive devices or apparatus have been developed on the other hand, with some of them being already commercially available.

Concerning the above-mentioned PC samples, those are briefly summarized as types of PCs in the next section. As for the common and fundamental features of PBS, and applications or PC-based devices, see the next chapter.

1.2 Types of Photonic Crystals

A PC is defined as a crystal such that the dielectric constant (ε) periodically varies spatially in specific directions. In the case where the variation is along one direction, we call it a 1D PC. Similarly, we define 2D and 3D PCs, corresponding to the respective cases where ε varies along two and three independent directions. We survey their typical or representative examples below without referring to the respective literatures except for a few cases: those are presented in the related chapters in more detail.

In Fig. 1.1 are shown schematics of the respective examples. For a 1D PC there is the well-known example of the multi-layer dielectric film. In the case of a 2D PC as shown in Fig. 1.1b, intersections of air- or dielectric rod-axes with a perpendicular plane form a 2D lattice, so this plane is the 2D PC plane. Let us consider the simple case of a 1D PC. As will be shown later, the states of light inside this PC sample, equivalently, the eigenmodes of light with wavevector $\boldsymbol{k}$ parallel to this particular direction are quite different from

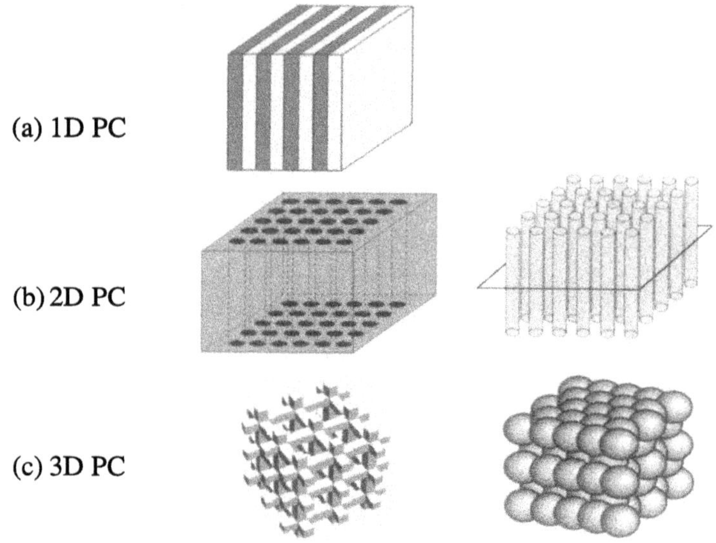

(a) 1D PC

(b) 2D PC

(c) 3D PC

Fig. 1.1. Schematics of representative 1D, 2D and 3D photonic crystals

those in a uniform dielectric. Similarly, the respective eigenmodes of light with $\boldsymbol{k}$ in the 2D plane in the 2D PC case and with $\boldsymbol{k}$ in an arbitrary direction in the 3D case become unique as compared again to homogeneous 2D and 3D dielectrics.

The situations for PC slabs shown in Fig. 1.2, where examples of 1D and 2D slabs are presented, greatly differ also from those for the corresponding 1D and 2D PCs, respectively. Namely, if the thickness of the slab is on the same order with, or smaller than the relevant wavelength of light, the situation, i.e., that for 1D (2D) PC slab is completely different from 1D (2D) PCs where the thickness of a sample is a priori assumed to be infinitely large. For the latter case, the plane wave propagating along one direction (1D PC) or in the 2D plane can be well defined, while for the former (slab) such a plane wave can no longer be defined, so that we need to treat a 2D PC problem, for example, three-dimensionally.

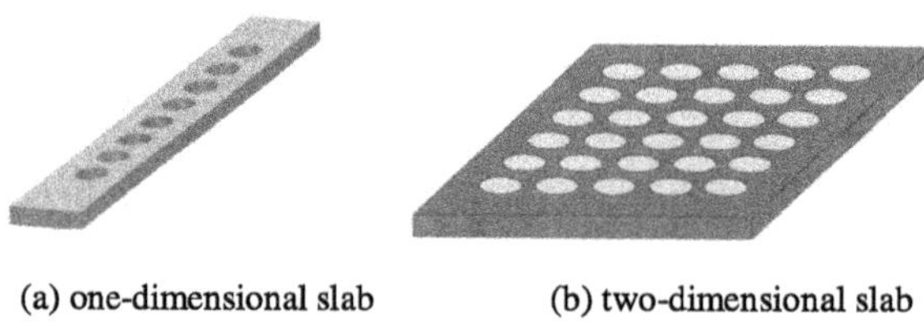

Fig. 1.2. Schematics of photonic crystal slabs

Fig. 1.3. A schematic drawing of an example of a photonic crystal fiber

Figure 1.3 shows a schematic of a PC fiber, another important example of PCs where the plane perpendicular to the fiber or the rods forms a 2D PC lattice except for the central portion [13]: light can propagate in the central portion along the fiber, as will be explained later in Chaps. 5 and 11.

For a 3D PC only two examples are schematically presented in Fig. 1.1c, but there are a variety of other 3D examples, some of which will be explained in detail in Chap. 7. By the way, the example on the right-hand side corresponds to the suspended polysterene spheres already described in the preceding section. There are other important examples of this array type of particles. Namely, these types of examples currently studied most intensively are a 3D array of opal particles [14, 15] and the so-called inversed opals [16].

Thus far we classify PCs according to their dimension. There are different kinds of PCs belonging to the other categories, which include a PC made of metallic material [17], a PC with quasi-crystalline structure [18], a PC with double periodicities [19, 20], and others, all of which will be explained in a bit more detail in Chap. 8.

As for how to fabricate those samples, a variety of methods have been invented or introduced up until now. Readers are asked to refer to the relevant chapters from 5 to 8.

1.3 Light States in a Photonic Crystal

1.3.1 Description of Light State in Vacuum

First let us consider a simple case of a uniform and transparent medium. In this case, the eigenstate (mode) of light is denoted by a set of indices ($\hbar\omega$, $\boldsymbol{k}$, $\boldsymbol{e}$) where $\hbar\omega$ and $\boldsymbol{k}$ refer to the photon energy and wavevector, respectively, and $\boldsymbol{e}$ denotes the polarization state. For each eigenstate a characteristic spatial pattern of electric ($\boldsymbol{E}$) and magnetic ($\boldsymbol{H}$) field is specified, which is called the mode. For each polarization, there is a relation between $\hbar\omega$ and $\boldsymbol{k}$; the dispersive property in a medium is reflected in this dispersion relation. For example, in a sample of infinitely large space of vacuum, the relation is expressed as $\omega = ck\,(k = |\boldsymbol{k}|)$ with c referring to the velocity of light in vacuum, where the eigenstate of a transverse plane wave of $\boldsymbol{k}$ with an arbitrary magnitude and direction is allowed to exist.

Next, as an example of the characteristic modes of electromagnetic waves in a closed space (vacuum), let us consider the case in a cube of side ℓ in each of the x, y and z directions. By adopting the periodic conditions on the surfaces of the cube of side ℓ, we have $k_i = (2\pi/\ell)n_i$ with $i = x, y, z$ where $n_i = 0, \pm 1, \pm 2, \pm 3, \cdots$. Therefore, the modes with a set of discrete values of $(2\pi/\ell)(n_x, n_y, n_z)$ for (k_x, k_y, k_z), are allowed with k expressed as

$$k^2 = (2\pi/\ell)^2(n_x^2 + n_y^2 + n_z^2). \tag{1.1}$$

Aside from the freedom of 2 for polarization, the number of points (n_x, n_y, n_z) in the sphere of radius R, corresponds to the number of modes with frequencies in a range from 0 to $\omega = (2\pi c/\ell)R$; notice that ℓ is by far larger than the wavelength of light $\lambda\,(= 2\pi/k)$, so $(n_x^2 + n_y^2 + n_z^2)$ is a very large quantity. Thus, from the ω derivative of twice the number, $2(4\pi/3)(\ell/2\pi c)^3\omega^3$, we have the density of electromagnetic states, or the density of states (DOS) of photon $\rho(\omega)$, i.e., the number of modes per unit volume with frequencies between ω and ω+dω as

$$\rho(\omega)\mathrm{d}\omega = [\omega^2/(\pi^2c^3)]\mathrm{d}\omega. \tag{1.2}$$

The mode density in a free space is also given by the above equation. It is important to note that the mode density completely different from that in the free space can be obtained in a PC; notice that the magnitude of any physical quantity is determined by the mode density.

1.3.2 Light State and Its Density for a Photonic Crystal

The eigenstates of light in a PC where $\varepsilon(\boldsymbol{r})$ varies periodically with respect to $\boldsymbol{r}$ with $\boldsymbol{r}$ being the position vector differ very much from that in a homogeneous medium. Suppose for simplicity the non-magnetic case, i.e., the dielectric (insulator) where the magnetic permeability $\mu(\boldsymbol{r}) = \mu_0$ with μ_0 being that in vacuum. The wave equation, e.g., for $\boldsymbol{H}(\boldsymbol{r})$ that can be derived by combining Maxwell's electromagnetic equations, is expressed as

$$\boldsymbol{\nabla} \times [\varepsilon(\boldsymbol{r})^{-1}\boldsymbol{\nabla} \times \boldsymbol{H}(\boldsymbol{r})] = \mu_0\omega^2\boldsymbol{H}(\boldsymbol{r})\,, \tag{1.3}$$

where the harmonic modes for $\boldsymbol{H}(\boldsymbol{r},t)$ and $\boldsymbol{E}(\boldsymbol{r},t)$ are assumed, i.e.,

$$\boldsymbol{H}(\boldsymbol{r},t) = \boldsymbol{H}(\boldsymbol{r})\exp(\mathrm{i}\omega t)\,, \quad \boldsymbol{E}(\boldsymbol{r},t) = \boldsymbol{E}(\boldsymbol{r})\exp(\mathrm{i}\omega t)\,. \tag{1.4}$$

Namely, the problem is reduced to solving an eigenvalue problem (1.3). Once we have a solution for $\boldsymbol{H}(\boldsymbol{r})$ together with the eigenvalue $\mu_0\omega^2$ by solving this eigenvalue equation, then $\boldsymbol{E}(\boldsymbol{r})$ is obtained from $\boldsymbol{H}(\boldsymbol{r})$ by using the following relation,

$$\boldsymbol{E}(\boldsymbol{r}) = (\mathrm{i}\omega)^{-1}[\varepsilon(\boldsymbol{r})^{-1}(\boldsymbol{\nabla} \times \boldsymbol{H}(\boldsymbol{r}))]\,. \tag{1.5}$$

How to solve (1.3) for a given PC is a key ploblem. This problem is explained in detail in Chaps. 3 and 4, together with several examples; for a simple example of a 1D PC we solve the equation in Chap. 2.

Genarally, the result, i.e., the eigenenergy $\hbar\omega$ thus obtained as a function of $\boldsymbol{k}$, is depicted in the reduced Brillouin zone (BZ), which will be explained in Appendix A. The dispersion curve of $\hbar\omega$ versus $\boldsymbol{k}$ is divided into bands called photonic bands (PBs); the entire structure of the bands is called the photonic band structure (PBS). A few examples of the PBS will be shown also in Chap. 2. The eigenstate has the specific spatial pattern of $\boldsymbol{E}(\boldsymbol{r})$ and $\boldsymbol{H}(\boldsymbol{r})$ reflecting the symmetry of the lattice structure. Most importantly, the DOS of photon for a PC varies very much with the frequency, and also remarkably, in general, with polarization, and thereby differs completely from the case for a homogeneous material including vacuum; in particular, the DOS vanishes over an energy region for a particular PC, which is called the PBG. Physically, the unique PBS including the PBG is the origin for controlling light.

References

1. E. Yablonovitch, Phys. Rev. Lett. **58**, 2059 (1987)
2. S. John, Phys. Rev. Lett. **58**, 2486 (1987)
3. K. Ohtaka, Phys. Rev. B**19**, 5057 (1979)
4. See, for example, P. S. Hiltner and I. M. Krieger, J. Phys. Chem. **73**, 2386 (1969)
5. K. Ohtaka and M. Inoue, Phys. Rev. B**25**, 677 (1982)
6. M. Inoue, K. Ohtaka, and S. Yanagawa, Phys. Rev. B**25**, 689 (1982)
7. M. Inoue and K. Ohtaka, Phys. Rev. B**26**, 3487 (1982); J. Phys. Soc. Jpn. **52**, 1457 (1983)
8. E. Yablonovitch, T. J. Gmitter and K. J. Leung, Phys. Rev. Lett. **67**, 2295 (1991)
9. J. D. Joannopoulos, R. D. Meade and J. N. Winn, *Photonic Crystals*, (Princeton University Press, Princeton, 1995)
10. K. Inoue, M. Wada, K. Sakoda, A. Yamanaka, M. Hayashi and J. W. Haus, Jpn. J. Appl. Phys. **33**, L1463 (1994)
11. S. Noda, K. Tomoda, N. Yamamoto, and A. Chutinan, Science, **289**, 604 (2000)
12. C. C. Cheng, V. Arbet-Engels, A. Scherer, and E. Yablonovitch, Phys. Scr. **T68**, 17 (1996)
13. J. C. Knight, J. Broeng, T. A. Birks, and P. St. J. Russell, Science **282**, 1476 (1998)
14. T. Holland, C. E. Blanford, and A. Stein, Science **281**, 538 (1998)
15. J. E. G. J. Wijnhoven and W. L. Vos, Science **281**, 802 (1998)
16. A. Blanco, E. Chomski, S. Grabtchak, M. Ibisate, S. John, S. W. Leonard, C. Lopez, F. Meseguer, H. Miguez, J. P. Mondia, G. A. Ozin, O. Toader, and H. M. van Driel, Nature (London) **405**, 437 (2000)

17. V. Kuzmiak and A. A. Maradudin, Phys. Rev. B**55**, 7427 (1997); *ibid*, B**58**, 7230 (1998)
18. M. E. Zoorob, M. D. B. Charlton, G. J. Parker, J. J. Baumberg, and M. C. Netti, Nature (London) **404**, 740 (2000)
19. S. Yano, Y. Segawa, J. S. Bae, K. Mizuno, H. Miyazaki, K. Ohtaka, and S. Yamaguchi, Phys. Rev. B**63**, 153316 (2001)
20. R. Shimada, T. Koda, T. Ueta, and K. Ohtaka, J. Phys. Soc. Jpn. **67**, 3414 (1998);
J. Appl. Phys. **90**, 3905 (2001)

2 Survey of Fundamental Features of Photonic Crystals

K. Inoue and K. Ohtaka

In this chapter we take a survey of our current research work as well as fundamental features of PCs for non-spcialists or newcomers including undergraduate students. By grasping those first, it should be, we believe, easier for them to understand the later chapters. For this purpose, in addition to surveying a few representative PBSs, their prominent features, and PC-based devices, we also present and explain basic concepts and fundamental laws or relations needed to understand the underlying physics, by considering mainly a simple 1D PBS. In some other cases, we use the band structures without explaining here how to obtain those.

2.1 One-Dimensional Photonic Crystal: Band Calculation

2.1.1 Bloch Theorem

To see the properties of photonic bands (PBs), a 1D periodic multilayer of dielectric films is illustrative; such films have long been used, for example, as a lossless mirror (Bragg reflecter) or a quater-wave plate. Let two kinds of layers of dielectric substances A and B be alternating in the z direction periodically, as shown in Fig. 2.1. The layers are assumed to have no internal structure in the x and y directions and thus form a 1D PC. Let $\varepsilon_A, \varepsilon_B$ and a_A, a_B be the dielectric constants and thickness of the two layers, respectively. The size of the unit cell is

$$a = a_A + a_B,$$

which defines the lattice constant of this PC. We confine ourselves to the wavevector directed in the z direction. Then, the electric field $\boldsymbol{E}(z)$ and magnetic flux density $\boldsymbol{B}(z)$, which are mutually orthogonal, are both polarized perpendicularly to the z direction. The magnetic flux density $\boldsymbol{B}(z)$ is simply related to the magnetic field $\boldsymbol{H}(z)$ through $\boldsymbol{B}(z) = \mu_0 \boldsymbol{H}(z)$, with μ_0 the vacuum permeability in our nonmagnetic PC. Let us take the direction of the polarization of $\boldsymbol{E}(z)$ as the x direction and that of $\boldsymbol{B}(z)$ in the y direction.

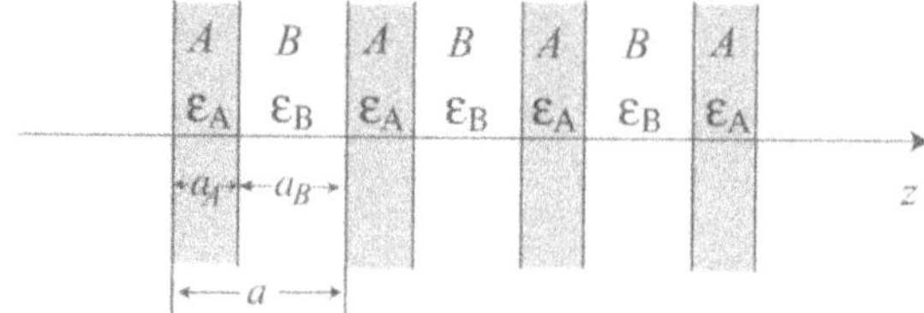

Fig. 2.1. One-dimensional stack of A and B films in the z direction. The length of the unit cell $a = a_A + a_B$ defines the periodicity of this 1D PC

Using the scalar symbol for the amplitudes, the electric field satisfies

$$\frac{\partial^2}{\partial z^2}E(z,t) - \frac{1}{c^2}\frac{\partial^2}{\partial t^2}\varepsilon(z)E(z,t) = 0 \tag{2.1}$$

with the periodic dielectric constant $\varepsilon(z)$ defined by

$$\varepsilon(z) = \begin{cases} \varepsilon_A & \text{if } z \text{ is in A} \\ \varepsilon_B & \text{if } z \text{ is in B.} \end{cases} \tag{2.2}$$

The periodic function $\varepsilon(z)$ $(-\infty < z < \infty)$ is expanded into a Fourier series:

$$\varepsilon(z) = \sum_{p=-\infty}^{\infty} \varepsilon_p \mathrm{e}^{\mathrm{i}\frac{2\pi p}{a}z}. \tag{2.3}$$

Here the basis functions $\{\mathrm{e}^{\mathrm{i}\frac{2\pi p}{a}z}\}_{p=-\infty}^{\infty}$ have periodicity of the lattice and

$$\frac{2\pi p}{a} \quad (p = 0, \pm 1, \pm 2, \cdots.)$$

defines the reciprocal lattice "vector". The Fourier coefficients are given by

$$\varepsilon_p = \frac{1}{a}\int_0^a \mathrm{d}z\varepsilon(z)\mathrm{e}^{-\mathrm{i}\frac{2\pi p}{a}z}, \tag{2.4}$$

the prefactor $1/a$ being the inverse of the unit cell "volume", over which the integral is made. For our unit cell of

$$\varepsilon(z) = \begin{cases} \varepsilon_A & 0 < z < a_A, \\ \varepsilon_B & a_A < z < a, \end{cases} \tag{2.5}$$

(2.4) yields

$$\varepsilon_p = \begin{cases} \varepsilon_A \frac{a_A}{a} + \varepsilon_B \frac{a_B}{a} & \text{for } p = 0 \\ \frac{\mathrm{i}}{2\pi p}(\varepsilon_A - \varepsilon_B)\left(\mathrm{e}^{-\mathrm{i}\frac{2\pi p}{a}a_A} - 1\right) & \text{for } p \neq 0. \end{cases} \tag{2.6}$$

We assume the harmonic oscillation in the temporal behavior

$$E(z,t) = \mathrm{e}^{-\mathrm{i}\omega t}E(z).$$

To find the solutions of wave equation (2.1), a Bloch form is substituted for $E(z)$, which is given by

$$E(z) = \sum_{p=-\infty}^{\infty} c_p \mathrm{e}^{\mathrm{i}(k+p\frac{2\pi}{a})z}. \tag{2.7}$$

This is a plane-wave expansion form of the PB solution. Inserting (2.3) and (2.7) into (2.1) and putting the prefactor of the plane wave $\mathrm{e}^{\mathrm{i}(k+\frac{2\pi p}{a})z}$ to zero, we find

$$\left(k+\frac{2\pi p}{a}\right)^2 c_p - \sum_{p'=-\infty}^{\infty} \left(\frac{\omega}{c}\right)^2 \varepsilon_{p-p'} c_{p'} = 0 \qquad (p = 0, \pm 1, \pm 2, \cdots). \tag{2.8}$$

These are linear coupled equations for the coefficients $\cdots, c_{-1}, c_0, c_1, c_2, \cdots$. The solutions exist only when ω satisfies the secular equation constructed from the coefficients of the coupled equations:

$$\det \begin{vmatrix} \cdots & \cdots & \cdots & \cdots & \cdots \\ \cdots & (k-\frac{2\pi}{a})^2 - (\frac{\omega}{c})^2\varepsilon_0 & -(\frac{\omega}{c})^2\varepsilon_{-1} & -(\frac{\omega}{c})^2\varepsilon_{-2} & \cdots \\ \cdots & -(\frac{\omega}{c})^2\varepsilon_1 & (k)^2 - (\frac{\omega}{c})^2\varepsilon_0 & -(\frac{\omega}{c})^2\varepsilon_{-1} & \cdots \\ \cdots & -(\frac{\omega}{c})^2\varepsilon_2 & -(\frac{\omega}{c})^2\varepsilon_1 & (k+\frac{2\pi}{a})^2 - (\frac{\omega}{c})^2\varepsilon_0 & \cdots \\ \cdots & \cdots & \cdots & \cdots & \cdots \end{vmatrix} = 0. \tag{2.9}$$

We note the regularity of the matrix elements; the diagonal elements are

$$\cdots, \quad \left(k-\frac{4\pi}{a}\right)^2, \quad \left(k-\frac{2\pi}{a}\right)^2, \quad (k)^2, \quad \left(k+\frac{2\pi}{a}\right)^2, \quad \left(k+\frac{4\pi}{a}\right)^2, \quad \cdots$$

and the Fourier components $\{\varepsilon_p\}$ of the dielectric function $\varepsilon(z)$ appear in each row in the order

$$\cdots, \quad \varepsilon_2, \quad \varepsilon_1, \quad \varepsilon_0, \quad \varepsilon_{-1}, \cdots$$

with ε_0 present at the diagonal position of the matrix.

That the solutions of (2.1) could be found in this way is wholly attributed to the Bloch form (2.7) assumed for the solution. It satisfies

$$E(z+a) = \mathrm{e}^{\mathrm{i}ka} E(z), \tag{2.10}$$

which is the Bloch theorem, to be met by any solution of a periodic system. It is the wavenumber k that specifies the eigensolutions. We note the periodicity of the solutions with respect to k; the solution for k of (2.8) or (2.9) and that for $k + p(2\pi/a)$ with an integer p are identical. In fact, these two k lead to the same form in (2.7), as seen by renumbering the coefficient c_p, and to the same eigenvalues for ω in (2.9), as seen by using the regularity of

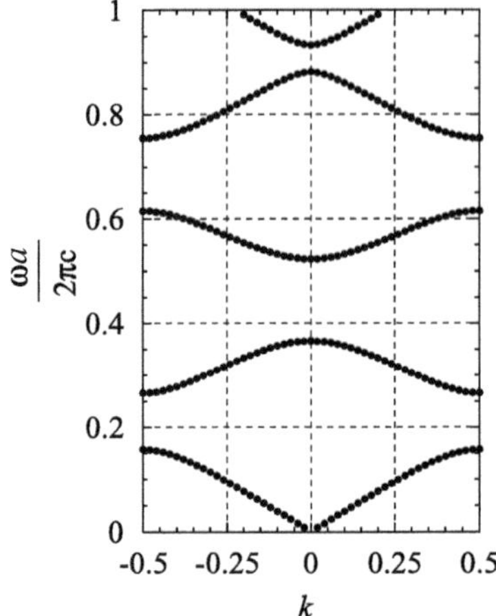

Fig. 2.2. Band structure of the 1D photonic crystal shown in Fig. 2.1 for A = Si and B = air. The case of $a_A = a_B = a/2$ with $\varepsilon_A = 12$ (dielectric constant of Si in the visible range) and $\varepsilon_B = 1$ is shown. Both the frequency ω and wavenumber k are shown in dimensionless units

the matrix elements. Since the identical solutions are obtained for the set of wavenumbers k, $k \pm \frac{2\pi}{a}, \cdots$, all possible normal modes of (2.9) are covered by letting k run only over the first Brillouin zone (BZ), which is for the present 1D system a region of k defined by

$$-\frac{\pi}{a} \leq k < \frac{\pi}{a}. \tag{2.11}$$

For a given value of k in the first BZ, the secular equation (2.9) gives the eigenvalues ω_1, ω_2, $\cdots$.

The number of the eigenvalues is fixed by the size of the matrix truncated in the secular equation. As k varies, each of the eigenvalues changes gradually. As a result, (2.9) gives a set of continuous functions of k, $\omega_1(k)$, $\omega_2(k)$, $\cdots$. They are the dispersion curves of the PBs, of our 1D system. The subscript n of $\omega_n(\boldsymbol{k})$ is usually called a band index. (We sometimes use n as a superscript as in $v_g^{(n)}$, the group velocity of nth band.) For the concrete case of A = Si ($\varepsilon_A = 12$), B = air ($\varepsilon_B = 1$), and $a_A = a_B = a/2$, the band structure is given in Fig. 2.2.

If we take account of a larger range of integer p in truncating the sum of (2.7), the matrix of the secular equation becomes bigger. In calculating Fig. 2.2 we have used 11 plane waves ($-5 \leq p \leq 5$) in (2.7). As the size of the secular determinant increases, so does the number of solutions for ω for a fixed k. With increase of the number of plane waves, the contributions of the shorter waves are incorporated, for the larger the wavenumber $|k + \frac{2\pi p}{a}|$ is, the shorter becomes the wavelength of $\exp \mathrm{i}(k + \frac{2\pi p}{a})z$. As a result, if we

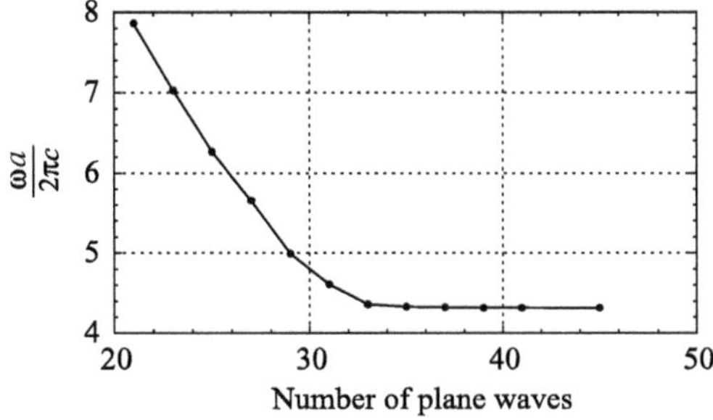

Fig. 2.3. Convergence of the band energy as a function of the number of plane waves used in the calculation. The result is given for the 20th band at the BZ edge. The parameters used in the band calculation are the same as those used in Fig. 2.2

attempt to obtain the PBS in a higher frequency range, we necessarily need to treat a larger secular matrix.

The actual trend that a PB of higher frequency is better represented by a larger number of plane waves, is seen in Fig. 2.3, which shows the calculated band frequency at $k = \pi/a$ to be improved progressively with the increase of the cutoff size of the secular matrix.

It is imortant here to understand the PBS of a system in terms of the dispersion relation of photons in the empty lattice. An empty lattice of a PC is a fictitous system which has only the periodicity of that PC but is identical to free space otherwise. In an empty lattice, therefore, $\varepsilon_{\pm 1}$, $\varepsilon_{\pm 2}$, ... all vanish identically except $\varepsilon_0 = 1$ in (2.9). The PBS of the empty lattice is obtained by folding the free space dispersion curve $\omega = ck$ at the edge of the first BZ. This is because what remains in the secular matrix of (2.9) are solely the diagonal elements

$$(k + 2\pi p/a)^2 - (\omega/c)^2.$$

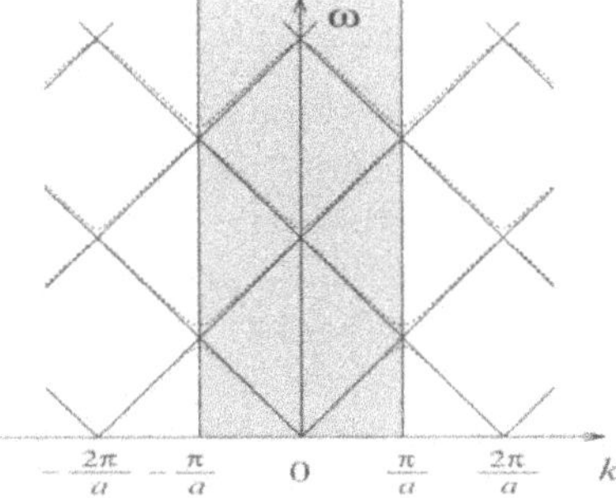

Fig. 2.4. Band structure of a 1D photonic crystal of weak periodicity. The straight lines show the dispersion curves of the empty lattice, $\omega = c|k + \frac{2\pi}{a}p|$ ($p = 0, \pm 1, \pm 2, \cdots$). The band structure with small band gaps is the one calculated for $a_A = a_B = a/2$ with $\varepsilon_A = 2$ and $\varepsilon_B = 1$

The empty-lattice test is especially useful in a PC of weak periodicity. See Fig. 2.4 for the PBS of a model system.

2.1.2 Scaling Property of Photonic Band Structure

From (2.1) and (2.8) we can deduce an important scaling property. We introduce, in place of k and ω, the dimensionless wavenumber k' and frequency ω', which are defined by

$$k = k'\left(\frac{2\pi}{a}\right), \qquad \omega = \omega'\left(\frac{2\pi c}{a}\right) \tag{2.12}$$

or

$$k' = \frac{ka}{2\pi}, \qquad \omega' = \frac{\omega a}{2\pi c}. \tag{2.13}$$

The new quantity k' is the wavenumber of photons measured in units of $2\pi/a$ and ω' is the frequency measured in units of $2\pi c/a$. In accordance with the new scale of the wavevector, the reciprocal lattice points are now simply given by $\cdots, -2, -1, 0, 1, 2 \cdots$. Thus the first BZ is defined in the k' axis to be $-0.5 \leq k' < 0.5$. When the secular equation (2.9) is rewritten using k' and ω', the size a can be removed everywhere. To see it, let us return to the Fourier component (2.4). We find

$$\begin{aligned} \varepsilon_p &= \frac{1}{a}\int_0^a \mathrm{d}z\varepsilon(z)\mathrm{e}^{-\mathrm{i}\frac{2\pi p}{a}z} \\ &= \int_0^1 \mathrm{d}z'\varepsilon(az')\mathrm{e}^{-\mathrm{i}2\pi pz'} \\ &= \int_0^1 \mathrm{d}z'\tilde{\varepsilon}(z')\mathrm{e}^{-\mathrm{i}2\pi pz'} \\ &\equiv \tilde{\varepsilon}_p \end{aligned} \tag{2.14}$$

where

$$\tilde{\varepsilon}(z') = \varepsilon(az') \tag{2.15}$$

is the dielectric function expressed using the new cordinate z' measured in units of a. The conclusion is that in terms of the dimensionless quantities of k', ω' and $\tilde{\varepsilon}_p$, (2.8) is transformed to

$$(k'+p)^2 c_p - \sum_{p'} \omega'^2 \tilde{\varepsilon}_{p-p'} c_p = 0. \tag{2.16}$$

This final result has nothing to do with the periodicity a of the PC under consideration. Therefore, if there are two PCs and their dielectric functions are identical when expressed using the lattice constant a as the unit of length, the PBSs of the two systems are the same, when exhibited using ω' as functions of k'. For example, two PCs of repeated A and B, the first having

$$a = 2\,\mu\mathrm{m}, \ a_A = 0.8\,\mu\mathrm{m}, \ a_B = 1.2\,\mu\mathrm{m}$$

and the second one having

$$a = 1\,\mathrm{mm},\ a_A = 0.4\,\mathrm{mm},\ a_B = 0.6\,\mathrm{mm},$$

have the same PBS when expressed using ω' and k', if ε_A and ε_B of the two cases are identical. In the scaled units, the dispersion curve $\omega = ck$ of a free photon gives $\omega' = 0.5$ at the BZ edge $k' = \pm 0.5$. Because of these scaling properties, PBSs are usually discussed and displayed in these scaled units. We should, however, note that in the above example, the BZ edge ± 0.5 corresponds to the wavelength of the PB of the order of microns and millimeters, respectively. In two systems having such a huge difference of scales, and hence involving totally different wavelengths of PBs, identical values of dielectric functions are hardly expected.

2.2 One-Dimensional Photonic Crystal: Various Concepts and Characteristic Features of Photonic Bands

In the calculated PBS, several characteristic features of BSs are seen, which are, in fact, common not only in 1D but also in 2D and 3D PCs. Let us again take a look at Fig. 2.2.

2.2.1 First Band at $k \simeq 0$

The first band, the lowest of all in frequency, has a linear dispersion relation near the center of the BZ. In view of the empty-lattice test, this feature is natural; since the state with $k \simeq 0$ in the first band has a wavelength much longer than the lattice constant, there are many periods involved within one wavelength of light and a photon in the state of $k \simeq 0$ sees effectively a homegeneous substance with uniform ε of a value between ε_A and ε_B.

2.2.2 Photonic Bands for k near the BZ Boundary

As k approaches the BZ edge, the dispersion curves are gradually curved. At the edge $k' = 0.5$, the PB state is an equal admixture of e^{ikx} and e^{-ikx} and propagates neither to right nor to left by forming a standing wave. This is why the group velocity v_g of the photon vanishes there:

$$v_g \equiv \frac{\partial}{\partial k}\omega_1(k) = 0 \qquad \text{at } k = 0.5\frac{\pi}{a}. \tag{2.17}$$

Beyond the BG, there is a partner of the standing-wave with $v_g = 0$ at the BZ edge.

2.2.3 Tendency of Photon Localization: Dielectric and Air Bands

To show the difference of the two states at the BZ edges, the intensity of the electric field is plotted in Fig. 2.5. We see that the state below the BG (band $n = 1$) has a higher intensity within the region of higher ε-value, while the reverse is true for the state above the BG. In this sense we often call the lower band of the pair a dielectric band and the higher band an air band, having in mind a periodic repetition of dielectric and air, which presents the contrasting behavior most vividly at the BZ edge. This type of distribution of the electric field is more or less seen in the pair of states at the BZ edge in the higher frequency range, though the contrast gets gradually obscured due to the nodes appearing both in dielectric and air regions.

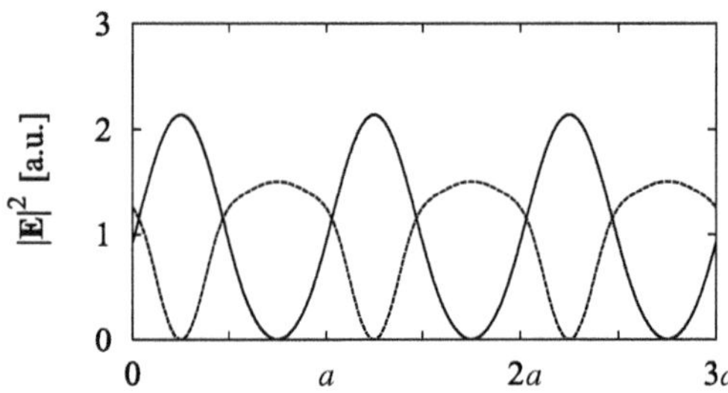

Fig. 2.5. Squared electric field as a funtion of z of the states at the BZ edge. The solid curve shows the first band (dielectric band) and the dashed curve shows the second band (air band) for the band structure shown in Fig. 2.2

The contrasting feature of the two bands stems from the general trend for photons that a higher ε induces a stronger localization of photons. To see it, let us return to the wave equation (2.1). For the stationary state of ω, we have $\partial^2/\partial t^2 \sim -\omega^2$. Adding $(\omega/c)^2 E(z)$ on both side of (2.1) then yields

$$-\frac{\partial^2}{\partial z^2}E(z) - \frac{\omega^2}{c^2}(\varepsilon(z) - 1)E(z) = \frac{\omega^2}{c^2}E(z). \tag{2.18}$$

We have rewritten the wave equation to compare it with the Schrödinger equation

$$-\frac{\hbar^2}{2m}\frac{\partial^2}{\partial z^2}\psi(z) + V(z)\psi(z) = E\psi(z) \tag{2.19}$$

for a paticle of mass m and energy eigenvalue E in the potential energy $V(z)$. By comparison, we see at once that

$$-\frac{\omega^2}{c^2}(\varepsilon(z) - 1)$$

plays the role of the potential $V(z)$ (the dimensions are not of potential, however). Therefore we see that a substance with ε higher than unity works as an attractive potential relative to free space, by having a negative potential. This is why the band having a higher electric-field intensity at A ($\varepsilon_A = 12$) than B ($\varepsilon_B = 1$) is found to have a lower frequency. Also note that in (2.18) the quantity ω^2/c^2 plays the role of an energy eigenvalue of the electron. Because ω^2/c^2 is always positive in the photonic case, there cannot be a bound state in an ordinary sense, i.e., the situation of a photon being localized completely in an attractive potential can never be achieved. The leakage of photons as an outflow of electromagnetic energy is thus inevitable in the photonic case. This causes trouble on the side of practical applications of PCs. At the same time, however, it provides interesting physical problems. For example, we shall often encounter in later chapters the virtual bound states of photons, which are bound states in the positive energy region. The topics related to the leak or how to avoid the leak in PCs is one of the topics of this book.

2.2.4 Slow Group Velocity

We have seen that photons propagate very slowly near the BZ edge due to the slowing down of v_g. The slow v_g is a feature generally observed in the PBS not only in the BZ edge region, but also in the vicinity of the Γ point, as seen in Fig. 2.2. At $k = 0$, v_g of each of the bands vanishes. In Fig. 2.4, we see a pair of dispersion curves, which cross each other in the empty lattice, repel each other to lead to a band gap at $k = 0$. In 2D and 3D structures, the band repulsion takes place more often than in 1D PCs and makes the band edges appear frequently inside the BZ. Since the states at band edges have vanishing v_g, we encounter the feature of small v_g more often in a 2D or 3D PC than in a simple 1D system.

2.2.5 Density of States

The density of states (DOS) of photons, $\rho(\omega)$, is defined so that the number of states $N(\omega)$ within the frequency region $[\omega, \omega + \Delta\omega]$ of an infinitesimally small interval $\Delta\omega$ is given by

$$N(\omega) = \rho(\omega)\Delta\omega. \tag{2.20}$$

Since k is quantized by the spacing $2\pi/L$, L being the size of the 1D system (see Appen. A), the number $N(\omega)$ should be equal to the number of the allowed k values within the interval Δk determined by the prescribed allowance $\Delta\omega$. From Fig. 2.6, which shows a small part of the dispersion curve of a band $\omega = \omega_n(k)$, we have, for an infinitesimal $\Delta\omega$,

$$\Delta\omega = \frac{\partial}{\partial k}\omega_n(k)\Delta k \qquad \text{i.e.} \qquad \Delta k = \left(\frac{\partial}{\partial k}\omega_n(k)\right)^{-1}\Delta\omega. \tag{2.21}$$

Dividing Δk by the spacing $2\pi/L$ of the quantization yields $N(\omega)$. Therefore, it follows that

$$\rho(\omega) = 2\frac{L}{2\pi}\left(\frac{\partial}{\partial k}\omega_n(k)\right)^{-1} . \tag{2.22}$$

The DOS is thus related to the inverse of v_g, the prefactor 2 of (2.22) being given to take account of the two polarizations of photons propagating in the z direction of the system shown in Fig. 2.1. That the DOS is inversely proportional to v_g simply reflects the fact that the allowance Δk for a given $\Delta\omega$ increases with the decrease of the curvature of the dispersion curve. This is shown in Fig. 2.6. At the frequencies of the bands at the BZ edge, where vanishing of v_g of the band 1, 2, ... takes place, DOS peaks arise one after another.

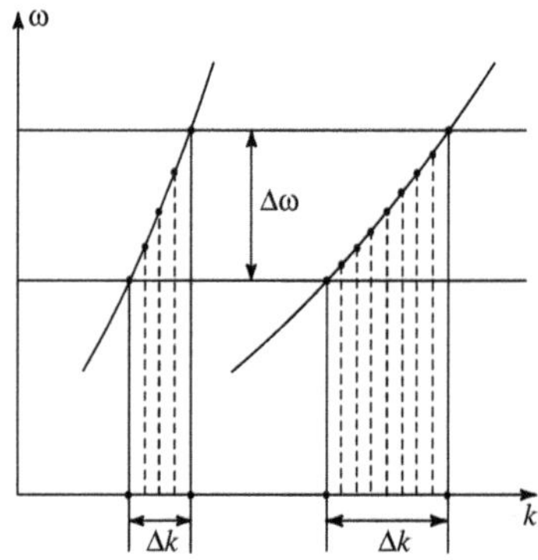

Fig. 2.6. DOS versus the curvature of photonic bands. As a band becomes flatter, the number of the states lying in a given range $\Delta\omega$ increases. The spacing between the quantized values of k is $2\pi/L$.

The feature of enhanced DOS is one of the characteristics of PCs. For example, the emission probability of photons of frequency ω from an atom in a PC is proportional to $\rho(\omega)$ (see Sect. 10.1) and it is natural to expect an enhanced light emission from the atoms at the frequency of the DOS peaks. In this sense the enhancement of the DOS of the PB is often called the group velocity anomaly.

Together with the enhancing behavior, the suppression of the DOS is spectacularly demonstrated by the existence of the region of zero DOS at the frequency regions of PBGs. This was indeed what let Yablonovitch first propose an artificial PC called Yablonovite [1, 2]. In the PBG regions we have the zero emission rate of photons from an atom. Control of the light emission rate by designing a PC to meet one demand and another is one of the ultimate goals of the technological application of PCs.

In essence, all the optical responses expected and exploited are related more or less to an enhanced or suppressed DOS due to the PBS. One reason for the increasing demand for high-quality 2D and 3D systems is that these characteristics of group velocity or DOS manifest themselves much more dramatically in a PC of higher dimension.

2.3 Concept of the Light Cone and Example of One-Dimensional Off-Axis Band

In this book, finite PCs, especially those with finite thickness in one direction and infinite extention in the other two directions, are treated repeatedly both theoretically and experimentally. In such a system, called in this book a slab type PC or simply a slab PC, it is important to consider whether light coming from an outside homogeneous medium can couple to the inside modes through the entrance boundary plane. If there is a coupling the mode is called a leaky mode because it is not an eigenmode in a strict sense because of its leakage to the outer space through the coupling. The leakage gives a finite lifetime to the mode.

From the physical point of view, a mode with wavevector $\boldsymbol{k}_1$ and frequency ω_1 $(= \omega(\boldsymbol{k}_1))$ of an outside region can be coupled to one with $\boldsymbol{k}_2$ and ω_2 in a PC. The spatial dispersion $\omega_2(\boldsymbol{k}_2)$ is determined by the PBS calculation. In this coupling process, the conservation of both energy and $\boldsymbol{k}_{\|}$, the wavevector component parallel to the surface, is established between the two outside fields, such that

$$\omega_1 = \omega_2 \quad \text{and} \; (\boldsymbol{k}_1)_{\|} = (\boldsymbol{k}_2)_{\|}. \tag{2.23}$$

The concept of light cone or line, used to divide the phase space $(\omega, \boldsymbol{k}_{\|})$ into two parts based on (2.23), is of fundamental importance in the study of PCs. In a homogeneous substance of uniform refractive index n the light cone is literally a cone in the $(\omega, \boldsymbol{k}_{\|})$ space defined by the relation

$$\omega = \frac{c}{n}|\boldsymbol{k}_{\|}|. \tag{2.24}$$

When the region is air with $n = 1$ and the x axis is chosen in the direction of $\boldsymbol{k}_{\|}$, it holds that

$$\omega = ck_x. \tag{2.25}$$

Having this case in mind, we often call the light cone as the light line. The concept of the light cone was first introduced as early as 1966 by K. L. Kliewer and R. Fuchs, as a concept used to distinguish the leaky and nonleaky modes of a dielectric slab [3, 4]. In PCs, this concept was first used in 1982 to examine the photonic modes of a periodic system of finite thickness, a slab PC in the present terminology, by Inoue and Ohtaka [5]. The classification between leaky and nonleaky modes is still an important fundamental in the technological application of PCs, because a PC of practical use is always bounded by surfaces, as in slab-type PCs and PC fibers.

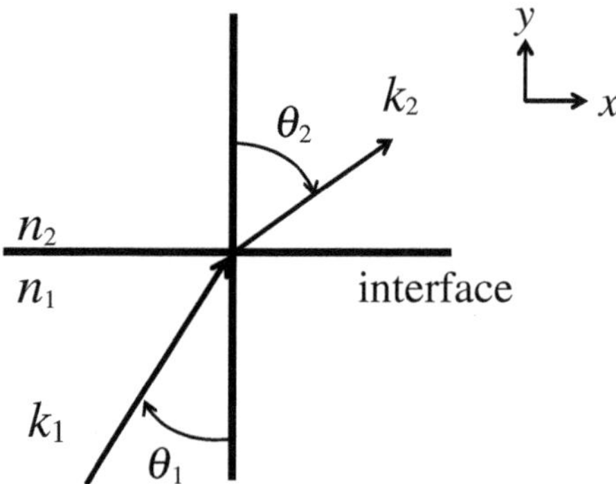

Fig. 2.7. A schematic drawing of refraction at a boundary

We briefly explain generally how to use this concept in a PC, taking a simple example of a 1D PC. We plot the energy of the mode in a homogeneous medium as a function of $\boldsymbol{k}_{\|}$, instead of $\boldsymbol{k}$. Use of the $\boldsymbol{k}_{\|}$ component corresponds to a change of incident angle at the interface. Before showing an example, let us consider first refraction in a well-known case where light is incident on the boundary plane between two homogeneous media of different refractive indices (n). We take the boundary to be the xy plane, with n_1 for $z < 0$, and n_2 for $z > 0$ with $n_1 > n_2$. The plane of incidence is taken as the xz plane with the incident angle of θ_1 (see Fig. 2.7).

In Fig. 2.8 we plot two straight lines, one corresponding to a light line (the dispersion relation) in medium 1, and the other in medium 2, respectively, which are expressed as $\omega = (c/n_i)k_{\|}$ with i referred to 1 or 2. These lines just correspond to the cases where light propagates in parallel to the plane in the respective media. Suppose in medium 1 one mode on the point B has

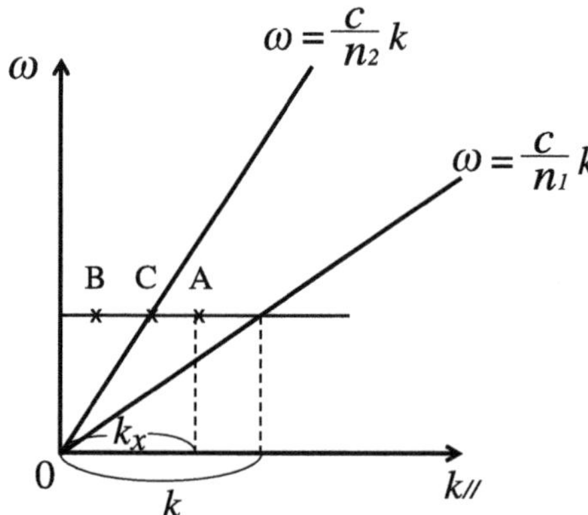

Fig. 2.8. A schematic drawing of an example explaining the concept of the light line

the same energy ω as that on A. Since the magnitude of $\boldsymbol{k}$ of the mode with the $\boldsymbol{k}_{\parallel}$ component is given by $\omega = (c/n_1)k$, the mode is nothing other than one propagating with an angle θ_1 such that

$$\sin\theta_1 = k_{\parallel}/k = k_x/(c^{-1}n_1\omega). \tag{2.26}$$

Therefore, all modes in the medium are presented on the left-hand side of the line: any light modes do not exist on the right-hand side. The same is also true with medium 2. As a consequence, light that is incident on the boundary plane with ω and $k_{\parallel}$ on the left-hand side of medium 2 can go inside medium 2 with the refraction angle θ_2 given by

$$\sin\theta_2 = k_{\parallel}/k = k_x/(c^{-1}n_2\omega). \tag{2.27}$$

On the other hand, light with ω and $k_{\parallel}$ on the right-hand side cannot go inside medium 2, but is totally reflected back into medium 1. Therefore, from the above two equations the critical condition is given by the light line of medium 2, such that

$$\sin\theta_c = n_2/n_1, \tag{2.28}$$

which is well known. In the case of total reflection $\theta > \theta_c$, there is no wave propagating at large distances from the boundary in medium 2. However, this does not indicate that the light field vanishes completely in medium 2 for $z > 0$. Namely, the evanescent wave can propagate along the boundary plane, the wavevector k_x of which is expressed as

$$k_x = (n_1/c)\omega\sin\theta_1 = n_1(2\pi/\lambda)\sin\theta_1. \tag{2.29}$$

Those features which descriminate the leaky modes and guided modes are shown in Fig. 2.9. See Sects. 3.4 and 4.3 for leaky modes.

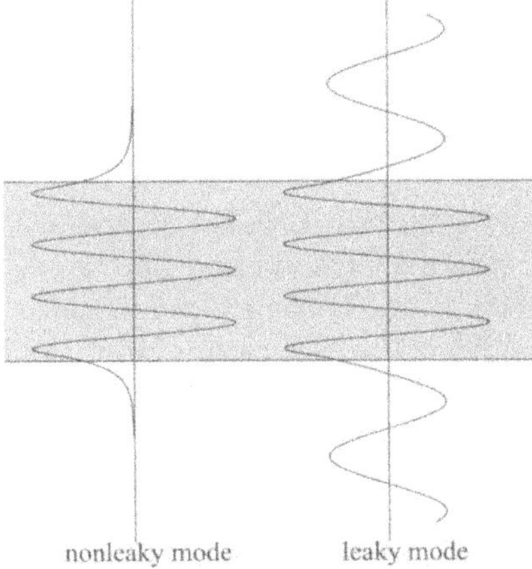

Fig. 2.9. Leaky modes and guided modes.The outside fields are those of plane-wave light in the leaky modes, while they are evanescent light in the guided modes

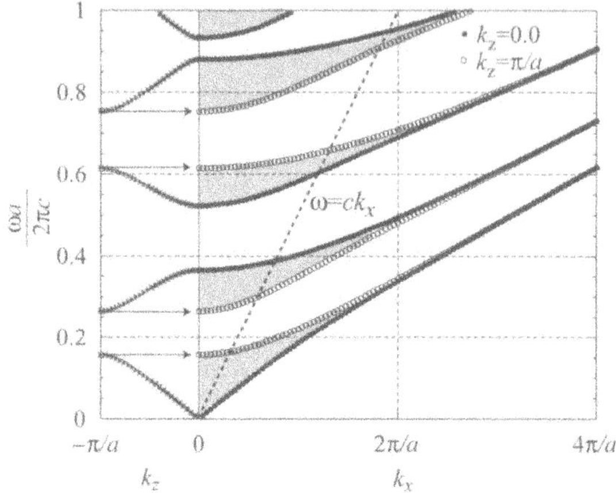

Fig. 2.10. The off-axis modes in a 1D multiple-layer film. Two layers A and B are stacked in the z direction. The dispersion curves of the modes of $k_y = 0$ and polarized in the y direction are plotted as functions of k_x, the boundary of the grey regions being given by the dispersion curve with $k_z = 0$ or $k_z = \pi/a$. From the points at $k_x = 0$, the dispersion curves of the modes of $k_x = 0$ are drawn on the left hand side in the first BZ $0 < k_z < \pi/a$

Now we are in a position to use this concept for the PC case. As the simplest case, let us consider the off-axis modes, i.e., the mode with $\boldsymbol{k}$ not perpendicular to the layers or film, in the multiple-layers film described in Sect. 2.1 and 2.2. The structure considered here is the same as that shown in Fig. 2.1, except that it is not infinitely long but semi-infinite by the presence of the surface. Let $\boldsymbol{k}_x$ be the wavevector component parallel to the surface between the semi-infinite 1D PC and air. Now the guided mode is specified by $\boldsymbol{k}_z$ and $\boldsymbol{k}_x$.

In Fig. 2.10 is shown the dispersion of the TM (s-polarized) guided modes of the 1D PC used in Fig. 2.2 The band structure with $k_x = 0$ is given on the left as a function of k_z, which is reproduced from Fig. 2.2. The shaded regions of the band structure on the right show the continuum of bands coming from the freedom of the k_z values. For a multilayer stack of finite thickness, bounded by a surface parallel to the xy plane, only the modes below the light line, shown by the dashed line in the figure, are true guided modes confined completely inside the PC, while those above the light line are actually the leaky modes which can escape from the PC into air.

2.4 Band Structures of Two- and Three-Dimensional Photonic Crystals

2.4.1 Examples of Two-Dimensional Photonic Band

As for PBS in 2D and 3D PCs, many examples will be presented in the subsequent chapters in relation to the concrete samples fabricated. So, here we show only a few typical examples. First, in Fig. 2.11 we show an example of a 2D PC with the square lattice of dielectric rods of circular cross-section; as for how to derive this PBS, refer to Sect. 3.1 in the next chapter. The first BZ corresponding to this structure is also a square lattice, where there exist two high symmetry points, i.e., X $(a/\pi,\ 0)$ and M $(a/\pi,\ a/\pi)$ other than the Γ point $(0,0)$ (for the naming of these special points, see Appendix A).

The eigenmodes for this structure can be specified by the polarization. That is, those are classified into the $E(H)$-polarized modes with E (H) parallel to the dielectric cylinders, which are called TM-(TE-)modes, respectively. It is seen that an ample variety of PBSs manifest themselves in the 2D PC as compared to 1D PCs (see Sect. 3.2 for PBS of 2D PCs).

First, each band belongs to the specific irreducible representation of the relevant point group that depends not only on polarization, but also on the $\boldsymbol{k}$ direction. For example, if the mirror plane is present for the lattice, the respective bands are classified to even or odd symmetry ones according as $\boldsymbol{E}$ becomes symmetric or antisymmetric with respect to the mirror plane. Importantly, in this case the odd modes, or odd-parity modes cannot be coupled to the external plane wave at normal incidence for symmetric reasons [6, 7]. This situation is met frequently in 2D and 3D PCs, although it does not occur in 1D PCs. Those modes are called the "uncoupled modes" [8] (see Sect. 4.2 for the group theory of PCs).

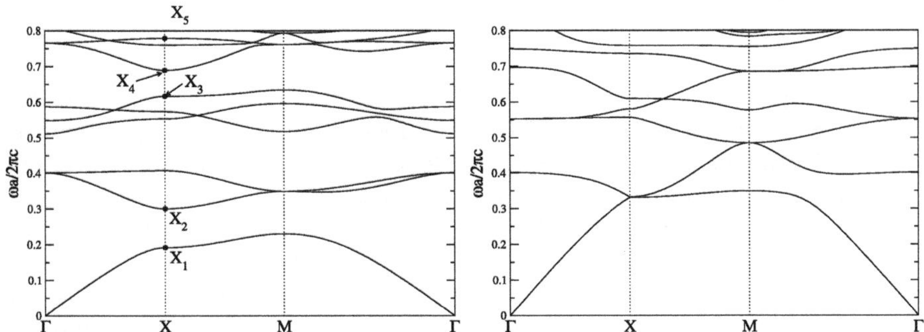

Fig. 2.11. Photonic band structure of a 2D PC of cylinders arrayed in a square lattice: TM bands (left) and TE bands (right). The cylinders of $\varepsilon = 12$ (Si) are arrayed in a lattice in the air. The ratio of the radius r of cylinders to a is $r/a = 0.3$. The states X_1 through X_5 at the X point are examined below

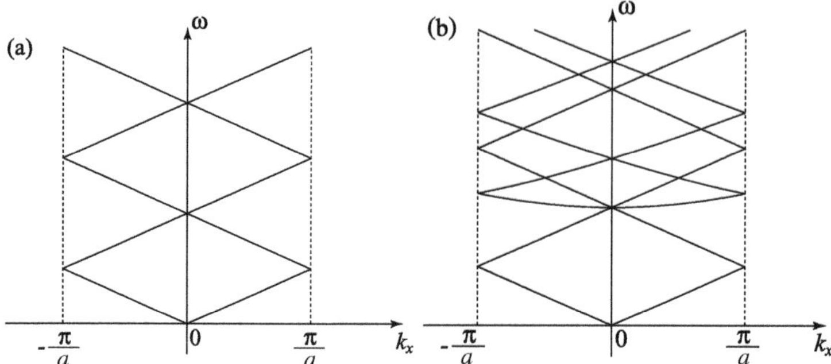

Fig. 2.12. Photonic band structure of empty lattice: (a) 1D photonic crystal and (b) 2D photonic crystal of square lattice ($\Gamma - X$) direction

Second, there exist three PBGs for TM-modes, although the PBG common to both polarized modes does not open in this case. Third, there are a number of very narrow bands and the v_g-anomalies occur not only at the zone edge but also in the interior of the first BZ. The appearance of so many narrow bands can be understood by the empty lattice test. Figure 2.12 shows the BS of the empty-lattice for $\boldsymbol{k}_{\|}$ along the Γ-X axis, i.e., $\boldsymbol{k}_{\|} = (k_x, 0)$ as a function of k_x. In obtaining it, we drew the dispersion curves for plane-wave light of wavevector $\boldsymbol{k}_{\|} + \boldsymbol{h}_{\|}$

$$\omega = c|\boldsymbol{k}_{\|} + \boldsymbol{h}_{\|}|, \tag{2.30}$$

with $\boldsymbol{h}_{\|}$ given by

$$\boldsymbol{h}_{\|} = \frac{2\pi}{d}(\pm 1, 0), \ \frac{2\pi}{d}(0, \pm 0), \ \frac{2\pi}{d}(\pm 1, \pm 1), \tag{2.31}$$

etc. and plotted the curves by putting $\boldsymbol{k}_{\|} = (k_x, 0)$.

In short, some of them originate from the flat bands in the empty 2D lattice; notice that this is not the case for 1D.

Next, the eigenvector of the PB of $(n\boldsymbol{k}_{\|})$ gives the intensity of the electric field, $|\boldsymbol{E}_{n\boldsymbol{k}_{\|}}(\boldsymbol{r})|^2$, associated with that Bloch state. Without specifying the normalization of the eigenvectors, Fig. 2.13 shows the intensity of the electric field $|\boldsymbol{E}_{n\boldsymbol{k}_{\|}}(\boldsymbol{r})|^2$ in arbitrary units, for the bands $n = 1$ and 2 of Fig. 2.11 at the X point of the first BZ. We see the typical feature of a dielectric band in the lower band-state X_1. The air-band characteristic of the X_2 state is less obvious than the dielectric-band feature of X_1, however. As for the band states with higher energy marked by X_3 to X_5, a similar plot of intensity reveals that the distinction between the dielectric and air bands becomes progressively vague, since the mixing among the increasing number of plane waves of shorter wavelengths comes in.

In Fig. 2.14 is shown another example of a 2D PBS for a triangular lattice of air holes; the 2D BZ corresponding to this structure is shown in Appendix

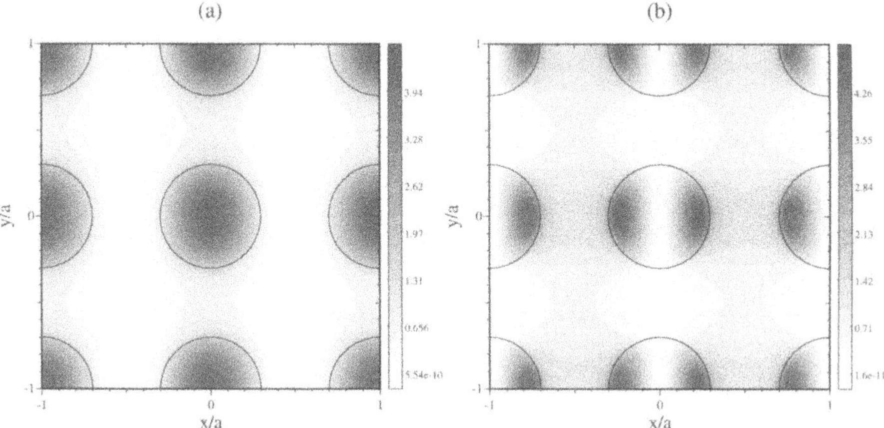

Fig. 2.13. Plot of intensity $|\boldsymbol{E}(\boldsymbol{r})|^2$ of the states X_1 and X_2 of Fig. 2.11 (arbitrary units). The state X_1 (left) has a typical dielectric-band character in (a) and band X_2 (right) shows an air-band character in (b).

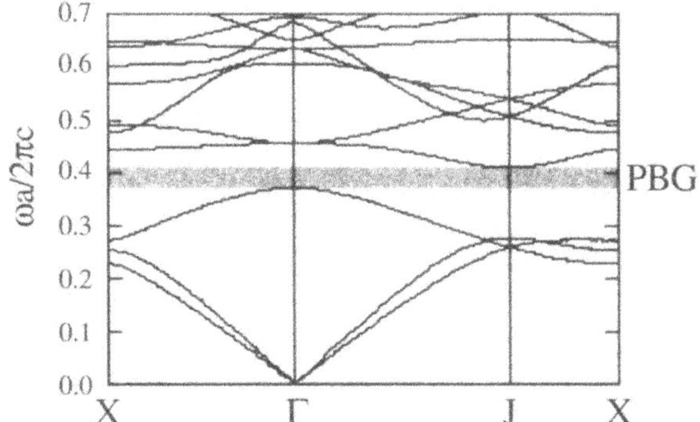

Fig. 2.14. An example of 2D photonic band structure with a 2D photonic band gap (a triangular air-hole structure)

A. It is well-known that the PBS for this structure exibits a 2D PBG, irrespective of polarization, as is seen in Fig. 2.14, if the difference of ε between the background material and air (holes) is large enough; in this example the difference is 12 used for Si against 1. Rather unusually, for this structure the gap does not open for E-polarized modes between the lowest (first) and the second-lowest (second) bands, but instead, it does between the higher-energy bands. This is because the first and second bands for E-polarized modes are degenerate at the K point [9]. Note that the notation X and J are sometimes used in the literature (also in this book) in place of M and K, respectively.

2.4.2 Example of Three-Dimensional Photonic Band

In Fig. 2.15 is shown an example of a calculated PBS with a full 3D BG in an inverse diamond lattice structure, which is composed of the respective tetra-bonds consisting of air-rods (diameter R) in the dielectric of $\varepsilon = 7$ (see also Sect. 13.4); $R/a = 0.1$ is adopted in calculation. The shape of the first BZ corresponding to this structure is the face-centered cubic (fcc) lattice, as shown in Appendix A. There, it is seen that a full BG opens in all directions.

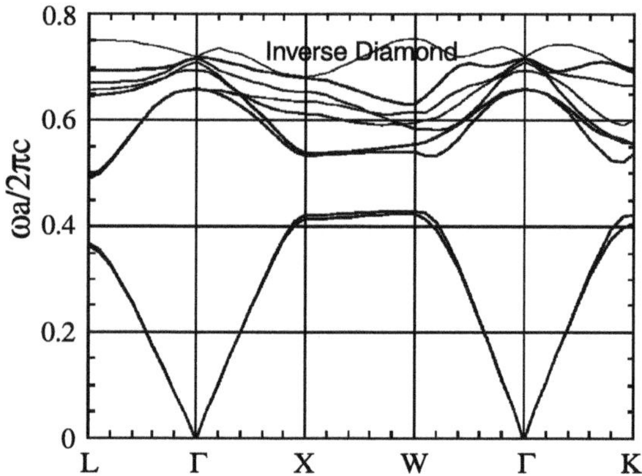

Fig. 2.15. An example of 3D photonic band structure with the diamond lattice structure (a sample of inverse diamond structure)

This PBG corresponds to the well-known BG of electron in semiconductors such as GaAs, which also has the same BZ structure of the fcc lattice. As is well known, a BG for GaAs exists between the valence and conduction bands [10]. Notice that the two band structures (photonic and electronic) resemble each other in many respects, which arises from the resemblance of the two different wave equations. In particular, the existence of the BG is common between the two. However, the two differ from each other in some respects. For example, in the photon case one can excite any state, while in the electron case one can do so only when the state is not occupied. This is because the photon is a boson governed by Bose–Einstein statistics, whereas the electron is a fermion governed by Fermi–Dirac statistics.

2.5 How to Experimentally Explore the Band Structure

In the preceding two sections we have described a few examples of representative PBSs. How can we experimentally get information about individual PBSs corresponding to a sample? We explain very briefly the methods below. For details readers are recommended to refer to Chap. 9.

A simple and reliable method is to observe either the transmittance (T) or reflectance (R) spectrum, or both in general, as a function of wavelength over a broad range. We need to do so with the propagation direction of incident light varied. This can be done by preparing samples with the surface normals directed in several high-symmetry directions such as $\Gamma - K$, $\Gamma - M$, etc. Therefore it needs rather a troublesome task. However, in some simple cases, information about such spectra only for the particular directions may suffice for the purpose.

For wavelengths corresponding to a stop band, T drops very much, whereas R should be unity, since there exist no photon modes inside in that direction for the external light to couple to, causing it to be reflected completely. However, it is important to note that the reverse is not necessarily true. In other words, observation that the external light is completely or totally reflected over a wavelength range does not necessarily indicate that the range corresponds to a stop band. This is because complete reflection or drastic attenuation of T occurs for a range of the uncoupled band, which has already been explained in the preceding section. Next, to what extent the incident light can couple to an individual band (coupled band) of a PC depends on the band itself, or the character of the band.

Finally, in the case of a practical sample T drops considerably even for a coupled band, whenever the quality of the sample under study is not good. Also, another problem exists from a practical point of view in that we have to make a measurement by using a sample with finite periods, although the PBS is calculated, with a few exceptions, for an infinitely long sample (see Sect. 4.3, for the treatment of PBS for a system of finite thickness). If the number of periods of the sample is too small, say, 5 or 10, the correspondence of the T-spectrum between the observed and the calculated PBS is not good. Except these points, the correspondence is very good between theories and experiments carried out for PCs. This is because the calculated PBS is reliable enough due primarily to the lack of photon–photon scattering; in contrast the one-electron approximation is often not good in the electronic case because of the existence of electron–electron and electron–phonon (vibration) interactions.

2.6 Defect Modes

Thus far we have described and discussed a regular or completely periodic PC lattice, i.e., one that can be called a bulk PC. It is rather easy to technically introduce a defect or disorder to a particular lattice point or place, or thereby. As a consequence, this kind of defect causes, generally speaking, breakdown of the symmetry that a PC has originally possessed. An eigen-mode due to the defect is possible to newly manifest itself in a PBG [11]. As an example of the zero-dimensional (0D) or point defect in a 1D and 2D PC of air cylinder or dielectric pillar type, we have only to enlarge or lessen the size (diameter), for

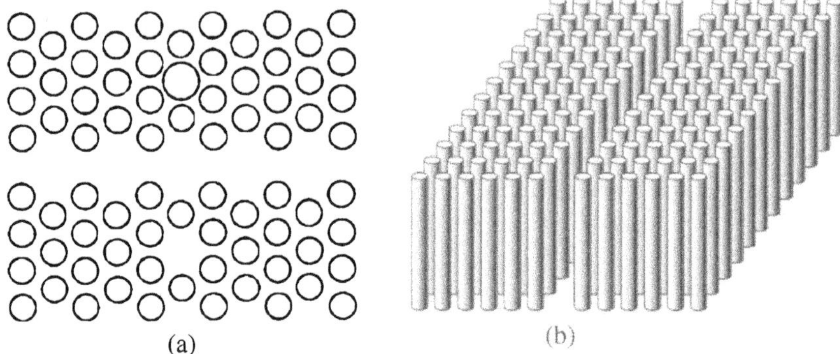

Fig. 2.16. A schematic drawing of examples of the defect modes in a 2D PC. (a) 0D (point) defect modes (top view) in an array of air holes; the acceptor (upper) and donor (lower) types, and (b) 1D (line) defect mode in an array of dielectric pillars

example, of only one air cylinder at the particular lattice point, as compared to others, or the surrounding ones; the special case of this is to leave it without any air cylinder (see Sect. 4.5 for the origin of defect modes). Such examples of an enlarged air cylinder and without (missing) air cylinder in a 2D PC are schematically shown in Fig. 2.16(a). Another typical example of a 0D defect in a 1D PC is to make the period at one lattice position different, as compared to the others. This kind of 0D defect mode corresponds to an impurity state, either the donor or the acceptor state of electrons in a semiconductor; in the case of a different size of air hole introduced, increasing (decreasing) the hole size that causes decrease (increase) of the dielectric constant there corresponds to an acceptor (donor) state for electrons.

In the case of PCs, in addition to the above-mentioned 0D defects, 1D and 2D types of defects, i.e., line and plane defects, can also be created, which are also important in controlling light or developing unique devices in optoelectronics. As an example of a 1D defect, we create it in a 2D PC of air cylinders or dielectric pillars by leaving a single line of air cylinder imperforated, or introducing a single line of air, as is schematically shown in Fig. 2.16(b). One can also create it in a 3D PC, e.g., of air-hole type in such a way that we leave a single line of holes unopened. A typical example of a 2D defect is such that one lattice plane is left homogeneous in a 3D PC.

Now, introducing this kind of defect into the bulk PC creates newly an eigenmode or an eigen band called a defect-mode or a defect-band, generally speaking, in a BG: the point-defect modes in a BG correspond to the impurity modes in the electronic case, or in semiconductor. It is noted that the line-defect band exhibits a $\omega - k$ dispersion along the direction, which will be repeatedly discussed in later chapters, e.g., in Chaps. 7, 11 and 12.

2.7 Common and Fundamental Features of Photonic Band Structure

In Sect. 2.2, based on an example we have already explained several basic features of the 1D PBS. This section is devoted to newly summarize the unique and outstanding features that 2D and 3D PCs generally exhibit. As will be shown in the next section, PCs are very well suited for controlling light, i.e., both the radiation field and light propagation characteristics. This important potential is primarily based on one or two of the following features:

1. Existence of photonic band gap
2. Existence of defect or local mode
3. Anomalous group velocity
4. Remarkable polarization dependence
5. Manifestation of peculiar band
6. Others

Let us make a survey of each item one by one, taking into account how to utilize those for observing new phenomena and developing PC-based devices.

2.7.1 Existence of Photonic Band Gap

A question arises as to in what cases a full PBG opens. There exist rather severe requisites to satisfy for a 3D PC to possess such a PBG. Those are related to the crystalline structure, the difference of the relevant ε-values, and the occupation ratio f between two constituted materials. It is generally recognized that the diamond lattice structure is the best in order to obtain a full PBG as well as a wide gap. This is primarily because the shape of the first BZ is close to the sphere (most spherelike) as compared to other structures such as the simple or body-centered cubic one. This feature is advantageous to opening a gap between the second- and third-lowest bands. However, this is not the case for the higher bands. In fact, it is known that a full PBG opens for such higer bands in a face-centered cubic lattice [12] and even in the simple cubic lattice [13]. In Fig. 2.17 is shown the BS. It is remarked that in this case the conditions regarding the contrast of ε as well as f are relatively severe, and more importantly, the relative gap width $\Delta\omega/\omega_0$ is not large enough, where $\Delta\omega$ and ω_0 refer to the width and the center frequency of the gap, respectively.

In contrast, it is rather easier to design a sample with a 2D PBG. In both 2D and 3D cases, in a given crystal structure there is, as far as the structure is adequate, a trend that as the ratio of $\varepsilon_A/\varepsilon_B$ is larger, a gap is easier to open, where A and B refer to the two constituted materials.

Here we mention how the term PBG is used. It is better to use the terminology "PBG" only for the complete BG already defined. However, it is true that in some cases people in this field use this term even for the case where

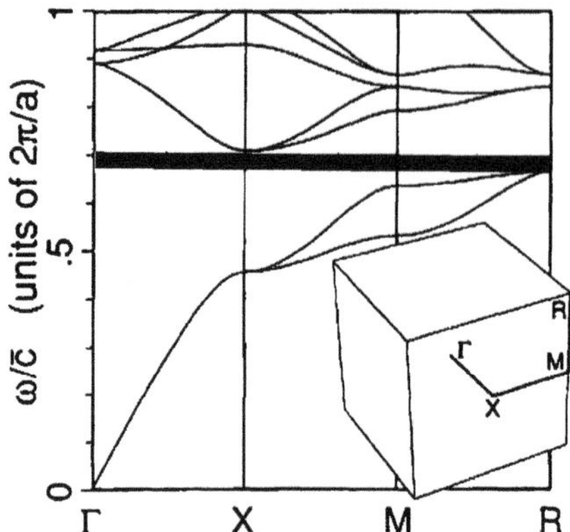

Fig. 2.17. Another example of 3D photonic band structure for square-shaped air rods in a simple cubic lattice: $\varepsilon_b = 13$, $\varepsilon_a = 1$ and $f = 0.83$, quoted from [13] by courtesy of Professor J. W. Haus

there is no such BG, or for an incomplete BG: in physics we have used the term of "stop band" for the latter case. Therefore, in order to avoid confusion, we will use hereafter in this book the term "a full PBG" only in a strict sense.

In this connection, it is important to note that the situation for a 2D PC slab (a sort of quasi-3D PC) is substantially different from the 2D case; it differs also from the 3D case. Namely, because of the existence of the so-called light cone, already explained in Sect. 2.3, any complete BG should not exist within the 2D slab plane: inside the light cone there are, in principle, continuous extended states (not eigenmodes for the PC slab) for arbitrary wavelength. However, we define a 2D PBG even for this case in such a manner that any guided modes (eigenmodes) of the PC slab are missing over a specific energy range in all 2D directions [14]. This will be presented in more detail in Chap. 6.

Now we mention briefly a few examples of direct application of a full BG. First, unique mirrors for complete reflection without loss can evidently be developed, and in fact, some of those have already been commercially available. Next, it can be utilized for controlling the radiation from matter. For example, an emission from atoms placed inside a PC with a full 3D PBG is inhibited when the photon energy is in an energy range of the PBG. This is because the emission probability is zero, since the DOS for photons within the PBG vanishes; the probability in the radiative process is proportional to the DOS. In order to completely suppress the spontaneous emission from atoms one needs to use a 3D PC with a full 3D PBG [1] (see also the literature [15]).

On the other hand, whether the probability of the spontaneous emission becomes enhanced or not at the edge of a full 3D PBG, as compared to that in the homogeneous case with the same effective refractive index is not clear. First of all, the radiative lifetime of an oscillating electric dipole placed inside a PC depends not only on the DOS per unit volume, but also on a few other factors. Namely, it depends on whether the dipole is placed in the constituent material with the higher or lower dielectric constant, and on the oscillation direction, i.e., the orientation of polarization. Furthermore, the probability of the spontaneous emission depends also on in what direction the emission is observed. So, generally speaking, the probability is possible to be enhanced or reduced in the case of a 3D PC. In particular, the probability is expected to be enhanced to some extent around the band edge depending on the case. The same is also true in the case for a 2D BG. In contrast, there exists the case where the probability should increase in a divergent way at the edge of a BG for a 1D PC. This problem in quantum electrodynamics will be discussed in a bit more detail in Chap. 10.

It is important to note that in many cases we can control the radiation field and light propagation properties with use of a 3D PC without such a full PBG, except for some cases including complete suppression of the spontaneous emission described above. For the same reason, 2D PCs also serve for controlling light. Thus, we should be able to develop a variety of devices by using such a PC, irrespective of the 2D or 3D nature.

2.7.2 Existence of Defect or Local Modes

The existence of the defect modes has already been stated in the preceding section, with special emphasis placed on how to create those. Here we discuss the physical property to some extent. By exciting this mode, light can be localized around the defect or the region with disorder. Therefore, the defect mode is localized both energetically and spatially, as stated in Sect. 2.6. In the 0D case, where the mode is localized in a small spatial region, the mode does not have a definite wavevector $\boldsymbol{k}$ because of the uncertainty principle concerning $\boldsymbol{k}$ versus $\boldsymbol{r}$ in quantum mechanics. This feature is very similar to that for the impurity state, i.e., the donor or acceptor state, for electrons in a BG, as already described in Sect. 2.6. In contrast to the point-defect (0D) case, the line-defect mode shows a particular dispersion in a BG. Namely, with $\boldsymbol{k}_\mathrm{l}$ defined to be the wavevector in the direction along the line defect, the mode frequency ω varies with $\boldsymbol{k}_\mathrm{l}$ in the gap. An example of such a line-defect mode in a 2D PC of a triangular array of dielectric pillars in air is shown in Fig. 2.18. The present line-defect is composed of one row of missing pillars along the $\Gamma - K$ direction (adopted as the x axis), so the band structure for the TM-modes with $\boldsymbol{k}_\mathrm{l}$ parallel to the x axis is shown in Fig. 2.18 with k_x used for $|\boldsymbol{k}_\mathrm{l}|$; the x and y axes are taken within the 2D plane, whereas the z axis is in the direction of the pillar axis. In Fig. 2.18 the hatched areas are obtained by projecting all the bands $\omega_n(k_x, k_y)$ in the first BZ onto the $\omega - k_x$ plane, these bands being often called the slab bands in the literature. The dispersion relation of the line-defect bands are shown by the thin curves

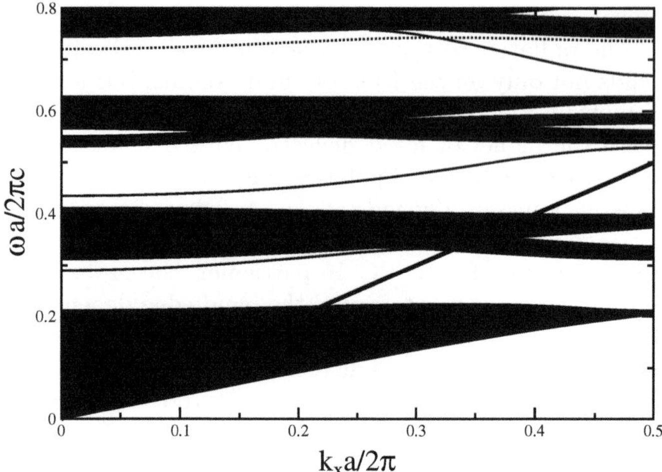

Fig. 2.18. Dispersion feature of the line-defect modes in a 2D PC of dielectric pillers of $\varepsilon = 12$ arrayed in the triangular lattice in free space. The electric fields of the defect modes and the projected bands are both polarized in directions of the pillar axis. The thick line is the air light-line $\omega = ck_x$

between the hatched regions. The solid and dotted curves show the even and odd defect modes, with the parity of the defect modes defined according as E_z is symmetric and antisymmetric with respect to the mirror plane (xz plane). The parameters are $r/a = 0.3$ and $\varepsilon = 12$, r being the radius of pillars and a the lattice constant. Needless to say, the defect modes shown here provide the light propagation along the x direction.

Let us consider in a bit more detail the significance of localization of the electric (magnetic) field due to the defect or disorder mode. That light without real mass can be localized, e.g., around a particular point in space, just like the case for an electron, is very important from the physical point of view; this phenomenon of light localization should be discriminated from the case where light is confined with use of the specific boundaries such as metal walls. Anyway, the 0D defect mode can be utilized, for example, either as an extremely small optical cavity, or as a very narrow band-pass filter. On the other hand, the line-defect mode already described is considered to serve as a novel type of waveguide without any propagation loss even at an abruptly bent corner [16, 17]. This type of waveguide, called a PC waveguide, becomes at present very important [18] (see also the literature in Chaps. 11 and 12, where this PC waveguide is discussed in more detail).

2.7.3 Anomalous Group Velocity

Both v_g and group velocity dispersion (GVD) play an important role in light propagation. Theoretically, v_g is given by the slope ($\mathrm{d}\omega/\mathrm{d}k$) of a band, where

k refers to the magnitude of $\boldsymbol{k}$ in the relevant direction of light propagation; usually, the magnitude of v_g is estimated for the band calculated for an infinitely long or large sample. In general, there appear in a PC a lot of band portions or positions where the band slope is flat, which makes v_g anomalous. At the band edge v_g should become extremely small without fail [19–21]. Such a flat energy position is also encountered at both the maximum and minimum points of the lower and upper bands resulting from an anti-crossing between two bands with the same symmetry, as already described in Sect. 2.2. A photon with such a small v_g at those positions is sometimes called a heavy photon in analogy with the heavy electron (see Sect. 3.6).

It is also important to note that very frequently a relatively flat band, i.e., a very small v_g, is encountered in a 2D and 3D PC [21, 22], as the origin of the flat band was already explained in Sect. 2.2. Moreover, there are many bands where v_g becomes negative. We call all these anomalous group velocity.

That v_g is very small indicates that the electric field strength becomes very large there; the fact that the Poynting vector must be constant everywhere requires that E^2 becomes large for a small v_g. This fact is very important, since the effective interaction length between light and matter becomes short enough by using such a small v_g. That is, the radiation signal in many physical phenomena such as second-harmonic generation (SHG) can be very effectively generated from a PC, or equivalently, can be greatly enhanced in a PC sample as compared to a conventional sample with the same length [23–25]. This also makes sense in the case of a laser, because the threshold value of electrical current or fluence for pumping can be greatly reduced by utilizing a small v_g [26–29].

As for the GVD in PCs, it is very large in general; the line-defect mode already described can be designed to also have a large GVD [30]. This is because the GVD in a PC originating from the periodicity can be artificially altered unlike the GVD in a transparent wavelength region of a material which is governed by the dispersion of the refractive index, i.e., essentially by the intrinsic absorption.

2.7.4 Remarkable Polarization Dependence

Characteristically, in many PCs propagation of light incident from the outside through a sample is markedly influenced or governed by the polarization; of course this property arises from the fact that the PBS under study is polarization-dependent. This property applies generally to all PCs irrespective of the dimension, with the exception of on-axis light propagation in a 1D multilayer film; notice that even in this case the BS for the off axis propagation already described in Sect. 2.3 becomes polarization-dependent. If we focus on only the guided modes, the polarization dependence is particularly remarkable even around the first BG in 2D PCs and 2D PC slabs. In contrast, it is, generally speaking, less remarkable for 3D PCs. This is particularly true with the diamond-lattice structure, as compared to the cases for a simple cubic lattice [13] and a face-centered cubic lattice [31].

As a consequence, a unique polarizer can be developed by utilizing a pair of specific bands for two different polarized light beams with the same frequency. A simple example is the case that there exists a stop band for one polarized light beam, while there is a band for the other. Even for the case where there are respective bands for two different polarizations, the magnitude of v_g should be generally significantly different from each other, except for an accidental case. Based on this feature, a different kind of polarizer can also be developed for an ultra-fast light pulse. Namely, such a pulse is easily split into two, depending on the polarization, after passing through a thin PC sample [32]. Thus, utilizing time delay together with a gate, one can pick up only a pulse with the desired polarization.

2.7.5 Manifestation of Peculiar Bands

There are a variety of peculiar PBs showing anomalous or singular points [33]. Roughly speaking, such an anomaly is likely to manifest itself in the higher energy region where many bands run. It has already been revealed that well established crystalline optics no longer apply to the case of a PC; it is again remarked that a PC is not a natural crystal, but an artificially fabricated one. Consequently, we need to introduce new crystalline optics or concepts, called "photonic crystal optics" to cover the case of PCs. Let us show the simplest example below to understand this fact. It is well known that all crystals except for quasi-crystals in an optical region are classified into three cases, i.e., uniform crystalline, uniaxial crystalline, and biaxial crystalline, corresponding, respectively, to

$$\varepsilon_1 = \varepsilon_2 = \varepsilon_3, \quad \varepsilon_1 = \varepsilon_2 \neq \varepsilon_3, \quad \varepsilon_1 \neq \varepsilon_2 \neq \varepsilon_3, \tag{2.32}$$

where $\varepsilon_1, \varepsilon_2$, and ε_3 refer to the dielectric constants along the optic axes. This classification no longer applies to a PC; for detail, see Sect. 4.7.

First of all, Snell's law of refraction does not generally apply to the boundary plane between a PC and a normal material. Anyway, propagation characteristics of light at such a singular point are very complicated on one hand, and unique on the other hand. As a result, several peculiar phenomana have already been found, such that the phenomenon of trirefringence occurs [34], and v_g cannot be determined uniquely [33]. From the view point of application, the superprism phenomenon [35] is very important, which manifests itself at a specific equi-energy surface portion of a particular band. There, the propagation direction of light going inside a PC through the interface varies remarkably with small change of wavelength of the incident light. This phenomenon showing anmalously large dispersion of wavelength versus incident angle enables one to develop, in principle, a new type of device of wavelength dispersion with the size much reduced as compared to the conventional ones.

Similarly, a phenomenon called supercollimation has also been found in a PC [36]: a light wave with a finite cross-section can propagate in a PC with the size unchanged, which occurs also at a specific band.

When we formally apply Snell's law for an anomalous band or at wavelengths exhibiting negative v_g, n becomes negative. Utilizing this sort of band, one can create a superlens (an ideal lens in a sense) [37]. However, in order to implement this kind of lens, v_g must be negative over a wide range of wavelengths for all incident angles. This requisite is very severe, but it is reported that this is still possible [38].

2.8 Application of Photonic Crystals

Because of their unique properties, PCs are very attractive for exploring new physical phenomena and developing novel and important devices in the field of optoelectronics as well. In this survey, we focus rather on the latter. Concerning the latter, some simple passive optical components such as narrow-band filters, unique polarizers and wonderful superprisms described in the preceding section have already been developed, and some others are expected to be developed one after another from now on and to be commercially avalable in the near future.

Let us mention some applications below, other than the above-mentioned simple components. First, we describe applications to nonlinear optical phenomena, or the related active components. By utilizing the unique PBS in either 2D or 3D PCs, the phase-matching (PM) can be easily fulfilled at several wavelengths; it is well known that the PM is very important in nonlinear optical phenomena, particularly in coherent phenomena such as second harmonic generation (SHG). Furthermore, a novel type of PM can be expected to be utilized in PCs including 1D in this case [39]. That is, an umklapp process can be utilized, since the reciprocal vector $\boldsymbol{q}$ is on the same order in magnitude with the relevant wavevectors involved in the phenomenon under study; it has already been proved that this process is useful in SHG for a 1D periodic system (1D PC). We describe below the SHG case as an example,

$$\boldsymbol{k}_s = 2\boldsymbol{k}_i \pm \boldsymbol{q}, \qquad \omega_s = 2\omega_i, \tag{2.33}$$

where $\boldsymbol{k}_i$ and $\boldsymbol{k}_s$, and ω_i and ω_s are the wavevectors, and angular frequencies of incident (fundamental) and second-harmonic light waves, respectively.

As for other active components, several kinds of unique 2D PC lasers have been successfully developed. Those are categorized into two types, i.e., one [40] utilizing an ultra-small PC cavity, or a point defect already mentioned in Sect. 2.7.2, and the other [26–28] utilizing a small v_g. So, some of them may be in practical use soon. The first type of laser has prominent characteristic of being extremely small. The latter type of lasers include a sort of vertically emitted laser excited by an electrical current [27], a 2D laser with a low threshold, and an small size of laser. The same is also true with a light emitter embedded in a PC slab, where the extraction efficiency of light has already been demonstrated to be greatly improved compared to the record obtained thus far [41].

PC fibers, both of PBG type and refractive-index confined, are also very attractive. The latter type of optical fiber is considered to be an ideal transmission line for light, so it will be in practical use, depending on the case. Independently of the above application, it has been demonstrated that light can be generated over an extremely broad range of wavelengths by illuminating the latter type, i.e., with the silica core in the central portion, using an ultra-short, i.e., subpicosecond light pulse. This phenomenon should be utilized as a white-light source for spectroscopic use [42].

Next, the Smith–Purcell radiation with use of a PC is also attractive and important, because it has the possibility of developing a compact wavelength-tunable coherent light source in a range from far- to near-infrared; the Smith–Purcell effect is such that coherent light is radiated due to interaction between high-speed traveling charged particles and a periodic system such as a grating. It has been predicted theoretically [43] and has already been verified very recently that use of a PC instead of a grating enables one to observe the radiation much more efficiently than the previous cases [44]. This topic is treated in Sect. 10.7.

Finally, we would like to mention one of the most important applications of a PC, i.e., development of ultra-fast and ultra-compact integrated light circuits; such a development is crucially important for future telecommunications. It is considered that such a circuit based on PCs may be developed in the future. For this purpose, there are a few key components to be developed. One is a novel type of light waveguide that is capable of being bent sharply. This can be achieved by utilizing a PC-based waveguide. The PC-based waveguide is constituted of line defect modes in a 2D or 3D PC; in the case of a 2D PC of triangular lattice of air hole type, for example, a line-defect is produced by leaving a single line of air holes imperforated along one of six equivalent $\Gamma - K$ directions. Importantly, in the PC-based waveguide the light wave is confined due to the PBG, so that it can be bent sharply without any propagation loss, while light is confined by total reflection in the lateral plane in the conventional dielectric waveguide; in the case of PC-slab waveguides, the vertical confinement of light is achieved by the total reflection. Another is the compact demultiplexer, but we do not go into details here (see Chaps. 11 and 12).

From the practical point of view, the planar type of integrated circuit, in particular, made of PC slabs, has the advantage over the 3D type. This is because it is much easier to fabricate the planar type than a 3D one. So, up until now many investigations have been devoted to this type, both thoretically and experimentally. As a result, it may be possible to finally develop such an ultra-small integrated circuit. The details will be presented also in Chap. 12.

References

1. E. Yablonovitch, Phys. Rev. Lett. **58**, 2059 (1987)
2. E. Yablonovitch, T. J. Gmitter, and K. M. Leung, Phys. Rev. Lett. **67**, 2295 (1991)
3. K. L. Kliewer and R. Fuchs, Phys. Rev. **144**, 495 (1966)
4. K. L. Kliewer and R. Fuchs, Phys. Rev. **150**, 573 (1966)
5. M. Inoue and K. Ohtaka, Phys. Rev. B**26**, 3487 (1982)
6. W. M. Robertson, G. Arjavalingham, R. D. Meade, K. D. Brommer, A. M. Rappe, and J. D. Joannopoulos, Phys. Rev. Lett. **68**, 2023 (1992)
7. M. Wada, K. Sakoda and K. Inoue, Phys. Rev. B**52**, 16297 (1995)
8. K. Sakoda, Phys. Rev B**51**, 4672 (1995)
9. K. Inoue, M. Wada, K. Sakoda, A. Yamanaka, M. Hayashi, and J. W. Haus, Jpn. J. Appl. Phys. **33**, L1463 (1994)
10. M. L. Cohen and T. K. Bergstresser, Phys. Rev. **141**, 789 (1966)
11. E. Yablonovitch, T. J. Gmitter, and K. M. Leung, Phys. Rev. Lett. **67**, 3380 (1991)
12. H. S. Sozuer, J. W. Haus, and R. Inguva, Phys. Rev. B**45**, 13,962 (1992)
13. H. S. Sozuer and J. W. Haus, J. Opt. Soc. Am. B **10**, 296 (1993).
14. S. Johnson, S. Fan, P. R. Villeneuve, and J. D. Joannopoulos, Phys. Rev. B**60**, 5751 (1999)
15. S. Noda, M. Imada, M. Okano, S. Ogawa, M. Mochizuki, and A. Chutinan, IEEE J. Quantum Electronics, **38**, 726 (2002)
16. A. Makis, S. Fan and J. D. Joannopoulos, Phys. Rev. B**58**, 4809 (1998)
17. S. Johnson, P. R. Villeneuve, S. Fan, and J. D. Joannopoulos, Phys. Rev. B**62**, 8212 (2000)
18. T. Baba, N. Fukaya, and J. Yonekura, Electron. Lett. **27**, 654 (1999)
19. M. Scalora, R. J. Flynn, S. B. Reinhardt, R. L. Fork, M. D. Tocci, M. J. Bloemer, C. M. Bowden, H. S. Ledbetter, J. M. Bendickson, J. P. Dowling, and R. P. Leavitt, Phys. Rev. E**54**, R1078 (1996)
20. A. Imhof, W. L. Vos, R. Sprik, and Ad. Lagendijk, Phys. Rev. Lett. **83**, 2942 (1999)
21. K. Inoue, N. Kawai, Y. Sugimoto, N. Ikeda, N. Carlsson, and K. Asakawa, Phys. Rev. B**65**, 121308R (2002)
22. K. Sakoda, Opt. Express, **4**, 167 (1999) [http://epubs.osa.org/opticsexpress]
23. M. Scalora, M. J. Bloemer, A. S. Manka, J. P. Powling, C. M. Bourden, R. Viswanathan, and J. W. Haus, Phys. Rev. A**56**, 3166 (1999)
24. M. Centini, C. Sibilia, M. Scalora, G. D'Aguanno, M. Bertolotti, M. J. Bloemer, C. M. Bowden, and I. Nefedov, Phys. Rev. E**60**, 4891 (1999)
25. Y. Dumeige, I. Sagnes, P. Monnier, P. Vidakovic, I. Abram, C. Meriadec, and A. Levenson, Phys. Rev. Lett., **89**, 043901 (2002)
26. K. Inoue, M. Sasada, J. Kawamata, K. Sakoda, and J. W. Haus, Jpn. J. Appl. Phys., **38**, L157 (1999)
27. M. Imada, S. Noda, A. Chutinan, T. Tokuda, M. Murata, and G. Sasaki, Appl. Phys. Lett., **75**, 316 (1999)
28. M. Notomi, H. Suzuki, and T. Tamura, Appl. Phys. Lett. **78**, 1325 (2001)
29. K. Sakoda, K. Ohtaka, and T. Ueta, Opt. Express, **4**, 481 (1999)
30. M. Notomi, K. Yamada, A. Shinya, J. Takahashi, C. Takahashi, and I. Yokohama, Phys. Rev. Lett. **87**, 253902 (2001)

31. K. M. Ho, C. T. Chan, and C. M. Soukoulis, Phys. Rev. Lett. **65**, 3152 (1990)
32. M. C. Netti, C. E. Finlayson, J. J. Baumberg, M. D. B. Charlton, M. E. Zoorob, J. S. Wilkinson, G. J. Parker, Appl. Phys. Lett. **81**, 3927 (2002)
33. K. Ohtaka, T. Ueta, and Y. Tanabe, J. Phys. Soc. Jpn. **65**, 3068 (1996)
34. M. Netti, A. Harris, J. J. Baumberg, D. M. Whittakar, M. B. D. Charlton, M. E. Zoorob, and G. J. Parker, Phys. Rev. Lett. **86**, 1526 (2001)
35. H. Kosaka, T. Kawashima, A. Tomita, M. Notomi, T. Tamamura, T. Sato, and S. Kawakami, Phys, Rev. B**58**, R10096 (1998)
36. H. Kosaka, T. Kawashima, A. Tomita, M. Notomi, T. Tamamura, T. Sato, and S. Kawakami, Appl. Phys. Lett. **74**, 1212 (1999)
37. M. Notomi, Phys. Rev. B**62**, 10696 (2000)
38. C. Luo, S. G. Johnson, and J. D. Joannopoulos, Appl. Phys. Lett. **81**, 2352 (2002)
39. N. Bloembergen and A. J. Sievers, Appl. Phys. Lett. **17**, 483 (1970)
40. O. Painter, R. K. Lee, A. Yariv, A. Scherer, J. D. O'Brian, P. D. Dapkus, and I. Kim, Science, **284**, 1819 (1999)
41. T. Baba, K. Inoshita, H. Tanaka, J. Yonekura, M. Ariga, A. Matsutani, T. Miyamoto, F. Koyama, and K. Iga, J. Lightwave Technol., **17**, 2113 (1999)
42. J. K. Ranka, R. S. Windeler, and A. J. Stentz, Optics Lett. **25**, 25 (2000)
43. K. Ohtaka and S. Yamaguti, Optical and Quantum Electronics, **34**, 135 (2002); S. Yamaguti, J. Inoue, O. Haeberlé and K. Ohtaka, Phys. Rev. B **66**, 195202 (2002)
44. K. Yamamoto et al., Abstract of International Workshop on Photonic and Electromagnetic Crystal Structures, ed. S. Yu Lin (UCLA, 2002), p. 36

3 Theory I: Basic Aspects of Photonic Bands

K. Ohtaka

This chapter is devoted to presenting the fundamental aspects of photonic crystals (PCs), especially for newcomers to the field of PCs, who are hoping to gain a basic understanding of it. We have in mind mainly readers who are already equiped with the knowledge of the band structure of electrons. Although this chapter is intended to be as self-contained as possible, I would like to cite two articles which look useful to readers of this section. They are the monographs by Joannopoulos et al. [1] for earlier work and by Sakoda [2] for recent studies. A number of books and journals editing the work of early studies are also useful, among which are the book edited by Soukoulis [3] and the feature issues of Journal of Modern Optics [4] and Journal of The Optical Society of America [5]. For the band structure of electrons, the book, e.g., by Ibach and Lüth [6] will be useful among others.

If you are fairly well familiar with photonic crystals, you may skip this chapter and jump directly to the next chapter, which treats the basic but advanced topics of PCs. Some of the topics of this chapter will appear once again there in a little more detail. This chapter should thus serve as an introduction to the next chapter.

3.1 2D or 3D Photonic Band Structure

3.1.1 Full Maxwell's Equations

Since a detailed description was already given for 1D PCs in Chap. 1, we concentrate mainly on PCs of higher dimension. In a 2D or 3D PC, its dielectric function $\varepsilon(\boldsymbol{r})$ is Fourier expanded using 2D or 3D reciprocal lattice vectors, which characterizes the periodicity of the PC:

$$\varepsilon(\boldsymbol{r}) = \sum_{\boldsymbol{h}} \varepsilon_{\boldsymbol{h}} \mathrm{e}^{\mathrm{i}\boldsymbol{h}\cdot\boldsymbol{r}}, \tag{3.1}$$

with

$$\varepsilon_{\boldsymbol{h}} = \frac{1}{v_{\mathrm{c}}} \int_{\text{unit cell}} \varepsilon(\boldsymbol{r}) \mathrm{e}^{-\mathrm{i}\boldsymbol{h}\cdot\boldsymbol{r}}, \tag{3.2}$$

v_{c} being the volume of the unit cell. Photons suffer periodic scattering and the photonic band structure (PBS) appears in the dispersion relations of photons.

Scattering between electromagnetic waves takes place among the set of plane waves $\boldsymbol{k}$, $\boldsymbol{k}+\boldsymbol{h}_1$, $\boldsymbol{k}+\boldsymbol{h}_2$, $\cdots$, where $\boldsymbol{h}_1$, etc. are reciprocal lattice vectors of the photonic crystal. Thus, $\boldsymbol{k}$ specifies a set of plane waves. The plane waves in the two different sets $\boldsymbol{k}_1$ and $\boldsymbol{k}_2$ in the first Brillouin zone (BZ) never mix; the wavevector $\boldsymbol{k}$ within the first BZ thus exhausts all possible photonic band states. In App. A, we have given a brief description of reciprocal lattice vectors and some illustrations of the first BZ for typical lattice structures of 2D and 3D PCs.

The scattering of the waves is governed by Maxwell's equations

$$\begin{aligned} \nabla \times \boldsymbol{E}(\boldsymbol{r},t) &= -\frac{\partial}{\partial t}\boldsymbol{B}(\boldsymbol{r},t), \\ \nabla \times \boldsymbol{B}(\boldsymbol{r},t) &= \mu_0 \frac{\partial}{\partial t}\boldsymbol{D}(\boldsymbol{r},t). \end{aligned} \tag{3.3}$$

Here $\boldsymbol{E}(\boldsymbol{r},t)$ is the electric field and $\boldsymbol{B}(\boldsymbol{r},t)$ is the magnetic flux density at the point $\boldsymbol{r}$ and time t. μ_0 is the permeability of vacuum ($\varepsilon_0\mu_0 = 1/c^2$ with ε_0 the dielectric constant of vacuum). Throughout this book, we consider nonmagnetic PCs, where the relation

$$\boldsymbol{B}(\boldsymbol{r}) = \mu_0 \boldsymbol{H}(\boldsymbol{r})$$

holds between $\boldsymbol{B}(\boldsymbol{r})$ and the magnetic field $\boldsymbol{H}(\boldsymbol{r})$. The symbols $\boldsymbol{B}(\boldsymbol{r})$ and $\boldsymbol{H}(\boldsymbol{r})$ are both used, however. The vector field $\boldsymbol{D}(\boldsymbol{r},t)$ is the displacement field, which is related to $\boldsymbol{E}(\boldsymbol{r},t)$ through $\varepsilon(\boldsymbol{r})$.

In a stationary state the fields of frequency ω may be written as

$$\boldsymbol{E}(\boldsymbol{r},t) = \boldsymbol{E}(\boldsymbol{r},\omega)\mathrm{e}^{-\mathrm{i}\omega t} \tag{3.4}$$

etc. Equation (3.3) then becomes

$$\begin{aligned} \nabla \times \boldsymbol{E}(\boldsymbol{r}) &= \mathrm{i}\omega \boldsymbol{B}(\boldsymbol{r}), \\ \nabla \times \boldsymbol{B}(\boldsymbol{r}) &= -\mathrm{i}\varepsilon_0\mu_0\omega\varepsilon(\boldsymbol{r})\boldsymbol{E}(\boldsymbol{r}). \end{aligned} \tag{3.5}$$

Here and in what follows we often drop the frequency argument ω in various quantities and denote them simply as $\boldsymbol{E}(\boldsymbol{r})$, $\varepsilon(\boldsymbol{r})$, etc., which should be taken as $\boldsymbol{E}(\boldsymbol{r},\omega)$, $\varepsilon(\boldsymbol{r},\omega)$, etc.

Using this convention, the fields $\boldsymbol{D}(\boldsymbol{r})$ and $\boldsymbol{E}(\boldsymbol{r})$ are linked by

$$\boldsymbol{D}(\boldsymbol{r}) = \varepsilon_0\varepsilon(\boldsymbol{r})\boldsymbol{E}(\boldsymbol{r}). \tag{3.6}$$

This is called a local relation connecting $\boldsymbol{D}(\boldsymbol{r})$ and $\boldsymbol{E}(\boldsymbol{r})$ of the same position by way of $\varepsilon(\boldsymbol{r})$ of that point. An unabbreviated form of this equation

$$\boldsymbol{D}(\boldsymbol{r},\omega) = \varepsilon_0\varepsilon(\boldsymbol{r},\omega)\boldsymbol{E}(\boldsymbol{r},\omega) \tag{3.7}$$

will be worth reproducing to emphasize the ω dependence of $\varepsilon(\boldsymbol{r})$ in a strict sense. Note that for the relation (3.7) to hold in Fourier space, $\boldsymbol{D}(\boldsymbol{r},\, t)$ and $\boldsymbol{E}(\boldsymbol{r},\, t)$ are connected through the nonlocal relation in the t domain given by

$$\boldsymbol{D}(\boldsymbol{r}, t) = \varepsilon_0 \int_{-\infty}^{t} \mathrm{d}t' \varepsilon(\boldsymbol{r}, t-t') \boldsymbol{E}(\boldsymbol{r}, t')$$

using the nonlocal (in t) dielectric function $\varepsilon(\boldsymbol{r}, t-t')$. The nonlocality in the t domain is a prerequisite for the relations (3.6) or (3.7).

For a PC of metal, where the optical properties are determined basically by conduction electrons, we often use the equation for $\nabla \times \boldsymbol{B}(\boldsymbol{r})$ in the form

$$\nabla \times \boldsymbol{B}(\boldsymbol{r}) = \mu_0 \boldsymbol{J}(\boldsymbol{r}) - \mathrm{i}\omega\varepsilon_0\mu_0\varepsilon_{\mathrm{others}}(\boldsymbol{r})\boldsymbol{E}(\boldsymbol{r}), \tag{3.8}$$

where the electric current density $\boldsymbol{J}(\boldsymbol{r})$ is defined by

$$\boldsymbol{J}(\boldsymbol{r}) = \sigma(\boldsymbol{r})\boldsymbol{E}(\boldsymbol{r}) \tag{3.9}$$

with the conductivity $\sigma(\boldsymbol{r})$ of conduction electrons, which is ω dependent. The dielectric constant $\varepsilon_{\mathrm{others}}(\boldsymbol{r})$ in (3.8) incorporates the polarizations other than conduction electrons, such as those of core and valence electrons of atoms, lattice vibrations and so on. We may collect the contributions of conduction electrons into the dielectric function $\varepsilon(\boldsymbol{r})$ by rewriting

$$\sigma(\boldsymbol{r})\boldsymbol{E}(\boldsymbol{r}) - \mathrm{i}\omega\varepsilon_0\varepsilon_{\mathrm{others}}\boldsymbol{E}(\boldsymbol{r}) = -\mathrm{i}\omega\varepsilon_0\varepsilon(\boldsymbol{r})\boldsymbol{E}(\boldsymbol{r}), \tag{3.10}$$

with

$$\varepsilon(\boldsymbol{r}) = \varepsilon_{\mathrm{others}}(\boldsymbol{r}) - \frac{\sigma(\boldsymbol{r})}{\mathrm{i}\omega\varepsilon_0}. \tag{3.11}$$

The dielectric function $\varepsilon(\boldsymbol{r})$ now involves all the contributions including that of conduction electrons. To summarize, if we use the dielectric function $\varepsilon(\boldsymbol{r})$ defined by (3.11), we can recover (3.5) and (3.7). It is important, however, to note that real $\sigma(\boldsymbol{r})$ makes the dielectric function $\varepsilon(\boldsymbol{r})$ complex by adding in (3.11) an imaginary part. Since the resisitivity, the inverse of $\sigma(\boldsymbol{r})$, causes an energy dissipation in the form of Joule heat, this should be so.

Let us return to (3.3). The two equations are combined to give

$$\nabla \times \nabla \times \boldsymbol{E}(\boldsymbol{r}) - \frac{\omega^2}{c^2}\varepsilon(\boldsymbol{r})\boldsymbol{E}(\boldsymbol{r}) = 0, \tag{3.12}$$

or

$$\nabla \times \left(\frac{1}{\varepsilon(\boldsymbol{r})}\nabla \times \boldsymbol{B}(\boldsymbol{r})\right) - \frac{\omega^2}{c^2}\boldsymbol{B}(\boldsymbol{r}) = 0, \tag{3.13}$$

using $\varepsilon_0\mu_0 = 1/c^2$. Equation (3.13) turns to the equation for $\boldsymbol{H}(\boldsymbol{r})$, if $\boldsymbol{B}(\boldsymbol{r})$ is replaced by $\boldsymbol{H}(\boldsymbol{r})$. From (3.7) and (3.12), we obtain the equation for $\boldsymbol{D}(\boldsymbol{r})$:

$$\nabla \times \left(\nabla \times (\frac{1}{\varepsilon(\boldsymbol{r})}\boldsymbol{D}(\boldsymbol{r}))\right) - \frac{\omega^2}{c^2}\boldsymbol{D}(\boldsymbol{r}) = 0. \tag{3.14}$$

These three equations are the starting equations employed ordinarily in the study of PCs.

They were derived by considering only two of the full set of Maxwell's equations. The other two, not explicit so far, are

$$\begin{aligned} \nabla \cdot \boldsymbol{D}(\boldsymbol{r}) &= 0, \\ \nabla \cdot \boldsymbol{B}(\boldsymbol{r}) &= 0, \end{aligned} \tag{3.15}$$

the second equation being equivalent to $\nabla \cdot \boldsymbol{H}(\boldsymbol{r}) = 0$. Therefore, the question should be answered whether the solution of (3.12) through (3.14) is consistent with (3.15). The answer is simple: if we operate with $\nabla \cdot$ on (3.12) through (3.14), the identity $\nabla \cdot (\nabla \times) = 0$ leads to

$$\frac{\omega^2}{c^2} \nabla \cdot \varepsilon(\boldsymbol{r}) \boldsymbol{E}(\boldsymbol{r}) = 0, \quad \text{and} \quad \frac{\omega^2}{c^2} \nabla \cdot \boldsymbol{B}(\boldsymbol{r}) = 0$$

the former yielding the first and the latter giving the second of (3.15). Therefore, any solution of $\omega \neq 0$ of (3.12) through (3.14) automatically satisfies the full set of Maxwell's equations; the PB solutions from them are exact solutions of Maxwell's equations. However, when we consider the completeness of the vector fields, we should take into account the solutions of $\omega = 0$ in addition to the PB solutions of $\omega \neq 0$. This topic will be treated in Sect. 4.5.1.

3.1.2 Plane-Wave Expansion Method

To obtain PBS, let us concentrate on the solution $\boldsymbol{E}_{\boldsymbol{k}}(\boldsymbol{r})$ of (3.12). We express it as a superposition of plane waves. This is the plane-wave expansion method used widely. We express $\boldsymbol{E}_{\boldsymbol{k}}(\boldsymbol{r})$ in the form of the Bloch type

$$\boldsymbol{E}_{\boldsymbol{k}}(\boldsymbol{r}) = \sum_{\boldsymbol{h}} \boldsymbol{e}_{\boldsymbol{k}}(\boldsymbol{h}) \exp[\mathrm{i}(\boldsymbol{k} + \boldsymbol{h}) \cdot \boldsymbol{r}]. \tag{3.16}$$

Here $\boldsymbol{e}_{\boldsymbol{k}}(\boldsymbol{h})$ is an unknown vector amplitude of the plane wave $(\boldsymbol{k}+\boldsymbol{h})$, which is to be determined so that $\boldsymbol{E}_{\boldsymbol{k}}(\boldsymbol{r})$ is a solution of Maxwell's equations (3.12). Note that the form given by (3.16) satisfies the Bloch theorem

$$\boldsymbol{E}_{\boldsymbol{k}}(\boldsymbol{r} + \boldsymbol{R}) = \mathrm{e}^{\mathrm{i}\boldsymbol{k} \cdot \boldsymbol{R}} \boldsymbol{E}_{\boldsymbol{k}}(\boldsymbol{r}), \tag{3.17}$$

for an arbitrary lattice translation vector $\boldsymbol{R}$, because

$$\boldsymbol{h} \cdot \boldsymbol{R} = 2\pi \times (\text{integer})$$

by definition.

Naturally, the reciprocal lattice vectors $\boldsymbol{h}$ and the geometry of the BZ change depending on the lattice structure. For example, in a 2D PC like the regular array of cylinders directed in the z direction, the vectors $\boldsymbol{h}$ are 2D vectors defined in the xy plane and the z dependence of the field of (3.12) is expressed by a single plane wave $\mathrm{e}^{\mathrm{i}k_z z}$. Once the electric field is obtained, the magnetic field is derived from it using (3.5).

Substituting (3.1) and (3.16) into (3.12) and putting the coefficients of each of the plane waves $\{e^{i(\boldsymbol{k}+\boldsymbol{h})\cdot\boldsymbol{r}}\}$ to zero, we find

$$\sum_{\boldsymbol{h}'}\left(-[(\boldsymbol{k}+\boldsymbol{h})\times(\boldsymbol{k}+\boldsymbol{h}')\times]\delta_{\boldsymbol{h}\boldsymbol{h}'}-\frac{\omega^2}{c^2}\varepsilon_{\boldsymbol{h}-\boldsymbol{h}'}\right)\boldsymbol{e}_{\boldsymbol{k}}(\boldsymbol{h}')=0, \tag{3.18}$$

where we have used

$$\nabla\times\boldsymbol{e}_{\boldsymbol{k}}(\boldsymbol{h})e^{i(\boldsymbol{k}+\boldsymbol{h})\cdot\boldsymbol{r}}=i(\boldsymbol{k}+\boldsymbol{h})\times\boldsymbol{e}_{\boldsymbol{k}}(\boldsymbol{h})e^{i(\boldsymbol{k}+\boldsymbol{h})\cdot\boldsymbol{r}}.$$

Equation (3.18) consists of three equations for the three cartesian components x, y and z. For concreteness, let us rewrite the term involving vector products. Using the identity

$$\boldsymbol{a}\times(\boldsymbol{b}\times\boldsymbol{c})=(\boldsymbol{a}\cdot\boldsymbol{c})\boldsymbol{b}-(\boldsymbol{a}\cdot\boldsymbol{b})\boldsymbol{c},$$

we find

$$\begin{aligned}&-(\boldsymbol{k}+\boldsymbol{h})\times((\boldsymbol{k}+\boldsymbol{h})\times\boldsymbol{e}_{\boldsymbol{k}}(\boldsymbol{h}))\\&\qquad=|\boldsymbol{k}+\boldsymbol{h}|^2\boldsymbol{e}_{\boldsymbol{k}}(\boldsymbol{h})-(\boldsymbol{k}+\boldsymbol{h})\left((\boldsymbol{k}+\boldsymbol{h})\cdot\boldsymbol{e}_{\boldsymbol{k}}(\boldsymbol{h})\right).\end{aligned}$$

The term coming from $\nabla\times\nabla\times$ of (3.18) then becomes

$$\left[\begin{pmatrix}|\boldsymbol{k}_{\boldsymbol{h}}|^2 & 0 & 0\\ 0 & |\boldsymbol{k}_{\boldsymbol{h}}|^2 & 0\\ 0 & 0 & |\boldsymbol{k}_{\boldsymbol{h}}|^2\end{pmatrix}-\begin{pmatrix}(\boldsymbol{k}_{\boldsymbol{h}})_x(\boldsymbol{k}_{\boldsymbol{h}})_x & (\boldsymbol{k}_{\boldsymbol{h}})_x(\boldsymbol{k}_{\boldsymbol{h}})_y & (\boldsymbol{k}_{\boldsymbol{h}})_x(\boldsymbol{k}_{\boldsymbol{h}})_z\\ (\boldsymbol{k}_{\boldsymbol{h}})_y(\boldsymbol{k}_{\boldsymbol{h}})_x & (\boldsymbol{k}_{\boldsymbol{h}})_y(\boldsymbol{k}_{\boldsymbol{h}})_y & (\boldsymbol{k}_{\boldsymbol{h}})_y(\boldsymbol{k}_{\boldsymbol{h}})_z\\ (\boldsymbol{k}_{\boldsymbol{h}})_z(\boldsymbol{k}_{\boldsymbol{h}})_x & (\boldsymbol{k}_{\boldsymbol{h}})_z(\boldsymbol{k}_{\boldsymbol{h}})_y & (\boldsymbol{k}_{\boldsymbol{h}})_z(\boldsymbol{k}_{\boldsymbol{h}})_z\end{pmatrix}\right]$$
$$\times\begin{pmatrix}e_{\boldsymbol{k}}(\boldsymbol{h})_x\\ e_{\boldsymbol{k}}(\boldsymbol{h})_y\\ e_{\boldsymbol{k}}(\boldsymbol{h})_z\end{pmatrix},$$

using the simplified notation

$$\boldsymbol{k}_{\boldsymbol{h}}=\boldsymbol{k}+\boldsymbol{h}.$$

From (3.18) the secular equation for the eigenvalues ω^2 is given by

$$\det\left|-\delta_{\boldsymbol{h}\boldsymbol{h}'}[(\boldsymbol{k}+\boldsymbol{h})\times(\boldsymbol{k}+\boldsymbol{h}')\times]-\frac{\omega^2}{c^2}\varepsilon_{\boldsymbol{h}-\boldsymbol{h}'}\boldsymbol{I}_{\boldsymbol{h}\boldsymbol{h}'}\right|=0. \tag{3.19}$$

The matrix of the determinant consists of block matrices, each of dimension 3×3, and this form shows the $(\boldsymbol{h}\boldsymbol{h}')$ block of 3×3 matrix using unit matrix

$$(\boldsymbol{I}_{\boldsymbol{h}\boldsymbol{h}'})_{ij}=\delta_{ij}.$$

Explicitly it is given by

$$\delta_{\boldsymbol{h}\boldsymbol{h}'}\left[\begin{pmatrix} |\boldsymbol{k_h}|^2 & 0 & 0 \\ 0 & |\boldsymbol{k_h}|^2 & 0 \\ 0 & 0 & |\boldsymbol{k_h}|^2 \end{pmatrix} - \begin{pmatrix} (\boldsymbol{k_h})_x(\boldsymbol{k_h})_x & (\boldsymbol{k_h})_x(\boldsymbol{k_h})_y & (\boldsymbol{k_h})_x(\boldsymbol{k_h})_z \\ (\boldsymbol{k_h})_y(\boldsymbol{k_h})_x & (\boldsymbol{k_h})_y(\boldsymbol{k_h})_y & (\boldsymbol{k_h})_y(\boldsymbol{k_h})_z \\ (\boldsymbol{k_h})_z(\boldsymbol{k_h})_x & (\boldsymbol{k_h})_z(\boldsymbol{k_h})_y & (\boldsymbol{k_h})_z(\boldsymbol{k_h})_z \end{pmatrix}\right]$$

$$-\frac{\omega^2}{c^2}\begin{pmatrix} \varepsilon_{\boldsymbol{h}\boldsymbol{h}'} & 0 & 0 \\ 0 & \varepsilon_{\boldsymbol{h}\boldsymbol{h}'} & 0 \\ 0 & 0 & \varepsilon_{\boldsymbol{h}\boldsymbol{h}'} \end{pmatrix}, \tag{3.20}$$

using the simplified symbol

$$\varepsilon_{\boldsymbol{h}\boldsymbol{h}'} = \varepsilon_{\boldsymbol{h}-\boldsymbol{h}'}.$$

The matrix and eigenvectors of (3.18) are both labeled by the reciprocal lattice vector $\boldsymbol{h}$ and the cartesian index j ($j = 1, 2$ and 3 for x, y and z, respectively). If we take into account the N reciprocal lattice vectors, the matrix involved in the seculer equation (3.19) has the dimension of $3N \times 3N$.

In place of the electric field, we can similarly obtain the secular equations for the field $\boldsymbol{B_k}(\boldsymbol{r})$, $\boldsymbol{H_k}(\boldsymbol{r})$ or $\boldsymbol{D_k}(\boldsymbol{r})$ by returning to (3.13) and (3.14), respectively. Sometimes they are easier to handle than that for $\boldsymbol{E_k}(\boldsymbol{r})$. As is obvious in (3.13) and (3.14), we need in these equations the Fourier components of the inverse of the dielectric constant $\varepsilon^{-1}(\boldsymbol{r})$

$$\varepsilon^{-1}(\boldsymbol{r}) = \sum_{\boldsymbol{h}} \varepsilon^{-1}(\boldsymbol{h})\mathrm{e}^{\mathrm{i}\boldsymbol{h}\cdot\boldsymbol{r}}, \tag{3.21}$$

with

$$\varepsilon^{-1}(\boldsymbol{h}) = \frac{1}{v_{\mathrm{c}}}\int_{\text{unit cell}} \varepsilon^{-1}(\boldsymbol{r})\mathrm{e}^{-\mathrm{i}\boldsymbol{h}\cdot\boldsymbol{r}}. \tag{3.22}$$

The eigenvalues for the PBs are the solutions for ω^2 of the secular equation (3.19). If we plot them as functions of wavevector $\boldsymbol{k}$ in the first BZ, we obtain PBS. The secular equations for $\boldsymbol{E_k}(\boldsymbol{r})$, $\boldsymbol{B_k}(\boldsymbol{r})$, $\boldsymbol{H_k}(\boldsymbol{r})$ and $\boldsymbol{D_k}(\boldsymbol{r})$ yield the same PB eigenvalues.

3.2 Parity of Mirror Reflection of 2D PBS

We have given typical exmples of PBS in Chap. 2 for prototypical 1D, 2D and 3D PCs. We can speak about the dimension of a PC by the dependence of $\varepsilon(\boldsymbol{r})$ on x, y, z of $\boldsymbol{r}$. In this section we describe in a little more detail the band calculation of a 2D PC of a square lattice of dielectric rods of circular cross-section. This is because 2D PCs are treated repeatedly in this book.

With the x and y axes chosen parallel to the sides of the square unit-cell, the center position of the cylinder of the (n, m) site is

$$\boldsymbol{R}_{nm} = a(n, m),$$

a being the lattice constant. We restrict ourselves to the case of $\boldsymbol{k} = (k_x, k_y, 0)$. In this section, therefore, the wavevector and reciprocal lattice vectors are all two-dimensional. We denote them as $\boldsymbol{k}_\| = (k_x, k_y)$ and $\boldsymbol{h}_\|$ for 2D wave vectors, for simplicity.

The system has a mirror symmetry in the xy plane. When a Bloch wave propagates in the xy plane with $k_z = 0$, $\boldsymbol{E}_{\boldsymbol{k}_\|}(\boldsymbol{r})$ and $\boldsymbol{B}_{\boldsymbol{k}_\|}(\boldsymbol{r})$ involved in it reflect this mirror symmetry and we can classify the photonic bands according to their parities.

An even parity mode is a mode which has an electric field invariant under mirror reflection, namely $\boldsymbol{E}_{\boldsymbol{k}_\|}(\boldsymbol{r})$ and the vector plane-waves involved are polarized within the xy plane. Its parity is even, because a vector in the xy plane is invariant by the mirror in the xy plane. The magnetic field is directed in the z direction, perpendicularly to the electric field. Therefore a plane wave $\mathrm{e}^{\mathrm{i}(\boldsymbol{k}_\|+\boldsymbol{h}_\|)\cdot\boldsymbol{r}}$ involved in the even-parity PB is

$$\boldsymbol{e}_{\boldsymbol{k}_\|}(\boldsymbol{h}_\|)\mathrm{e}^{\mathrm{i}(\boldsymbol{k}_\|+\boldsymbol{h}_\|)\cdot\boldsymbol{r}} \qquad \boldsymbol{e}_{\boldsymbol{k}_\|}(\boldsymbol{h}_\|) \text{ polarized in the } xy \text{ plane}$$

$$\boldsymbol{b}_{\boldsymbol{k}_\|}(\boldsymbol{h}_\|)\mathrm{e}^{\mathrm{i}(\boldsymbol{k}_\|+\boldsymbol{h}_\|)\cdot\boldsymbol{r}} \qquad \boldsymbol{b}_{\boldsymbol{k}_\|}(\boldsymbol{h}_\|) \text{ polarized in the } z \text{ direction,}$$

where $\boldsymbol{b}_{\boldsymbol{k}_\|}(\boldsymbol{h}_\|)$ is the amplitude of the plane-wave expansion of $\boldsymbol{B}_{\boldsymbol{k}_\|}(\boldsymbol{r})$. The polarization directions specified here applies to any $\boldsymbol{h}_\|$. Note that we use "even" and "odd", referring to the mirror property of the electric field. We often call the even-parity modes the transverse electric (TE) modes, because the electric field is polarized transverse to the reference axis of z.

The odd-parity modes have $\boldsymbol{E}_{\boldsymbol{k}_\|}(\boldsymbol{r})$ in the z direction. In this case the mirror reflection turns a z-directed polarization vector to a $-z$ directed vector. This is the transverse magnetic (TM) mode. For any $\boldsymbol{h}_\|$, $\boldsymbol{e}_{\boldsymbol{k}_\|}(\boldsymbol{h}_\|)$ and $\boldsymbol{b}_{\boldsymbol{k}_\|}(\boldsymbol{h}_\|)$ are directed parallel to the z axis and xy plane, respectively. Very often the TE and TM modes are called H mode and E mode, respectively. In summary, for $\boldsymbol{k}_\| = (k_x, k_y)$, two types of PBs arise:

$$\text{TE mode (H mode): } \boldsymbol{E}_{\boldsymbol{k}_\|}(\boldsymbol{r}) \parallel \hat{x},\, \hat{y}, \qquad \boldsymbol{B}_{\boldsymbol{k}_\|}(\boldsymbol{r}),\, \boldsymbol{H}_{\boldsymbol{k}_\|}(\boldsymbol{r}) \parallel \hat{z},$$
$$\text{TM mode (E mode): } \boldsymbol{E}_{\boldsymbol{k}_\|}(\boldsymbol{r}) \parallel \hat{z}, \qquad \boldsymbol{B}_{\boldsymbol{k}_\|}(\boldsymbol{r}),\, \boldsymbol{H}_{\boldsymbol{k}_\|}(\boldsymbol{r}) \parallel \hat{x},\, \hat{y}\,.$$

This classification of modes is shown in Fig. 3.1. Note that in, e.g., TE mode, we can only claim that the vector $\boldsymbol{E}_{\boldsymbol{k}_\|}(\boldsymbol{r})$ is lying in the xy plane; whether or not it is perpendicular to $\boldsymbol{k}_\|$ depends on the direction of $\boldsymbol{k}_\|$. Only when $\boldsymbol{k}_\|$ is parallel to the sides of the 2D square, we can assert that $\boldsymbol{E}_{\boldsymbol{k}_\|}(\boldsymbol{r})$ is indeed perpendicular to $\boldsymbol{k}_\|$ (see the discussion given below for the mirror symmetry in the xz plane).

The PBs of TE modes are obtained from the scalar wave equation when $\boldsymbol{B}_{\boldsymbol{k}_\|}(\boldsymbol{r})$ or $\boldsymbol{H}_{\boldsymbol{k}_\|}(\boldsymbol{r})$ is used, while the TM modes are derived from the scalar equation for $\boldsymbol{E}_{\boldsymbol{k}_\|}(\boldsymbol{r})$, both polarized in the z direction. Noting that $\boldsymbol{k}_\| + \boldsymbol{h}_\|$ is now directed within the xy plane and that the fields depend only on x and y, we see that

$$\boldsymbol{\nabla} \times \boldsymbol{\nabla} \times = -\left[\frac{\partial^2}{\partial x^2} + \frac{\partial^2}{\partial y^2}\right]$$

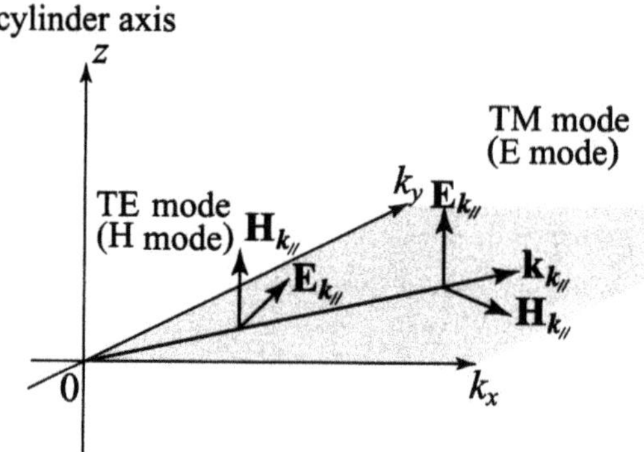

Fig. 3.1. Directions of the fields $\boldsymbol{E}_{\boldsymbol{k}_\parallel}(\boldsymbol{r})$ and $\boldsymbol{H}_{\boldsymbol{k}_\parallel}(\boldsymbol{r})$ of a 2D PC for $\boldsymbol{k}_\parallel$ in the xy plane. The z axis is parallel to the axes of the cylinders

when operated on a z directed field, because the term $\boldsymbol{\nabla}\boldsymbol{\nabla}\cdot$ of

$$\boldsymbol{\nabla} \times \boldsymbol{\nabla} = -\Delta + \boldsymbol{\nabla}\boldsymbol{\nabla}\cdot$$

vanishes due to the inner product between $(\boldsymbol{k}_\parallel + \boldsymbol{h}_\parallel)$ and a z-directed vector. Using the scalar symbols $B(\boldsymbol{r})$ and $E(\boldsymbol{r})$, we then find

$$\begin{aligned} &-\left[\frac{\partial}{\partial x}\frac{1}{\varepsilon(\boldsymbol{r})}\frac{\partial}{\partial x} + \frac{\partial}{\partial y}\frac{1}{\varepsilon(\boldsymbol{r})}\frac{\partial}{\partial y}\right] B(\boldsymbol{r}) - \left(\frac{\omega}{c}\right)^2 B(\boldsymbol{r}) = 0 \quad \text{(TE)} \\ &-\left[\frac{\partial^2}{\partial x^2} + \frac{\partial^2}{\partial y^2}\right] E(\boldsymbol{r}) - \left(\frac{\omega}{c}\right)^2 \varepsilon(\boldsymbol{r})E(\boldsymbol{r}) = 0 \quad \text{(TM)}, \end{aligned} \tag{3.23}$$

for TE and TM modes, respectively. Now assuming the Bloch form for the solution $B_{\boldsymbol{k}_\parallel}(\boldsymbol{r})$ and $E_{\boldsymbol{k}_\parallel}(\boldsymbol{r})$ using 2D wavevectors $\boldsymbol{k}_\parallel$ and $\boldsymbol{h}_\parallel$ in the same way as (3.16), the eigenvalue equation is straightforwardly obtained from (3.23). In our square lattice of cylinders, the PBs are specified by the 2D wavevector $\boldsymbol{k}_\parallel$ in the first BZ of the 2D square lattice:

$$-\frac{\pi}{a} \le k_x,\, k_y < \frac{\pi}{a}.$$

We can express the Fourier component $\varepsilon_{\boldsymbol{h}_\parallel \boldsymbol{h}'_\parallel}$ analytically using the Bessel function but this is not an essential point; for a regular array of cylinders of arbitrary cross-section, numerical evaluation of the Fourier components does not cause any serious trouble.

In this way we can get the PBS of a 2D PC numerically. The TM and TE bands of arrayed dielectic cylinders of circular cross-section are given in Fig. 2.11.

Finally we briefly discuss the case of nonzero k_z. For the state of

$$\boldsymbol{k} = (k_x, 0, k_z),$$

a mirror symmetry exists in the xz plane, because wavevector $\boldsymbol{k}$ as well as our PC itself remains unaltered by this mirror reflection. They are thus classified by the parity of this mirror reflection: even-parity modes which have their electric field vectors polarized in the xz plane and odd parity modes polarized in the direction y will then result. Consequently, for $\boldsymbol{k} = (k_x, 0, k_z)$ there appear two kinds of PBs, one which is capable of responding to incident light which has a polarization vector in the xz plane and the others responding to y-polarized light. Note that we have chosen the x and y axes parallel to the two sides of the 2D unit cell.

For a $\boldsymbol{k}$ off the xz plane, i.e., $\boldsymbol{k} = (k_x, k_y, k_z)$, this selection rule no longer applies, because $\boldsymbol{k}$ is now changed by the mirror reflection in the xz plane. Any photonic band of general $\boldsymbol{k}$ may be excited by light of arbitrary polarization. A consideration outlined here will suffice to show that the symmetry of PCs plays an important role not only in the PB calculation but also in understanding the optical response. Group theoretical properties of photonic crystals will appear in many ways in this book. The details of group theory will be given in the next chapter.

3.3 Light Transmission and Reflection

So far we have studied the PBSs of infinitely large PCs. To observe experimentally a PBS, however, we have to use a system of finite thickness. Transmission and reflection of light through a PC of finite thickness is an experiment most widely used for this purpose.

3.3.1 Transmission through a 1D Photonic Crystal

To grasp the features of light transmission, let us see the transmission of light in the slab of a 1D PC, whose PBS was examined in Sect. 2.1. Let the electric field of polarization in the x direction be incident normally on the system of Fig. 2.1 . We assume that the 1D PC of period a occupies the region $0 < z < Na$, Na being the thickness. Using the scalar notation again, the incident light of unit strength and wavenumber k is given by

$$E(z) = \exp(\mathrm{i}kz).$$

Scattering by the slab gives rise to reflected and transmitted light. Therefore, we have

$$E(z) = \begin{cases} \mathrm{e}^{\mathrm{i}kz} + r\mathrm{e}^{-\mathrm{i}kz} & : z < 0 \\ t\mathrm{e}^{\mathrm{i}kz} & : z > Na. \end{cases} \tag{3.24}$$

The complex amplitudes r and t of the reflected and transmitted light determine the reflectance $R(\omega)$ and transmittance $T(\omega)$ through

$$R(\omega) = |r(\omega)|^2 \qquad T(\omega) = |t(\omega)|^2. \tag{3.25}$$

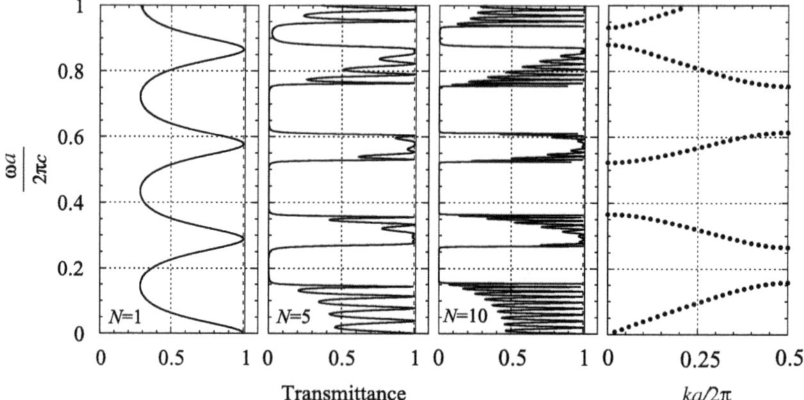

Fig. 3.2. Transmittance $T(\omega)$ as compared with the band structure shown in Fig. 2.1. The number of stacked units cells is N. Transmittance is calculated for the thickness of $N = 1$, 5 and 10. The parameters are the same as those used in Fig. 2.2, from which the PBS of the rightmost panel is reproduced. Inhibition of transmission at the band-gap regions and the interference fringes in the band continuum are two striking features of a larger N

They are, respectively, the reflection and transmission coefficients of energy flow, relative to the energy inflow of the incident light. They are both functions of $\omega = ck$ of the incident light. Flux conservation claims that the energy outflow from the slab should be equal to the energy inflow, implying

$$R(\omega) + T(\omega) = 1. \tag{3.26}$$

The amplitudes r and t in (3.24) are determined by matching the inside and outside fields.

Figure 3.2 shows the result for $T(\omega)$ for the case of A = Si and B = air and $a_A = a_B = a/2$ of the 1D PC used in Sect. 2.1. We have plotted $T(\omega)$ for $N = 1$, 5 and 10. The PBS for $N = \infty$, is reproduced from Fig. 2.2, for comparison.

In the case of $N = 1$, there is only a weak correlation between $T(\omega)$ and the band structure. As N increases, we see the stop bands develop, where the inhibition of transmission takes place. We find that roughly a thickness $N = 10$ is sufficient for a PC to present the essence of the band structure of $N = \infty$. This is indeed a rough, but quantitatively reasonable, estimate, which is often used as a rule of thumb for the low-lying PBs of 2D and 3D PCs.

Together with that, the density of the interference fringes increases with N. The appearence of the interference fringes may be viewed as indicating the existence of special photonic modes, which are favorably confined in a slab. This is another way of saying that the wavenumber k in the direction of wave propagation is effectively quantized and the interference fringes are caused by the resonant excitations of the quantized levels. If we assume that the photonic modes are confined completely inside the slab, the quantization is given by

$$k = p\frac{\pi}{Na} \quad (p = 1, 2, \cdots) \tag{3.27}$$

because $E(z)$ will be proportional to $\sin qz$ and vanish at the external sufaces at $z = 0$ and $z = Na$. Because of this quantization rule, the density of the special modes will increase near the band edges, just as the DOS enhancement, seen in Sect. 2.2.5. Also, (3.27) explains why the density of the fringes increases with N or the thickness of the system.

The width of an interference fringe is a direct measure of the finesse in optics, which determines the quality factor Q of the resonance of the optical response, which is defined by

$$Q = \frac{\Delta\omega}{\omega_0}, \tag{3.28}$$

where $\Delta\omega$ is the FWHM (full width at half maximum) of a resonant Lorentzian peak whose center is ω_0. The width $\Delta\omega$ will later be viewed as the measure of the lifetime of the special modes set up in a slab, which is caused by the leak of the electromagnetic energy out of a PC, neglected in (3.27).

3.3.2 Transmission through a 2D or 3D Photonic Crystal

We turn to a 2D or 3D PC. We consider plane-polarized light incident on a PC of a finite thickness. The direction of the thickness is again taken to be z and the lateral extension of the PC is assumed to be infinite in order for the system to have translational invariance in the x and y directions. We may thus regard the system to be composed of the layers stacked in the z direction, each layer having a 2D translational invariance in the x and y directions. Again, let N be the number of of stacked layers.

Let us consider a fcc PC, for example, which is bounded in the z direction by two (001) surface layers. Let the wavevector of the incident light have the wavevector component $\boldsymbol{k}_{\parallel}$, parallel to the surface. Figure 3.3 shows the typical situation of the light transmission and reflection. The incident light is usally taken to be either s- or p-polarized, as indicated. The polarizations of the transmitted and reflected light are not necessarily s or p but mixtures of both in general, except the case where the incident plane agrees with the mirror plane of the PC like the (100) or (010) plane. The direction and magnitude of the lateral wavevector $\boldsymbol{k}_{\parallel}$ is altered freely in the (k_x, k_y) plane by changing the incident plane and incident angle of light relative to the PC. The frequency ω of the light and the incident angle θ_0 determines $\boldsymbol{k}_{\parallel}$ through

$$|\boldsymbol{k}_{\parallel}| = \frac{\omega}{c}\sin\theta_0. \tag{3.29}$$

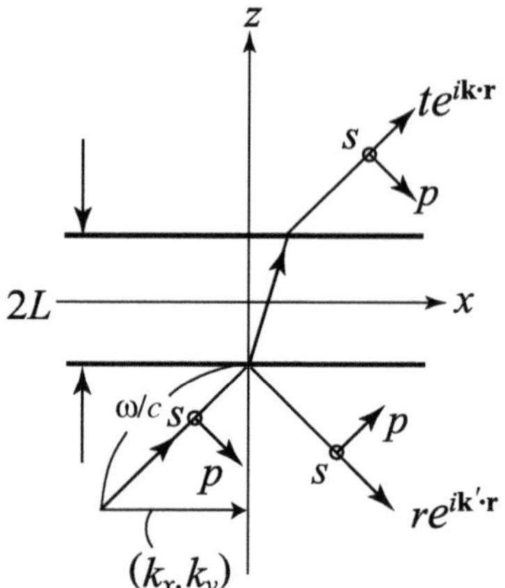

Fig. 3.3. Transmission and reflection of s- and p-polarized light. The s-light is polarized perpendicular to the incident plane and the p-light is polarized in the incident plane, with incident angle θ_0

Complex amplitudes $\boldsymbol{r}(\omega)$ and $\boldsymbol{t}(\omega)$ are those of the reflected and transmitted plane waves with wavevectors $(\boldsymbol{k}_{\parallel}, -k_z)$ and $(\boldsymbol{k}_{\parallel}, k_z)$, respectively. Here

$$k_z = \sqrt{(\frac{\omega}{c})^2 - |\boldsymbol{k}_{\parallel}|^2}$$

is determined by the dispersion relation of light in free space. Note that $\boldsymbol{r}$ and $\boldsymbol{t}$ are now 3D vectors. Namely, for the incident light of the electric field of unit amplitude

$$\boldsymbol{e}_0 \exp\left(\mathrm{i}(\boldsymbol{k}_{\parallel}, k_z) \cdot \boldsymbol{r}\right) \qquad (|\boldsymbol{e}_0| = 1),$$

the reflected and transmitted light have the forms

$$\begin{aligned} &\boldsymbol{r} \exp\left(\mathrm{i}\boldsymbol{k}^- \cdot \boldsymbol{r}\right), \\ &\boldsymbol{t} \exp\left(\mathrm{i}(\boldsymbol{k}^+ \cdot \boldsymbol{r}\right), \end{aligned}$$

respectively, with $\boldsymbol{k}^{\pm} = (\boldsymbol{k}_{\parallel}, \pm k_z)$. Note that the phases of the x, y and z components of $\boldsymbol{r}(\omega)$ and $\boldsymbol{t}(\omega)$ are different in general. Also, the s- and p-incidences have different coefficients $\boldsymbol{r}(\omega)$ and $\boldsymbol{t}(\omega)$. The reflectance $R(\omega)$ and transmittance $T(\omega)$ are again given by

$$R(\omega) = |\boldsymbol{r}(\omega)|^2, \qquad T(\omega) = |\boldsymbol{t}(\omega)|^2. \tag{3.30}$$

Hence in general s-light and p-light have different reflection and transmission properties. The flux conservation expressed by

$$R(\omega) + T(\omega) = 1 \tag{3.31}$$

always holds, when $\varepsilon(\boldsymbol{r})$ is real, or when there is no absorption of light. Neither of the reflected nor transmitted amplitudes are determined unless we know the light inside the PC, where there is a PBS.

3.3.3 Diffraction

Diffracted light appears associated with each of the 2D reciprocal lattice vectors. When a PB of 2D wavevector $\boldsymbol{k}_\parallel$ is excited inside the PC, it involves in its Bloch sum (3.16) various plane waves of the form

$$\exp\left(\mathrm{i}(\boldsymbol{k}_\parallel + \boldsymbol{h}_\parallel)\right)$$

for arbitrary 2D reciprocal lattice vectors $\boldsymbol{h}_\parallel$. These waves emerge from the entrance and exit surfaces of the PC as diffracted light. When ω of the incident light is such that the relation

$$\frac{\omega^2}{c^2} > |\boldsymbol{k}_\parallel + \boldsymbol{h}_\parallel|^2 \tag{3.32}$$

holds for a $\boldsymbol{h}_\parallel$, the inside plane-wave of $\boldsymbol{h}_\parallel$ of an excited PB gives rise to diffracted lights, having a spatial dependence given by

$$\begin{aligned} &\exp\left(\mathrm{i}(\boldsymbol{k}_\parallel + \boldsymbol{h}_\parallel)\cdot \boldsymbol{r}_\parallel + \mathrm{i}\Gamma_{\boldsymbol{k}_\parallel}(\boldsymbol{h}_\parallel)z\right) \quad \text{transmitted,} \\ &\exp\left(\mathrm{i}(\boldsymbol{k}_\parallel + \boldsymbol{h}_\parallel)\cdot \boldsymbol{r}_\parallel - \mathrm{i}\Gamma_{\boldsymbol{k}_\parallel}(\boldsymbol{h}_\parallel)z\right) \quad \text{reflected,} \end{aligned} \tag{3.33}$$

where

$$\Gamma_{\boldsymbol{k}_\parallel}(\boldsymbol{h}_\parallel) = \sqrt{\frac{\omega^2}{c^2} - |\boldsymbol{k}_\parallel + \boldsymbol{h}_\parallel|^2}, \tag{3.34}$$

defines the z component of the outside wavevector. The condition (3.32) guarantees that $\Gamma_{\boldsymbol{k}_\parallel}(\boldsymbol{h}_\parallel)$ is real and the diffracted wave propagates outside the PC in the $\pm z$ directions as a plane wave. We often refer to this diffraction channel $\boldsymbol{h}_\parallel$ as an open channel. When the incident frequency ω is short for (3.32) to hold, the outside $\boldsymbol{h}_\parallel$ wave is evanescent, due to an imaginary number $\Gamma_{\boldsymbol{k}_\parallel}(\boldsymbol{h}_\parallel)$ which causes an exponential decay. This is the case of a closed $\boldsymbol{h}_\parallel$ channel. The closed channels do not carry the electromagnetic energy away from the slab and they play no role in flux conservation. Note that as ω increases with a fixed $\boldsymbol{k}_\parallel$, the number of open channels increases abruptly, each time ω exceeds a threshold value of channel opening.

Suppose that the incident ω is in the range where there are several open diffraction channels, $\boldsymbol{h}_{\parallel 1}, \boldsymbol{h}_{\parallel 2}, \cdots$. Let their vector amplitudes of the plane waves be

$$\boldsymbol{r}_0, \quad \boldsymbol{r}_1 \quad \boldsymbol{r}_2, \quad \cdots \quad \text{and} \quad \boldsymbol{t}_0, \quad \boldsymbol{t}_1 \quad \boldsymbol{t}_2, \quad \cdots .$$

The flux conservation is now expressed by

$$\left(|\boldsymbol{r}_0|^2\cos\theta_0 + |\boldsymbol{r}_1|^2\cos\theta_1 + \cdots\right) + \left(|\boldsymbol{t}_0|^2\cos\theta_0 + +|\boldsymbol{t}_1|^2\cos\theta_1 + \cdots\right) = \cos\theta_0, \tag{3.35}$$

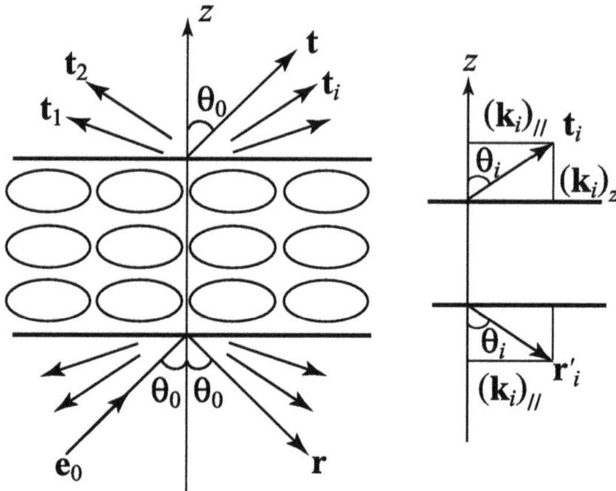

Fig. 3.4. Case of several diffraction channels open. In reflection and transmission, diffracted light waves are induced with amplitudes $\boldsymbol{r}_1$, $\boldsymbol{r}_2$, etc. by the incident light of amplitude $\boldsymbol{e}_0$. The amplitudes $\boldsymbol{r}$ and $\boldsymbol{t}$ are of specularly reflected and directly transmitted light, respectively. The angle θ_i of the channel i is defined in the right panel, with $\boldsymbol{k}_{i\|} = \boldsymbol{k}_\| + \boldsymbol{h}_{i\|}$

which is a natural extension of (3.31). Here θ_i is the angle of the ith diffracted light wave defined in terms of their wavevector $\boldsymbol{k}_\| + \boldsymbol{h}_{i\|}$ and $\Gamma_{\boldsymbol{k}\|}(\boldsymbol{h}_{i\|})$ (Fig. 3.4)

$$\cos\theta_i = \frac{|\Gamma_{\boldsymbol{k}\|}(\boldsymbol{h}_{i\|})_z|}{\omega/c}.$$

The cosine factors enter because what matters in the flux conservation is the z component of the Poynting vectors of each of diffracted plane waves.

3.3.4 Transmittance and Reflectance versus PBS

We concentrate on transmittance and reflectance, $T(\omega)$ and $R(\omega)$ of plane wave light in the frequency range of no diffraction. As in the case of a slab of homegeneous substance, the formulation of multiple scattering to calculate $T(\omega)$ or $R(\omega)$ is a nontrivial problem for a slab of PC. It is especially so when we wish to take into account exactly the energy transport caused by the external fields. In the x and y directions, however, the waves in the slab should be specified by $\boldsymbol{k}_\|$ of the incident light, so that $T(\omega)$ and $R(\omega)$ are expressed more properly as

$$T_{\boldsymbol{k}\|}(\omega) \quad \text{and} \quad R_{\boldsymbol{k}\|}(\omega).$$

Let us increase ω, with $\boldsymbol{k}_\|$ kept fixed. To understand how $T(\omega)$ or $R(\omega)$ changes, the band structure drawn as a function of k_z with a fixed $\boldsymbol{k}_\|$ is very helpful. By band structure we mean an ordinary PBS calculated for a fictitious system of infinite size. The $\omega - k_z$ relation thus provides us with a PBS along a straight line in the first BZ with a prescribed $\boldsymbol{k}_{\parallel}$ value in the $k_x\,k_y$ plane.

Figure 3.5 shows how the ω dependence of $R(\omega)$ is interpreted in terms of PBS. The system treated in this figure is a fcc array of dielectric spheres of $\varepsilon = 3.2^2$. The ratio of the radius r of spheres to the fcc cube size a is 0.3. The PBS is calculated by the vector KKR method and reflectance is obtained by the layer-doubling method, both to be explained in Chap. 4, but how they are obtained is not our chief concern here.

Panels (a) and (b) show the reflectance of light of oblique incidence to the entrance surface parallel to the xy plane, the x, y, z axes being parallel to the edges of the cube of the fcc lattice. The vertical axis shows $R(\omega)$ for the slab of 32 stacked (100) planes of spheres. Panel (c) shows the PBS calculated

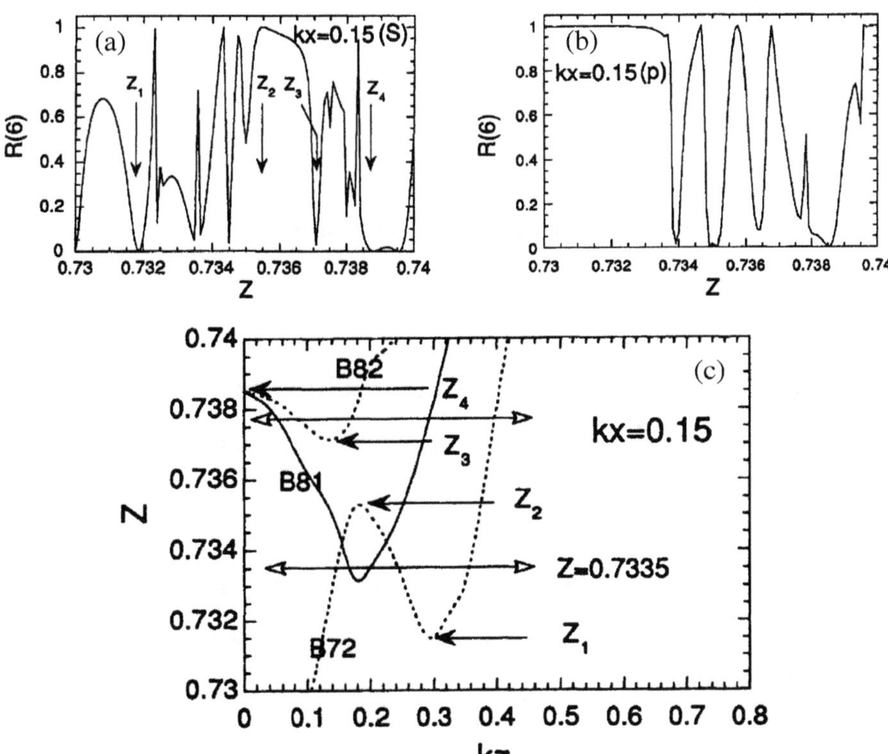

Fig. 3.5. Reflectivity of plane wave light of oblique incidence (a) for s-polarized light and (b) for p-polarized light. Z and k_x are the dimensionless frequency and wavevector component, normalized by $2\pi c/a$ and $2\pi/a$, respectively. The related band structure is given in (c) for wavevector $\boldsymbol{k} = (0.15, 0, k_z)$, as a function of k_z. The p-active (s-active) bands are shown by solid (dashed) curves. See [8] or Fig. 4.2 for the band index

for the infinite PC, as functions of k_z with $\boldsymbol{k}_\parallel$ kept fixed. The solid band is active to the p-light and the dashed bands to the s-light. The two horizontal lines with open arrows are drawn to cross the band dispersion curves, one at $Z = 0.7335$ and the other at a little higher Z than this. The crossing gives the values of k_z excited in PC by the lights of respective frequencies.

When the frequency of s-incident light increases to exceed one of the band edges marked as $Z_1, Z_2, \cdots$, the interference pattern changes abruptly. An example is a drastic change in (a), observed above and below Z_1. We also note that there are obviously two ranges of Z, one between Z_1 and Z_2 and the other between Z_3 and Z_4, where the horizontal line in (c) crosses s-active bands three times, i.e., three waves of different k_z are excited at a time by the s-polarized light. In these frequency ranges transmission and reflection are naturally complex, as exhibited in (a), because of possible beats caused by the interference among the excited waves. This is a typical example of triple refraction occurring in the transmission of the s wave.

If we add a small but nonzero k_y to the incident wavevector, the bands shown by the solid curves in (c), which were inactive to the s-polarized light and appeared only in panel (b) of the p-reflectance, will become active to the s-light [7]. At $Z = 0.7335$, for example, five waves (three on the dashed curve and two on the solid curves) will then be excited at a time.

Figure 3.5 thus provides an example of double, triple, ... refraction. Indeed, PCs are systems which provide a birefringence, so far familiar only in a uniaxial or biaxial anisotropic substance, in a far more striking way (see Sect. 4.7).

3.4 Photonic Crystals of Finite Thickness

We consider here the PBS of a PC of finite thickness. We have given in Chap. 1 the definitions of leaky PB and nonleaky PB. We start by recapitulating the classification.

3.4.1 Light Cone Dividing Leaky and Nonleaky regions

In free space, a plane-wave photon has the dispersion relation given by

$$\omega = c\sqrt{k_x^2 + k_y^2 + k_z^2}. \tag{3.36}$$

For a PC which is finite in the z direction and infinite in the x and y directions, photonic modes of the entire system, composed of the PC and surrounding free space, is specified by the lateral wavevector $\boldsymbol{k}_\parallel \equiv (k_x, k_y)$ due to the lateral translational invariance. Outside the PC, therefore, the plane-wave photons of $\boldsymbol{k}_\parallel$ can fill the phase-space region

$$\omega > c\sqrt{k_x^2 + k_y^2} \tag{3.37}$$

because $k_z^2 > 0$ in (3.36). This implies that a PB mode of $(\boldsymbol{k}_{\parallel}, \omega)$ set up in a PC of finite thickness can be connected to the plane-wave photons of the outside region, when its ω and $\boldsymbol{k}_{\parallel}$ satisfies (3.37). In other words, the PBs found in this region of the phase space $(\boldsymbol{k}_{\parallel}, \omega)$ inevitably have a finite lifetime due to the leakage of their energies carried away to $z = \pm\infty$ by the exterior plane-waves, while PBs existng outside this region of the phase space are confined to the PC. The lifetime due to the leakage of the modes of the first category is thus radiative in nature. The PBs of the second category which have a negative k_z^2 are connected only to the evanescent photons in the exterior. The evanescent photons decay exponentially away from the surfaces of the PC. The former PB modes are the leaky modes (radiative modes) and the latter PB modes are called the guided PBs . The boundary between them given by

$$\omega = c\sqrt{k_x^2 + k_y^2} \tag{3.38}$$

is the light cone. For the PB modes of $k_y = 0$, in particular, the leaky and nonleaky modes are divided by the light line $\omega = ck_x$.

3.4.2 Formation of Photonic Band Structure in a Slab

A rough idea of the PB modes set up in a PC of finite thickness is obtained if we start from the modes in a slab of no internal periodicity and afterwards give them the modifications due to the periodicity of actual PCs.

Let us consider a slab of homegeneous dielectric of dielectric constant ε_s . We temporarily neglect the leak and confine the fields completely to the slab region of $0 < z < L$, L being the slab thickness. The modes of wavevector $\boldsymbol{k}_{\parallel} = (k_x, 0)$ polarized in, say, the y direction will then have the amplitude

$$y(z) = \sin\sqrt{(\frac{\omega}{c})^2\varepsilon_s - k_x^2 z},$$

apart from a normalization factor and the x dependence given by $e^{ik_x x}$. The sine function is chosen by the boundary condition $y(z = 0) = 0$. Another boundary condition at $z = L$ then leads to the quantization rule

$$\sqrt{\frac{\omega^2}{c^2}\varepsilon_s - k_x^2 L} = p\pi,$$

from which we obtain

$$\omega_p(k_x) = \frac{c}{\sqrt{\varepsilon_s}}\sqrt{k_x^2 + \left(p\frac{\pi}{L}\right)^2}, \tag{3.39}$$

where the integer $p = 1, 2, \cdots$ specifies the quantized wavelength in the z direction. Thus, a series of modes appear in a slab, which have the k_x dependence shown in (3.39). The light line $\omega = ck_x$ introduces no distinction between leaky and nonleaky modes in the case of perfect confinement. Obviously, similar quantization occurs for the modes polarized in the xz plane.

Actually, the leakage of the modes out of the slab is important in the leaky region even in a homogeneous slab. The essence of relaxing the assumption of perfect confinement is to introduce the lifetime effect to the above dispersion relation in the leaky region. Therefore, the eigenvalues of quantized PBs are complex in the region $\omega > ck_x$, while those in the region $\omega < ck_x$ have real eigenvalues essentially given by (3.39). Strictly speaking, the dispersion relations of the nonleaky modes, too, are altered, for the incomplete confinement lets the modes go out of the slab as evanescent waves.

The precise theoretical treatment of the modes in the leaky region is not straightforward even for a homogeneous system, because we should take into account the energy dissipation by treating an open system consisting of the slab and surrounding free space. The secular equation yielding the eigenvalues should not be hermitian because the eigenvalues are complex. An example of the precise treatment of the leaky modes of a homegeneous slab will be given in Sect. 4.3.1 with some numerical results.

The internal periodicity of the lattice translation of a PC introduces a number of modifications to the above picture. In the case of an infinite lateral extension, a perfect 2D periodicity exists in the PC in the directions of x and y and the series of the dispersion curves obtained above are all folded back at the boundary of the 2D BZ. This is a rough but essential picture of the PBS of a slab PC. Thus the band gaps appear at the boundary of the 2D BZ.

Figure 3.6 is an example of the appearence of the band gaps in the nonleaky region of a slab of $\varepsilon_{\mathrm{s}} = 12$. A homogeneous slab of a finite thickness is given a periodicity a in the x direction by opening air gaps of width $0.1a$ periodically. In calculating the dispersion curves shown, the leak as evanescent light to the exterior is taken into account exactly. To see the positions of the gaps referring to the homogeneous slab, the solid circles show the slab modes of average dielectric constant of

$$\varepsilon_{\mathrm{av}} = 12 \times 0.9 + 1 \times 0.1,$$

taking into account the volume fraction 0.9 of the dielectric matrix.

When the air gap of the slab widens, the system gradually deviates from the homogeneous slab and larger band gaps develop at the BZ boundary. When the thickness L of the slab PC becomes larger, the density of band population increases in accordance with a finer quantization spacing, which is inversely proportiona to L. This feature is seen by comparing Fig. 3.6 with Fig. 3.7, which shows the result of a thinner slab. This is the same feature as we have remarked in relation to the density of fringes of Fig. 3.2

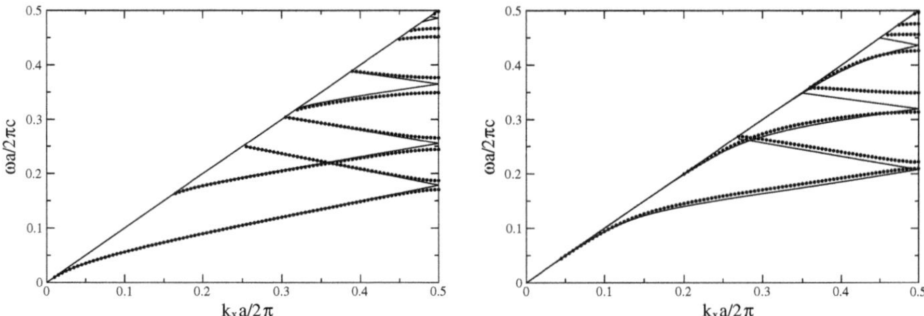

Fig. 3.6. Nonleaky photonic modes of a dielectric slab and appearence of band gaps due to periodic air gaps introduced to the slab. The dielectric constant of the slab is $\varepsilon_s = 12$. The solid curves show the nonleaky modes of a dielectric slab of thickness L. The left panel shows the modes polarized in the y direction (y-polarized modes) and the right panel shows those polarized in the xz plane (xz-polarized modes). Solid circles show the opening of gaps at the BZ when air gaps with period a are introduced to the slab. The width of the gap is $0.1a$. The case of $L = a$ is shown

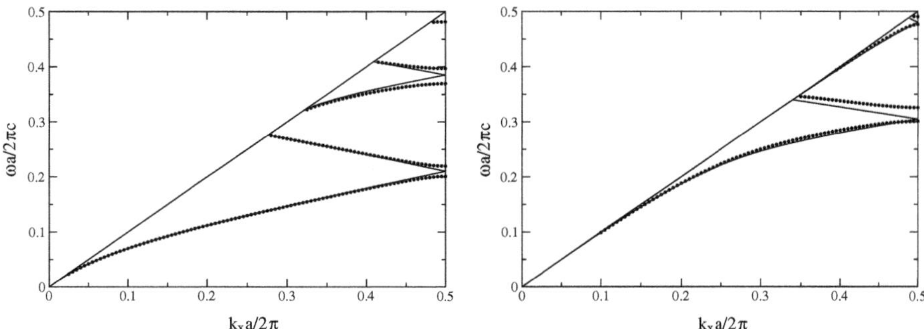

Fig. 3.7. Nonleaky photonic modes of a dielectric slab and opening of band gaps due to air gaps introduced periodically: y-polarized modes (left) and xz-polarized modes (right). The slab thickness is $L = 0.5a$, i.e., one half used in Fig. 3.6

In this way the BS set up in a slab PC has the dispersion relation of the form

$$\omega = \omega_p(k_x, k_y), \tag{3.40}$$

where the 2D wavevector (k_x, k_y) is defined within the first 2D BZ.

In the leaky region, $\omega_p(k_x, k_y)$ becomes complex in general due to the lifetime effect, while it is real in the nonleaky region. These are the essence of the PBs in a PC slab. How to treat the leaky PBs to obtain their precise dispersion relations and lifetime will be the theme of Sect. 4.3.3.

When we consider in addition a periodic structure in the thickness direction, the quantization due to the finite thickness is further modified. However, the photonic bands are still specified by the 2D dispersion relations. In a PC slab of thickness sufficiently large, the BS should be more or less similar to the PBS of the PC of infinite size in all three dimensions.

3.4.3 Thick Slab of Photonic Crystal

To understand the band structure in a very thick PC slab, there is another approach that starts from the PBS of an infinite system: we can regard the normal modes set up in a thick system as those obtained by quantizing the PBs of the PBS calculation. In this picture, we first imagine a PB of band index n of an infinite system,

$$\omega = \omega^{(n)}(k_x,\, k_y,\, k_z),$$

obtained by band calculation (in this subsection the band frequency $\omega_{\boldsymbol{k}}^{(n)}$ of the state $(n,\, \boldsymbol{k})$ is denoted as $\omega^{(n)}(k_x,\, k_y,\, k_z)$). We next quantize k_z considering the finiteness of the slab. In the slab PC of thickness L, we would have the original band n split into many subbands, each specified by a quantized value $k_z = p\pi/L$, if the leakage could be neglected. Each subband would then have a dispersion relation as a function of $(k_x,\, k_y)$ just as in (3.40). In a thick slab, therefore, we shall have a bundle of quantized bands specified by p in the frequency region where the band n would be present in an infinite system. In the frequency region of another band n', another bundle of subbands appears. Of course they are lifetime broadened in the leaky region of the phase space. As the thickness increases, the quantization becomes finer and the population of the subbands increases. This picture no longer applies to a very thin slab of PC of one or two unit-cell thickness.

An important problem should then be answered: whether or not the PBs of the dispersion relation $\omega_p^{(n)}(k_x, k_y)$ continue to have their characters as well-defined modes in the leaky region. If the lifetime of the modes in the leaky region is very short, they can hardly be called normal modes. This is why the lifetime of the leaky modes is so important. How to calculate the lifetime of such bands will be treated in Sect. 4.3.3.

3.5 Whispering Gallery Modes and Mie Resonances

As an atom binds electrons, a sphere binds photons but the leak is inevitable. The bound states of photons accompanying the leak have long been called Mie modes and, recently, whispering gallery modes (WGMs). By the name WGM, we have in mind a closed space of circular shape like an art gallery or something like a round corridor, where we often hear, strangely enough, other people's low conversation very clearly. Here, the whisper comes close by being transported by a bound state of sound set up in the closed area. A WGM is thus a sonic or photonic analog of an atomic bound state of electrons.

To see the origin more clearly in an isolated sphere of dielectric constant $\varepsilon_<$, we start from (3.12). Using

$$\nabla \times \nabla \times \boldsymbol{E} = -\Delta \boldsymbol{E} + \nabla\nabla \cdot \boldsymbol{E}, \tag{3.41}$$

we obtain

$$-\Delta \boldsymbol{E}(\boldsymbol{r}) - \frac{\omega^2}{c^2}\varepsilon(\boldsymbol{r})\boldsymbol{E}(\boldsymbol{r}) + \nabla\nabla \cdot \boldsymbol{E}(\boldsymbol{r}) = 0. \tag{3.42}$$

Adding $(\omega^2/c^2)\boldsymbol{E}(\boldsymbol{r})$ to both sides, as done to (2.18) for the 1D PC, we obtain

$$-\Delta \boldsymbol{E}(\boldsymbol{r}) - \frac{\omega^2}{c^2}(\varepsilon(\boldsymbol{r}) - 1)\boldsymbol{E}(\boldsymbol{r}) + \nabla\nabla \cdot \boldsymbol{E}(\boldsymbol{r}) = \frac{\omega^2}{c^2}\boldsymbol{E}(\boldsymbol{r}). \tag{3.43}$$

The similarity to or difference from the Schrödinger equation

$$-\frac{\hbar^2}{2m}\Delta\psi(\boldsymbol{r}) + V(\boldsymbol{r})\psi(\boldsymbol{r}) = E\psi(\boldsymbol{r}) \tag{3.44}$$

is obvious. If we discard the $\nabla\nabla \cdot \boldsymbol{E}(\boldsymbol{r})$ in (3.43), the similarity is complete with the correspondence

$$-\frac{\omega^2}{c^2}(\varepsilon(\boldsymbol{r}) - 1) \leftrightarrow V(\boldsymbol{r}). \tag{3.45}$$

As in Sect. 2.2.3 for the 1D PC, therefore, a dielectric sphere with $\varepsilon_<$ larger than unity works in the free space as an attractive potential for a photon. In (3.43), ω^2 plays the role of the energy E of the electron: the bound states of photons, if any, correspond to the electronic bound states of a *positive* eigen-energy. These types of bound states of electron can actually exist if the potential is a short-range attractive potential, in contrast to the long-range Coulomb potential which leads to the bound states of a hydrogen atom only in the region $E < 0$. Let us briefly see why this is so.

Due to the spherical symmetry, the modes in the sphere are classified by the angular momentum index (l, m) (the terms of quantum mechanics are used here associated with the spherical harmonics $Y_{lm}(\theta, \phi)$, θ, ϕ being the polar coordinates of $\boldsymbol{r} = (r, \theta, \phi)$). We thus need to examine a bound state in the effective potential, the sum of the centrifugal potential proportial to $l(l+1)$ and the original potential related to $\varepsilon(\boldsymbol{r})$. Let us denote the effective potential as $V_{\text{eff}}(r)$:

$$V_{\text{eff}}(r) = -\frac{\omega^2}{c^2}(\varepsilon(\boldsymbol{r}) - 1) + \frac{\ell(\ell+1)}{r^2}. \tag{3.46}$$

A rough sketch of $V_{\text{eff}}(r)$ is given in Fig. 3.8 for the dielectric function of

$$\varepsilon(\boldsymbol{r}) = \begin{cases} \varepsilon_< : r < r_0 \\ 1 \quad : r > r_0 \end{cases} \tag{3.47}$$

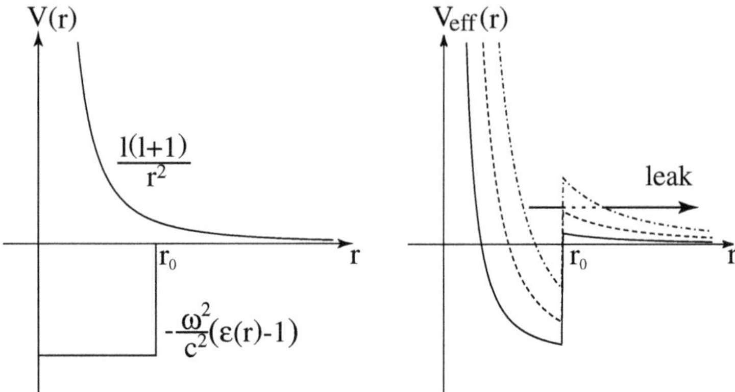

Fig. 3.8. $V_{\rm eff}(\boldsymbol{r})$ as a function of r. The radius of the sphere is r_0. In the right panel, curves are given for $V_{\rm eff}(\boldsymbol{r})$ for several ℓ. As ℓ increases, the barrier at $r = r_0$ becomes higher to lead to a better confinement of the photon

where $\varepsilon_< > 1$. The symbol r_0 is used here to stand for the radius of the sphere. The point of Fig. 3.8 is that $V_{\rm eff}(\boldsymbol{r})$ has a positive potential barrier at $r = r_0$ because of the abrupt vanishing of the attractive potential in the outside region $r > r_0$, so that a kind of bound states can certainly arise in the postive energy region, whose leak is hindered by the potential barrier at $r = r_0$. As shown in the right panel of Fig. 3.8, these bound states in the potential dip can finally leak by tunneling through the potential barrier out to the surrounding free space. If it were not for the tunneling, the sphere works as a perfect cavity for photons. In this sense, these bound states are often called virtual bound states, to emphasize that they are not true bound states but have finite lifetime due to the tunneling. Since the repulsive potential is proportional to $l(l+1)$, the binding is stronger for the modes of larger l by inducing a higher potential barrier at $r = r_0$. This means that the higher l makes the lifetime longer, or the quality of the mode as a bound state becomes higher. As ω becomes larger, the potential becomes more and more attractive. But the characters as bound states will get blurred when $(\omega/c)^2$ goes beyond the value $l(l+1)/r_0^2$, the height of the potential barrier at the sphere wall $r = r_0$.

This is the origin of the Mie modes or WGM of a sphere or dielectric cavity, so called in this field in place of virtual bound states familiar in electrons. They cause resonant light scattering from a single dielectric sphere. For each ℓ, i.e., for each $V_{\rm eff}$ of the modes of ℓ, there exist a series of WGM.

Since the degrees of freedom are twice as large as the Schrödinger equation, we have two series of bound states for each (l, m). The degrees of freedom of two (not three) comes from the transversality. Usually we classify the modes in the two series as M and N modes.

The existence of M and N modes corresponds just to the spin degrees of freedom of electrons. For a fixed (l, m), the frequencies of M and N modes are different, unlike the case of electrons, however, due to the difference of the accompanying polarizations; the N modes which have a finite radial component of the electric field satisfy the continuity of the radial component of the displacement field at $r = r_0$, while the M modes which have zero radial components have smooth variation of their electric fields at $r = a$.

Figure 3.9 shows the increase of the DOS of photons due to the presence of a dielectric sphere of $\varepsilon_< = 3.2^2$ in free space. The increase of the DOS is plotted only for the M type WGM as a function of the frequency $qr_0 = \omega r_0/c$. The DOS fomula necessary for this calculation will be given in Sect. 4.3.2. We note in Fig. 3.9 that the DOS profile is composed of approximate Lorentzian peaks at the frequencies of WGM. The FWHM of each peak is a measure of the inverse of the lifetime due to the leakage. The ω and ℓ dependences of the FWHM are well understood by what has been mentioned above. Note that optical responses of a dielectric sphere such as the light scattering of plane-wave incident light, reflect gross features of the DOS profile. A spectrum of any optical response of a dielectric sphere is characterized more or less by the series of resonant peaks. These features have stimulated the field of cavity quantum electronics (cavity QED), for the purpose of making use of the strong resonance of WGM in various optical processes, lasing oscillation, for example.

Photonic bands of a regular array of spheres are formed from the WGM of individual spheres. The important concept here is the hopping of the bound photons from sphere to sphere. A WGM of a sphere is transferred to the next

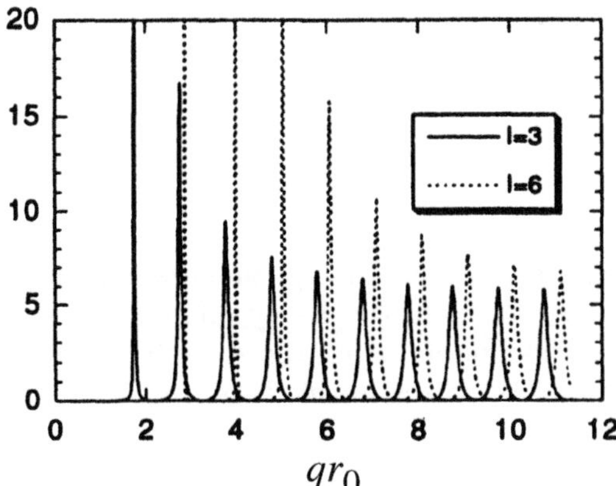

Fig. 3.9. DOS of M-type WGM of a sphere as a function of $qr_0(= \omega r_0/c)$. The radius of the sphere is r_0 and the dielectric constant is $\varepsilon_< = 3.2^2$ The curves of $\ell = 3$ and $\ell = 6$ are shown. For a fixed ℓ the sharpness of the DOS peaks decreases with increasing value of qr_0

sphere because two spheres are not completely isolated. The leak of WGM from a sphere wall is used in the hopping process to the neighboring spheres. In a periodic array of spheres, the cause of the finite lifetime in an isolated sphere thus turns out to be the origin of the coherent motion or band motion of photons, which is free from the lifetime effect. Roughly speaking, therefore, the band tructure of arrayed spheres is made of tight-binding bands.

3.6 Concept of Heavy Photons and Tight-Binding Bands

Heavy fermions are the name given to electrons whose motion in a substance is very coherent as a Bloch wave but has at the same time a very heavy mass, 500 or even 1000 times heavier than that of a free electron. The heavy mass of this type is a result of the very flat dispersion relation of those electrons. One of the remarkable consequences is a very high density of states of electrons, which leads to a huge electronic specific heat, for example. The origin is the virtual bound states of the d or f character of an atom, a heavy atom like U and Ce in, e.g., $CeAl_3$ and UBe_{13}. Their localization is virtual because their frequency lies within the continuum of the density of states of the s conduction electrons and the leak from the confined area is inevitable.

The d or f electrons in these states may hop to an adjacent atom making use of a finite tunneling probability and they finally have the character of a Bloch wave. During the electron motion they stay mostly in atomic states and hop to the neighbors only occasionally. This is why the mass is so big.

This scenario is completely identical to what we have seen in the hopping of photons via the WGM states in a lattice of dielectric spheres. Therefore, the PBs in the lattice of dielectric spheres or cylinders may well be considered to be a bosonic analog of the heavy Fermion. The name "heavy photon" is used in referring to these Bloch states. Near the Γ point, they are characterized by the dispersion curve of the form

$$\omega = A\sqrt{|\boldsymbol{k}|^2 + k_0^2}, \quad \text{or} \quad \omega = Ak_0 + \frac{A}{2k_0}|\boldsymbol{k}|^2 \tag{3.48}$$

with two positive constant A and k_0. The parabolic dispersion relation with a massive term Ak_0 characterizes heavy photons. In other words the heavy photons have a tight-binding band whose origin is the virtual bound states, i.e., WGM. We shall often see in this book the flat dispersion curves of the same origin, which have group velocities less than 1/10 of the light velocity in free space.

Since heavy photons induce a highly enhanced density of states, they are expected to induce a dramatic effect in many optical processes. The huge enhancement of the near-field intensity in the photonic crystal of arrayed spheres is related to the slow v_g of heavy photons. Also the light emission from a PC will be enhanced due to the effect of heavy photons, because it is directly proportional to the density of photons. Some numerical examples of these phenomena will be seen later.

References

1. J. D. Joannopoulos, R. D. Meade, J. N. Winn: *Photonic Crystals: Molding the Flow of Light*, (Princeton University Press, Princeton 1995)
2. K. Sakoda: *Optical Properties of Photonic Crystals*, (Springer, Berlin Heidelberg New York 2001)
3. C. M. Soukoulis (ed.): *Photonic Band Gaps and Localization*, (Plenum, New York 1993)
4. G. Kurizki and J. W. Haus (ed.): *Photonic Band Structures*, J. Mod. Opt. **41** (1994) pp.171–404
5. C. M. Bowden, J. P. Dowling and H. O. Everitt (ed.): *Development and Applications of Materials exhibiting Photonic Band Gaps*, J. Opt. Soc. Am. B **10** (1993) pp. 279–413
6. H. Ibach and H. Lüth: *Solid-State Physics*, 2nd edn (Springer, Berlin Heidelberg New York 1996) pp 35–63 and pp 129–154
7. K. Ohtaka and Y. Tanabe: J. Phys. Soc. Jpn. **65**, 2670 (1996)
8. K. Ohtaka and Y. Tanabe: J. Phys. Soc. Jpn. **65**, 3068 (1996)

4 Theory II: Advanced Topics of Photonic Crystals

K. Ohtaka

This chaper provides further theory which treats advanced topics of PCs, mostly from a theoretical point of view. It is intended for readers who are actively working in the field of PCs and hoping to gain deeper insight into the physics of PBs. Each section is intended to be as independent and self-contained as possible in order for you to read one section or another according to your interest. Most of the contents are related more or less to the work of mine and my colleagues and we have not attempted to make the references comprehensive. Including the monographs and reports cited in Chap. 3, a number of books and feature issures of journals are cited here for advanced studies of photonic crystals [1–10].

As a first topic of this chapter, we shall describe the methods of calculation of PBS other than the plane-wave expansion method. Then we give the group theory of PCs with some numerical applications. The third and fourth topics are related to the leak of PBs from a slab of PC. We describe a method of treating the lifetime of PBs caused by it. The completeness relation of the PB solutions is the next topic and we shall make use of it to construct the Green's function to solve the inhomogeneous Maxwell equations. We apply the Green's function to the calculation of band gap modes and to the analyses of a number of optical processes. As the final topic, we treat the optics of PCs.

4.1 Methods not Based on Plane-Wave Expansion

The study of the PBS of arrayed spheres and cylinders is useful in clarifying the basic properties of PCs of more sophisticated fabrication. By making use of the spherical and cylindrical symmetries, we can analyse these systems very precisely beyond the reach of the plane-wave expansion method. Including these special methods, there are a number of methods for the band structure calculation of PCs.

The first is the plane-wave expansion method, which is a topic of Chap. 3. Since it is a simple and straightforward method, it is used most widely of all. Less familiar than this method is to make use of the particular basis functions which fit the geometry of the PC under consideration. This method is exploited solely for systems of arrayed spheres or cylinders. The third one is to utilize the scattering data of a single layer of PC. By combining it with the Bloch theorem, we can treat an infinitely thick system as a system of repeated monolayers and obtain its PBS. The fourth one is methods making full use of numerical means. A typical example is the finite-difference time-demain (FDTD) calculation [11, 12]. We describe here the second and third methods.

4.1.1 Vector Korringa–Kohn–Rostoker (KKR) Method

In a group of PCs, which consist of arrayed spheres or cylinders of circular cross-section, we can make use of their spherical or cylindrical symmetry. The complementary PCs of arrayed air holes or air cylinders are treated similarly. These particular PCs are basic systems in the PC study, which provide precise information on the physics of PCs by enabling very compact and exact theoretical analyses.

The vector nature complicates not a little the formulation of the light scattering from these systems but, conceptually, the formuation is quite analogous to the electron case. The final result, too, is also very transparent. This method is often referred to as the vector KKR method after the names of Korringa and Kohn and Rostoker, who completed this method of electron band structure calculation [13]. We usually add “vector” to the acronym KKR, because we are dealing with Maxwell's fields.

The basic idea stems from the fact that the band structure in a periodic array of scatterers is determined by two factors: how a photon is scattered by an isolated scatterer and how the scatterers are distributed in the lattice. The scattering of a single scatterer is determined by the phase shifts of the S (scattering) matrix, while the distribution of the scatterers is incorporated into a factor called a structure factor (structure constant or form factor).

Suppose dielectric spheres of dielectric constant $\varepsilon_<$ are arrayed in a periodic lattice in free space. The quantity $\varepsilon_<$ may vary spatially within a sphere but the spherical symmetry of it is a crucial assumption. It may therefore depend only on the distance from the center of the sphere. Of course, the dependence on ω is allowed, as in metallic spheres (see Sect. 8.4). Let $\varepsilon(\boldsymbol{r})$ be the dielectric function for an arbitrary position $\boldsymbol{r}$ of this PC. Inside the spheres it takes the value $\varepsilon_<$. As in the Born series of electron scattering, Maxwell's equations for the PC are transformed to an integral equation. Our starting equation is (3.12)

$$\left(-\boldsymbol{\nabla}\times\boldsymbol{\nabla}\times+(\omega/c)^2\varepsilon(\boldsymbol{r})\right)\boldsymbol{E}(\boldsymbol{r})=0. \tag{4.1}$$

Then it follows that

$$E_i(\boldsymbol{r})=E_i^0(\boldsymbol{r})+\sum_j\int \mathrm{d}\boldsymbol{r}'G_{ij}(\boldsymbol{r},\boldsymbol{r}')V(\boldsymbol{r}')E_j(\boldsymbol{r}'), \tag{4.2}$$

where the potential $V(\boldsymbol{r})$, introduced in Sect. 3.5, is given by

$$V(\boldsymbol{r}) = -\frac{\omega^2}{c^2}(\varepsilon(\boldsymbol{r}) - 1) = \begin{cases} -\left(\frac{\omega}{c}\right)^2(\varepsilon_< - 1) & : \text{inside the sphere,} \\ 0 & : \text{outside the sphere.} \end{cases} \tag{4.3}$$

Here the first term $\boldsymbol{E}^0(\boldsymbol{r})$ is a solution of (4.1) for free space, i.e., for $V(\boldsymbol{r}) = 0$ everywhere. Usually we take it to be incident light when treating light scattering. In the PBS calculation, we put it to zero and the band energy is given by the self-consistent solution of

$$E_i(\boldsymbol{r}) = \sum_j \int \mathrm{d}\boldsymbol{r}' G_{ij}(\boldsymbol{r}, \boldsymbol{r}')V(\boldsymbol{r}')E_j(\boldsymbol{r}'). \tag{4.4}$$

In (4.2) the tensor Green's function $((\boldsymbol{G})_{ij} = G_{ij})$ is given by

$$G_{ij}(\boldsymbol{r}, \boldsymbol{r}') = \left(\delta_{ij} + \frac{c^2}{\omega^2}\frac{\partial^2}{\partial x_i \partial x_j}\right) G(\boldsymbol{r}, \boldsymbol{r}'), \tag{4.5}$$

with $i,\ j = x,\ y,\ z$ and $G(\boldsymbol{r}, \boldsymbol{r}')$ the Green's function of the Helmholtz equation, satisfying

$$\left(\Delta + \frac{\omega^2}{c^2}\right) G(\boldsymbol{r}, \boldsymbol{r}') = \delta(\boldsymbol{r} - \boldsymbol{r}'). \tag{4.6}$$

To check that the solution of (4.2) or (4.4) indeed satisfies (4.1), we first operate with $-\boldsymbol{\nabla} \times \boldsymbol{\nabla} \times +(\omega/c)^2$ on both sides of (4.2) and use the key relation for the tensor $\boldsymbol{G}(\boldsymbol{r}, \boldsymbol{r}')$, which reads

$$\left(-\boldsymbol{\nabla} \times \boldsymbol{\nabla} \times + \frac{\omega^2}{c^2}\right) \boldsymbol{G}(\boldsymbol{r}, \boldsymbol{r}') = \boldsymbol{I}\delta(\boldsymbol{r} - \boldsymbol{r}'). \tag{4.7}$$

Therefore, $\boldsymbol{G}$ may be said to be the Green's function of Maxwell's equation of free space. Both sides of this equation give a second rank tensor. The (ij) component of the first term of the left-hand side is

$$-\sum_p (\boldsymbol{\nabla}\times)_i(\boldsymbol{\nabla}\times)_p G_{pj}$$

and that of the right-hand side is

$$(\boldsymbol{I})_{ij}\delta(\boldsymbol{r} - \boldsymbol{r}') \equiv \delta_{ij}\delta(\boldsymbol{r} - \boldsymbol{r}'),$$

$\boldsymbol{I}$ being the unit tensor. To prove (4.7), we have only to use the identity $\boldsymbol{\nabla} \times \boldsymbol{\nabla}\times = -\Delta + \boldsymbol{\nabla}\boldsymbol{\nabla}\cdot$ and the property (4.6) of the scalar Green's function of the Helmholtz equation.

In the Born series for solving (4.2), the tensor Green's function $\boldsymbol{G}$ enters everywhere in place of the scalar Green's function G of the electronic case:

$$E_i(\boldsymbol{r}) = E_0(\boldsymbol{r})_i + \sum_j \int d\boldsymbol{r}' G_{ij}(\boldsymbol{r}, \boldsymbol{r}') V(\boldsymbol{r}') E_0(\boldsymbol{r}')_j + \sum_{p,q} \int \int d\boldsymbol{r}_1 d\boldsymbol{r}_2 (\boldsymbol{G} V \boldsymbol{G} V \boldsymbol{E}_0)_i + \cdots , \tag{4.8}$$

with an abbreviation in the third term

$$(\boldsymbol{G} V \boldsymbol{G} V \boldsymbol{E}_0)_i = G_{ip}(\boldsymbol{r}, \boldsymbol{r}_1) V(\boldsymbol{r}_1) G_{pq}(\boldsymbol{r}_1, \boldsymbol{r}_2) V(\boldsymbol{r}_2) \boldsymbol{E}_0(\boldsymbol{r}_2)_q .$$

In short, in the photonic case we can follow the electronic treatment by using $\boldsymbol{G}$ in place of G. Thus the Mie scattering of light from an isolated sphere, i.e., the light scattering accompanying the excitation of whispering gallery modes (WGM), can be handled by introducing the concepts of the t matrix, S matrix, partial waves, phase shift, etc., all familiar in the electronic case. For such a formulation of the light scattering from a spherical object, see, e.g. [14].

The situation is similar in the case of PCs. Consider plane-wave light entering a PC of arrayed spheres. By being scattered by a sphere the light aquires a phase shift relative to the incident phase. This is Mie scattering. After the Mie-scattering by one sphere, it propagates to the next sphere in the lattice and is once again Mie scattered. What is occurring in the PC amounts to this repetition. Therefore, the series of Mie scattering in the PC is a geometrical series which is expressed by

$$\boldsymbol{t} + \boldsymbol{tFt} + \boldsymbol{tFtFt} + \cdots = \boldsymbol{t}(\boldsymbol{I} - \boldsymbol{Ft})^{-1}, \tag{4.9}$$

where $\boldsymbol{t}$ is a t matrix of a sphere which incorporates the full Born series of the *single-sphere* scattering, and $\boldsymbol{F}$ is a matrix describing the light propagation between the spheres. The quantity $\boldsymbol{t}$ involves the phase shifts of the Mie scattering and $\boldsymbol{F}$ contains information on the distribution of the scatterers in the lattice. They are matrices both expressed using the basis functions of vector spherical waves. For their concrete expressions, we refer the readers to the original works of [14, 15]. If there is no incident light $\boldsymbol{E}^{(0)}$ in the Born series (4.8), the PBS is obtained from the poles of the inverse matrix $(\boldsymbol{I} - \boldsymbol{Ft})^{-1}$ of (4.9). We thus conclude that the secular equation to obtain the PBS is $\det(\boldsymbol{I} - \boldsymbol{Ft}) = 0$. For the array of cylinders, exactly the same procedure applies. In what follows, we concentrate on the secular equation in some detail.

We resolve the Bloch wave $E_{\boldsymbol{k}}(\boldsymbol{r})$ of wavevector $\boldsymbol{k}$ into a superposition of spherical waves in the region within the sphere whose center is chosen to be the origin of coordinates (called the reference sphere). Using the M and N partial waves of the anglular momentum $L = (l, m)$, it has the form

$$\boldsymbol{E}_{\boldsymbol{k}}^{<}(\boldsymbol{r})=\sum_{L}{}^{\prime}\left(\boldsymbol{E}_{L}^{\mathrm{M}}(\boldsymbol{r})\alpha_{L}^{\mathrm{M}}(\boldsymbol{k})+\boldsymbol{E}_{L}^{\mathrm{N}}(\boldsymbol{r})\alpha_{L}^{\mathrm{N}}(\boldsymbol{k})\right), \tag{4.10}$$

where we put the symbol $^{<}$ on $\boldsymbol{E}_{\boldsymbol{k}}(\boldsymbol{r})$ to emphasize that this expression holds inside the sphere. The primed sum over L means the removal of the space $l=0$ from the sum. The vector spherical waves $\boldsymbol{E}_{L}^{\mathrm{M}}(\boldsymbol{r})$ and $\boldsymbol{E}_{L}^{\mathrm{N}}(\boldsymbol{r})$ are the M and N fields set up in an isolated sphere discussed in Sect. 3.5. How they mix to form a Bloch state of wavevector $\boldsymbol{k}$ is given by the amplitudes $\alpha_{L}^{\mathrm{M}}(\boldsymbol{k})$ and $\alpha_{L}^{\mathrm{N}}(\boldsymbol{k})$ of (4.10). By summing the Born series (4.8) [14] or directly manipulating (4.4) [15], they are shown to satisfy

$$\begin{bmatrix}\delta_{LL'}+q^{-1}\tan\delta_{\ell}^{\mathrm{M}}\boldsymbol{A}^{\mathrm{MM}}(\boldsymbol{k}) & q^{-1}\tan\delta_{\ell}^{\mathrm{M}}\boldsymbol{A}^{\mathrm{MN}}(\boldsymbol{k})\\ q^{-1}\tan\delta_{\ell}^{\mathrm{N}}\boldsymbol{A}^{\mathrm{NM}}(\boldsymbol{k}) & \delta_{LL'}+q^{-1}\tan\delta_{\ell}^{\mathrm{N}}\boldsymbol{A}^{\mathrm{NN}}(\boldsymbol{k})\end{bmatrix}\begin{bmatrix}\boldsymbol{\alpha}^{\mathrm{M}}(\boldsymbol{k})\\ \boldsymbol{\alpha}^{\mathrm{N}}(\boldsymbol{k})\end{bmatrix}=0, \tag{4.11}$$

with $q=\omega/c$. Namely, the PBS is obtained by

$$\det\begin{bmatrix}\delta_{LL'}+q^{-1}\tan\delta_{\ell}^{\mathrm{M}}\boldsymbol{A}^{\mathrm{MM}}(\boldsymbol{k}) & q^{-1}\tan\delta_{\ell}^{\mathrm{M}}\boldsymbol{A}^{\mathrm{MN}}(\boldsymbol{k})\\ q^{-1}\tan\delta_{\ell}^{\mathrm{N}}\boldsymbol{A}^{\mathrm{NM}}(\boldsymbol{k}) & \delta_{LL'}+q^{-1}\tan\delta_{\ell}^{\mathrm{N}}\boldsymbol{A}^{\mathrm{NN}}(\boldsymbol{k})\end{bmatrix}=0. \tag{4.12}$$

This equation is in essence equal to

$$\det(\boldsymbol{t}^{-1}-\boldsymbol{F})=0,$$

if we use the quantities used in (4.9) to express the sequential scattering. In (4.11) and (4.12), there are two blocks labeled by M and N fields and each block is labeled by the indices L and L'. It is therefore seen that the two phase shifts δ_{l}^{β} ($\beta=$ M and N) which describe the Mie scattering of an isolated sphere and the structure constant $\boldsymbol{A}$ which is equal to $\boldsymbol{F}-iq\boldsymbol{I}$ are all that we need in the band structure calculation. Note that all these quantities are functions of $\boldsymbol{k}$ and ω.

From these equations, we find that the electron scattering (the KKR formula for electrons [13]) is described solely by the M block and the extension is carried out to the Maxwell fields by the enlarged degrees of freedom, twice as large due to the transversal vector nature. To accept the M nature of electrons, it will suffice to remember that the M modes are determined by the boundary condition of smoothness as electrons, as mentioned in Sect. 3.5; the phase shift $\delta_{\ell}^{\mathrm{M}}$ has an identical expression to the electronic phase shift. The similarity between electrons and photons is not limited to the form of the M-type phase shift, however. We can in fact show that the electromagnetic phase shift $\delta_{\ell}^{\mathrm{M}}$, $\delta_{\ell}^{\mathrm{N}}$ and the four types of structure factor $\boldsymbol{A}^{\mathrm{MM}}$, $\boldsymbol{A}^{\mathrm{MN}}$, $\boldsymbol{A}^{\mathrm{NM}}$ and $\boldsymbol{A}^{\mathrm{NN}}$ are all derived by a generalized linear transformation of the phase shift δ_{ℓ} and the structure factor $\boldsymbol{A}$ of electrons, respectively. Consequently, all that is required in coding the vector KKR band calculation is only the preparation of the transformation matrices (two vector matrices $\boldsymbol{P}^{\mathrm{M}}$ and $\boldsymbol{P}^{\mathrm{N}}$ each having three components, in the notation of [15]), which are to be applied on the electronic KKR coding. This is the point of the vector KKR method for arrayed spheres [15] and cylinders [16].

The WGM or Mie modes of an isolated sphere appear as singular behavior in the phase shift δ_ℓ^{M} and δ_ℓ^{N}, which, in the case of the constant dielectric constant $\varepsilon_<$ inside the sphere, are expressed by [15]

$$\begin{aligned}\cot\delta_l^{\mathrm{M}} &= \frac{q_< n_l(qr) j_l(q_< r)' - q n_l(qr)' j_l(q_< r)}{q_< j_l(qr) j_l(q_< r)' - q j_l(qr)' j_l(q_< r)}, \\ \cot\delta_l^{\mathrm{N}} &= \frac{q^2 n_l(qr)[q_< r j_l(q_< r)]' - q_<^2 [qr n_l(qr)]' j_l(q_< r)}{q^2 j_l(qr)[q_< r j_l(q_< r)]' - q_<^2 [qr j_l(qr)]' j_l(q_< r)},\end{aligned} \quad (4.13)$$

where $q_<$ is the wavenumber of light inside the sphere:

$$q_< = q\sqrt{\varepsilon_<} = \frac{\omega}{c}\sqrt{\varepsilon_<}. \quad (4.14)$$

The radius r of the sphere, which should not be confused with the radial variable, enters because the phase shifts are determined by matching the electric fields between the inside and outside regions at the surface of the sphere. A WGM mode M or N of the anglular momentum L appears as an abrupt increase of δ_ℓ^{M} or δ_ℓ^{N} as a function of q. These increases are the origin of the series of peaks of the DOS due to the existence of WGMs, an example of the DOS having been given in Fig. 3.9.

The example of the band structure calculation based on (4.12) is given in Fig. 4.1, for the array of spheres in a fcc lattice with $r/a = 0.3$, a being the cube size of the fcc lattice [17]. The dielectric constant $\varepsilon_< = 3.2^2$ is assumed. Figure 4.2 magnifies the details of Fig. 4.1. From these two figures, we notice at once that the bands are grouped into several bunches. This feature is understood in terms of the splitting of the degeneracy of the Mie modes; the $(2\ell+1)$-fold degeneracy associated with an angular momentum ℓ is partly lifted when the spheres form a lattice. How the degeneracy is lifted in the PC is only answered by group theory [18]. Also, in the case of such large $\epsilon_<$ as shown in Figs. 4.1 and 4.2, the mixing between Mie modes of different ℓ is relatively small and a typical picture of tight-binding bands or heavy photons applies well. This mechanism was discussed in Sect. 3.6.

The applications of the vector KKR formulations to obtain the band structures as given above have been actively pursued by many authors [20]. For the recent applications, see, e.g. [21].

4.1.2 Monolayer Scattering and Bloch Theorem

We now proceed to the second method of the calculation of PBS. Suppose we have somehow obtained information of scattering of a single monolayer. By assembling the monolayer data we can obtain the full scattering data of

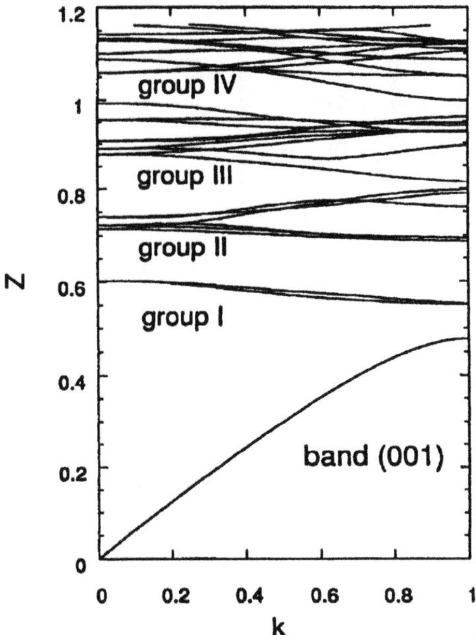

Fig. 4.1. Photonic band structure of the fcc lattice of dielectric spheres for $\boldsymbol{k}$ along the Δ axis. For the parameters, see the text. The vertical and horizontal axes are frequency and wavenumber normalized by $2\pi c/a$ and $2\pi/a$, a being the cube size of the fcc lattice

a system consisting of an arbitary number of stacked layers. The scattering data of many layers, however, involves the interference effects between the entrance and exit surfaces and is not appropriate to derive PBS of a PC of infinite size. For this purpose, we can use the Bloch therem to take into account the periodicity of an infinite system in the stacking direction.

Consider a periodic array of spheres whose centers are located in the xy plane. This is a representative layer comprising a PC of arrayed spheres. We have in mind a system of spheres, though the formulation is quite general, not at all limited to spheres, as mentioned at the end of this section. Below the layer there are series of waves propagating in the $+z$ and $-z$ directions

$$\boldsymbol{E}_n(\boldsymbol{r}) = \sum_{\boldsymbol{h}} \boldsymbol{a}_{\boldsymbol{h}}^{+} \exp(\mathrm{i}\boldsymbol{k}_{\boldsymbol{h}}^{+} \cdot \boldsymbol{r}) + \sum_{\boldsymbol{h}} \boldsymbol{a}_{\boldsymbol{h}}^{-} \exp(\mathrm{i}\boldsymbol{k}_{\boldsymbol{h}}^{-} \cdot \boldsymbol{r}). \tag{4.15}$$

Here $\boldsymbol{h}$ runs over all possible 2D reciprocal lattice vectors and

$$\boldsymbol{k}_{\boldsymbol{h}}^{\pm} = (k_x + h_x, k_y + h_y, \pm \boldsymbol{\Gamma}_{\boldsymbol{h}}), \tag{4.16}$$

where

$$\pm \boldsymbol{\Gamma}_{\boldsymbol{h}} = \pm\sqrt{(\omega/c)^2 - (k_x + h_x)^2 - (k_y + h_y)^2}, \tag{4.17}$$

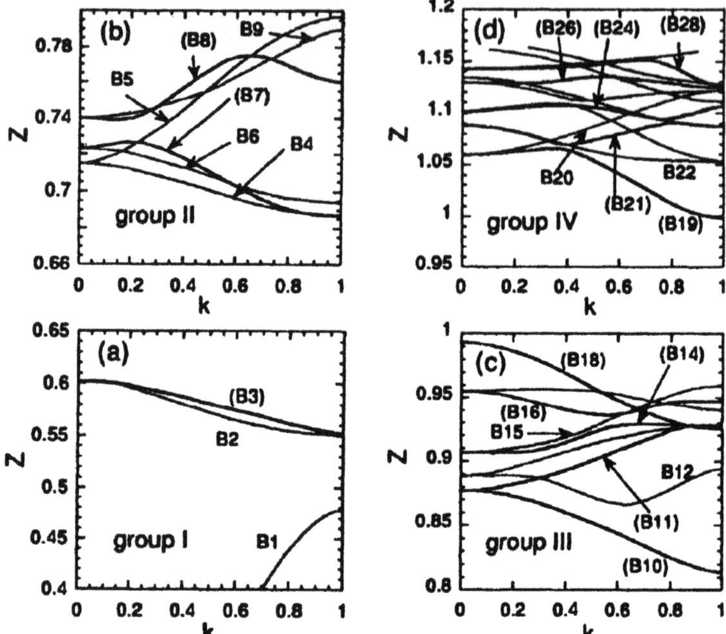

Fig. 4.2. Groups of photonic bands for $\boldsymbol{k}$ along the Δ axis. Each of the figures shows a magnified portion of the band structure shown in Fig. 4.1. Bands and groups of bands are labeled in increasing order of the band energy at the $\boldsymbol{\Gamma}$ point $\boldsymbol{k} = 0$

with the rule that $\sqrt{P} = \mathrm{i}\sqrt{|P|}$ when $P < 0$. Therefore (4.15) involves evanescent waves for closed channels at frequency ω (for the definition of open or closed channels, see Sect. 3.3.3). Above the layer there are waves of the form

$$\boldsymbol{E}_{n+1}(\boldsymbol{r}) = \sum_{\boldsymbol{h}} \boldsymbol{b}_{\boldsymbol{h}}^{+} \exp(\mathrm{i}\boldsymbol{k}_{\boldsymbol{h}}^{+} \cdot \boldsymbol{r}) + \sum_{\boldsymbol{h}} \boldsymbol{b}_{\boldsymbol{h}}^{-} \exp(\mathrm{i}\boldsymbol{k}_{\boldsymbol{h}}^{-} \cdot \boldsymbol{r}) \tag{4.18}$$

In (4.15) and (4.18), we have put the indices n and $n+1$ on the electric fields, to indicate that the layer in question is the nth layer and the two fields are those below the nth and $(n+1)$th layers, respectively (see Fig. 4.3). The amplitudes $\{\boldsymbol{a}_{\boldsymbol{h}}^{-}\}$, directed along $-z$ in the lower region, and $\{b_{\boldsymbol{h}}^{+}\}$, directed along $+z$ in the upper region, are those waves that are going away from the layer. They are brought about by the waves incoming to the layer, $\{\boldsymbol{a}_{\boldsymbol{h}}^{+}\}$ from below and $\{b_{\boldsymbol{h}}^{-}\}$ from above. Consider, e.g., the wave

$$\exp\left(\mathrm{i}\boldsymbol{k}_{\boldsymbol{h}}^{+} \cdot \boldsymbol{r}\right)$$

incident on the layer from below. It gives rise to a series of reflected waves by diffraction in the lower side and transmitted waves in the upper side. Let the process of the $\boldsymbol{h}$ wave being reflected back to an $\boldsymbol{h}'$ wave be described by

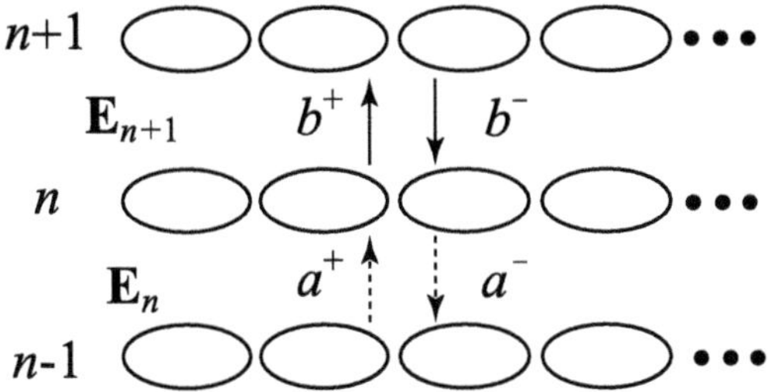

Fig. 4.3. Fields between the layers

the tensor reflection amplitude $\boldsymbol{r}_{\boldsymbol{h}'\boldsymbol{h}}^{(-+)}$, the superscript $^{(-+)}$ standing for the change of the propagation direction from $+z$ to $-z$. The transmission of the $\boldsymbol{h}$ wave is expressed similarly by the amplitide $\boldsymbol{t}_{\boldsymbol{h}'\boldsymbol{h}}^{(++)}$, etc. The assumption that we already have full knowledge of the monolayer scattering implies that we know $\{\boldsymbol{r}_{\boldsymbol{h}'\boldsymbol{h}}^{(-+)}\}$ and $\{\boldsymbol{t}_{\boldsymbol{h}'\boldsymbol{h}}^{(++)}\}$, etc., for all $\boldsymbol{h}$ and $\boldsymbol{h}'$ as a function of ω and k_x, k_y.

Combining the incoming waves coming from above the layer, it then follows that

$$\begin{pmatrix} \boldsymbol{b}^+ \\ \boldsymbol{a}^- \end{pmatrix} = \begin{pmatrix} \boldsymbol{t}^{(++)} & \boldsymbol{r}^{(+-)} \\ \boldsymbol{r}^{(-+)} & \boldsymbol{t}^{(--)} \end{pmatrix} \begin{pmatrix} \boldsymbol{a}^+ \\ \boldsymbol{b}^- \end{pmatrix}, \tag{4.19}$$

where we have used the symbols for the blocks and each of the two blocks $(+)$ and $(-)$ is composed of the channels specified by all possible 2D reciprocal lattice vectors $\boldsymbol{h}$, open or closed. The column vector on the right hand side represents the incoming waves. It gives rise to the outgoing waves of the left hand side through the 2×2 block matirix on the right hand side. This matrix is the scattering matrix of the monolayer scattering, which connects the outgoing waves with the incoming waves.

Now an infinite and periodic assembly of layers makes up a photonic crystal. There, the Bloch theorem should hold:

$$\boldsymbol{b}^{\pm} = \mathrm{e}^{\mathrm{i}k_z\ell_z}\boldsymbol{a}^{\pm} \tag{4.20}$$

ℓ_z being the lattice constant in the z direction, because $\boldsymbol{a}$ and $\boldsymbol{b}$ are the amplitudes involved in $\boldsymbol{E}(x, y, z)$ and $\boldsymbol{E}(x, y, z+\ell_z)$, respectively. Eliminating $\boldsymbol{b}^{\pm}$ from (4.19) and (4.20), we find

$$\begin{aligned} (1 - \mathrm{e}^{-\mathrm{i}k_z\ell_z}\boldsymbol{t}^{(++)})\boldsymbol{a}^+ - \boldsymbol{r}^{(+-)}\boldsymbol{a}^- &= 0, \\ -\boldsymbol{r}^{(-+)}\boldsymbol{a}^+ + (1 - \mathrm{e}^{\mathrm{i}k_z\ell_z}\boldsymbol{t}^{(--)})\boldsymbol{a}^- &= 0, \end{aligned} \tag{4.21}$$

from which we can construct the secular equation

$$\det \begin{vmatrix} 1 - \mathrm{e}^{-\mathrm{i}k_z\ell_z}\boldsymbol{t}^{(++)} & -\boldsymbol{r}^{(+-)} \\ -\boldsymbol{r}^{(-+)} & 1 - \mathrm{e}^{\mathrm{i}k_z\ell_z}\boldsymbol{t}^{(--)} \end{vmatrix} = 0. \tag{4.22}$$

Remember that the two blocks are composed of the matrices labeled by the channels and 1 in the diagonal element stands for the unit matrix. The solutions of (4.22) give the band energy for the wavevector (k_x, k_y, k_z), whose k_x and k_y are involved in the $\boldsymbol{r}$'s and $\boldsymbol{t}$'s and k_z appears in the two exponentials in the above secular equation.

So long as we know the $\boldsymbol{t}$'s and $\boldsymbol{r}$'s of a monolayer, we can get the band structure of any 3D PC by this method. For the cases of spheres and cylinders the expression of the monolayer data is found by [14] and [16]. In the general case, we may apply this method using the monolayer data obtained by plane-wave expansion method. If the composition geometry of the unit cell is complicated we have to employ numerical means to obtain the scattering data of the monolayer. The idea of this method is due to Pendry [22, 23].

4.2 Group Theory of Photonic Crystals

Except for some band assignments and related discussions, group theory has not been applied fully in the field of PCs. In the future, where more basic and detailed properties of PBs are expected to be used, knowledge of group theory will play a much more important role both theoretically and experimentally. Here we describe the basic aspects of the group theory of PBs, with their applications to some numerical data. The theoretical aspect given here is based upon the content of [18]. Other than this reference, many authors have treated group-theoretical topics. Among others, we cite some of the earlier works [24, 25]. See the book of Sakoda for illustrative descriptions [26]

4.2.1 Invariance agaist Lattice Symmetry Operations

Once the basic invariance properties are established in the Maxwell fields, group theory is applied literally to PBs just as in the band structures of phonons and electrons.

A PC of *infinite size* looks completely alike when a symmetry operation of the space group of that PC is applied. For example, an array of cylinders of circular cross-section which forms a 2D PC of a square lattice is invariant even if it is moved by an arbitrary lattice vector. This is the translational invariance. It also looks completely similar when we give a $\pi/2$ rotation about the cylinder axis (z axis) of any of the cylinders. Including this $\pi/2$ rotation, denoted as C_4, and the trivial identity operation I, it has a group of invariant operations composed of C_4^2 (rotation by π), C_4^3, σ_x, σ_y (mirror refections in the xz and yz planes, respectively, with x and y axes taken along two sides of a square), and σ_d, σ_d' (two mirror reflections in the two diagonal planes parallel to the z axis). These eight operations form a group, called the point group $\mathrm{C}_{4\mathrm{v}}$. Therefore, this 2D PC is invariant against any combined operation of a lattice translation and an operation of $\mathrm{C}_{4\mathrm{v}}$. The space group is defined to be the group of combined operations of them. The group composed solely of lattice translations is called the translation group. It is a subgroup of the space group. The Bloch theorem is established for the fields of PBs because of the translational invariance of PCs; any wavefunction of electrons, photons, phonons, etc. of a periodic system should be a basis function of the irreducible representation of the translation group. Besides the Bloch theorem coming from the translational invariance, the fields have additional important symmetry properties related to the rotations, mirror reflection and inversion of the space group.

Since the vector fields are the objects involved in PCs, we should begin by defining the transformation property of vector fields. Let R be a rotation operation of a point group. For an arbitrary vector $\boldsymbol{a}$, the transformed vector $R\boldsymbol{a}$ is defined to be a new vector generated by rotating the original one $\boldsymbol{a}$ by R. The operation R on a scalar field $F(\boldsymbol{r})$ is usually defined by the new function $F' = RF$, whose value at the position $\boldsymbol{r}'$ is

$$F'(\boldsymbol{r}') = F(\boldsymbol{r}), \tag{4.23}$$

where

$$\boldsymbol{r}' = R\boldsymbol{r}. \tag{4.24}$$

We may rewrite this as

$$F'(\boldsymbol{r}) = F(R^{-1}\boldsymbol{r}). \tag{4.25}$$

Therefore, the contour map of $F' = RF$ is that of F rotated by R.

In accordance with this, the operation R on a vector field $\boldsymbol{F}$ is defined by

$$\begin{aligned} \boldsymbol{F}'(\boldsymbol{r}') &\equiv (R\boldsymbol{F})(\boldsymbol{r}') \\ &= R\left(\boldsymbol{F}(\boldsymbol{r})\right), \end{aligned} \tag{4.26}$$

i.e., $\boldsymbol{F}'$ is obtained by rotating the vector $\boldsymbol{F}(\boldsymbol{r})$ by R at the original position $\boldsymbol{r}$ and then by transferring it to the rotated position $\boldsymbol{r}'$.

In the form equivalent to (4.25), this may be expressed as

$$\boldsymbol{F}'(\boldsymbol{r}) = R\left(\boldsymbol{F}(R^{-1}\boldsymbol{r})\right). \tag{4.27}$$

This means that when $\boldsymbol{F}$ is expressed as

$$\boldsymbol{F}(\boldsymbol{r}) = F_x(\boldsymbol{r})\boldsymbol{i} + F_y(\boldsymbol{r})\boldsymbol{j} + F_z(\boldsymbol{r})\boldsymbol{k} \tag{4.28}$$

with three cartesian unit vectors $\boldsymbol{i}$, $\boldsymbol{j}$, $\boldsymbol{k}$, the transformed vector field $R\boldsymbol{F}$ may be written as

$$(R\boldsymbol{F})(\boldsymbol{r}) = F_x(R^{-1}\boldsymbol{r})(R\boldsymbol{i}) + F_x(R^{-1}\boldsymbol{r})(R\boldsymbol{j}) + F_x(R^{-1}\boldsymbol{r})(R\boldsymbol{k}). \tag{4.29}$$

Namely, in the transformation of a vector field, the scalar field F_x, etc. is subject to the transformation of scalar fields and the vector $\boldsymbol{i}$, etc. is subject to that of vectors.

The invariance of a PC is usually meant to signify the invariance of Maxwell's equation against any operation of the space group. Using $\varepsilon(\boldsymbol{r})$, the periodic dielectric function of that PC, the invariance is defined to be

$$g(-\nabla\times\nabla\times+\omega^2\varepsilon(\boldsymbol{r}))g^{-1}=(-\nabla\times\nabla\times+\omega^2\varepsilon(\boldsymbol{r})) \tag{4.30}$$

for any operator g of the space group to which the PC belongs. Note that in the form $g(\cdots)g^{-1}$ the right operator g^{-1} works as terminating there the left operation g and does not let it operate further on what follows after g^{-1}. Let us see this by an example of an operator $g\varepsilon g^{-1}$ of the second term. Defining an operation ε on a vector field $\boldsymbol{f}$ to be

$$(\varepsilon\boldsymbol{f})(\boldsymbol{r})=\varepsilon(\boldsymbol{r})\boldsymbol{f}(\boldsymbol{r}),$$

we can understand the operation of $g\varepsilon g^{-1}$ on an arbitrary vector field $\boldsymbol{f}$ as follows:

$$\begin{aligned}(g\varepsilon g^{-1}\boldsymbol{f})(\boldsymbol{r})&=(g\varepsilon\boldsymbol{f}')(\boldsymbol{r})\\&=\varepsilon(g^{-1}\boldsymbol{r})g\left(\boldsymbol{f}'(g^{-1}\boldsymbol{r})\right)\end{aligned}$$

with $\boldsymbol{f}'=g^{-1}\boldsymbol{f}$. Since

$$\begin{aligned}\boldsymbol{f}'(g^{-1}\boldsymbol{r})&=(g^{-1}\boldsymbol{f})(g^{-1}\boldsymbol{r})\\&=g^{-1}\left(\boldsymbol{f}(gg^{-1}\boldsymbol{r})\right)\\&=g^{-1}\left(\boldsymbol{f}(\boldsymbol{r})\right),\end{aligned}$$

we find

$$(g\varepsilon g^{-1}\boldsymbol{f})(\boldsymbol{r})=\varepsilon(g^{-1}\boldsymbol{r})\boldsymbol{f}(\boldsymbol{r}). \tag{4.31}$$

Thus g^{-1} terminates the operation of g. Using this last relation, we find

$$(g\varepsilon g^{-1}\boldsymbol{f})(\boldsymbol{r})=\varepsilon(\boldsymbol{r})\boldsymbol{f}(\boldsymbol{r}) \tag{4.32}$$

because the value of the dielectric constant ε at the shifted position $g^{-1}\boldsymbol{r}$ on the right hand side of (4.31) is the same as the original position. Equation (4.32) proves

$$g\varepsilon g^{-1}=\varepsilon \tag{4.33}$$

i.e., the dielectric function is invariant.

Admitting this role of g^{-1}, we find

$$(g\nabla g^{-1}f)(\boldsymbol{r})=\frac{\partial f(\boldsymbol{r})}{\partial x'}g\boldsymbol{i}+\frac{\partial f(\boldsymbol{r})}{\partial y'}g\boldsymbol{j}+\frac{\partial f(\boldsymbol{r})}{\partial x'}g\boldsymbol{k}$$

with $g^{-1}\boldsymbol{r}=(x',y',z')$. That this last expression is indeed equal to

$$=\frac{\partial f(\boldsymbol{r})}{\partial x}\boldsymbol{i}+\frac{\partial f(\boldsymbol{r})}{\partial y}\boldsymbol{j}+\frac{\partial f(\boldsymbol{r})}{\partial x}\boldsymbol{k}\equiv(\nabla f)(\boldsymbol{r})$$

is obvious for a translational operator g. For a g involving a rotational operator, we need to express the transformed vector, $g^{-1}\boldsymbol{r}, g\boldsymbol{i}, g\boldsymbol{j}, g\boldsymbol{k}$, using the 3×3 rotational matrix. In this way we can show

$$g\boldsymbol{\nabla}g^{-1} = \boldsymbol{\nabla}.$$

The same invariant property holds for $\boldsymbol{\nabla}\cdot$ and $\boldsymbol{\nabla}\times$. The conclusion is thus that

$$g\boldsymbol{\nabla}g^{-1} = \boldsymbol{\nabla} \qquad g(\boldsymbol{\nabla}\cdot)g^{-1} = \boldsymbol{\nabla}\cdot, \qquad g(\boldsymbol{\nabla}\times)g^{-1} = \boldsymbol{\nabla}\times . \tag{4.34}$$

These lead to the invariance of each of the four Maxwell equations for $\boldsymbol{E}(\boldsymbol{r})$ and $\boldsymbol{B}(\boldsymbol{r})$.

¿From (4.34) it thus follows that

$$g\big(-\boldsymbol{\nabla}\times\boldsymbol{\nabla}\times + \omega^2\varepsilon(\boldsymbol{r})\big) = \big(-\boldsymbol{\nabla}\times\boldsymbol{\nabla}\times + \omega^2\varepsilon(\boldsymbol{r})\big)g. \tag{4.35}$$

The same thing holds for the magnetic field, displacement field, vector potential and so on. From this commutation relation, everything familiar in the group theory of the band structure of electrons applies to the PB structure. We refer readers unfamiliar with the group theory of solids to textbooks such as by Inui, Onodera and Tanabe [27].

4.2.2 Group of $\boldsymbol{k}$ and Basic Group-Theoretical Properties

Let us summarize some of the properties useful in PCs:

(1) The solutions of Maxwell's equations form the basis functions of an irreducible representation of the space group of the PC in question.

(2) A solution $\boldsymbol{E}(\boldsymbol{r})$ of a PC and its transform $g\boldsymbol{E}(\boldsymbol{r})$ by an operation g are degenerate, if g is an operator of the space group.
When g is a lattice translation, not involving a rotation, $g\boldsymbol{E}(\boldsymbol{r})$ is not indepenent of the original function $\boldsymbol{E}(\boldsymbol{r})$, because the Bloch theorem tells us that

$$\begin{aligned} g\boldsymbol{E}(\boldsymbol{r}) &= \boldsymbol{E}(g^{-1}\boldsymbol{r}) \\ &= \boldsymbol{E}(\boldsymbol{r}+\boldsymbol{R}) \\ &= \mathrm{e}^{\mathrm{i}\boldsymbol{k}\cdot\boldsymbol{R}}\boldsymbol{E}(\boldsymbol{r}), \end{aligned}$$

when $g = T(-\boldsymbol{R})$ is a pure lattice translation by a lattice vector $-\boldsymbol{R}$. This means that the solution $\boldsymbol{E}(\boldsymbol{r})$ is in itself a basis function of one-dimensional irreducible representation of the translation group. The exponent of the phase factor $\mathrm{e}^{\mathrm{i}\boldsymbol{k}\cdot\boldsymbol{R}}$ defines the wavevector of the Bloch state in the first BZ. Thus the wavevector $\boldsymbol{k}$ of a Bloch state specifies the irreducible representations of the translation group.
When g involves a rotation, $g = C_4$ for example, $g\boldsymbol{E}(\boldsymbol{r})$ belongs to the Bloch states of wavevector $g\boldsymbol{k}$. This is most straightforwardly seen if we

use the plane-wave expansion form of the PB state; for the Bloch state of wavevector $\boldsymbol{k}$, it holds that

$$g\boldsymbol{E}_{\boldsymbol{k}}(\boldsymbol{r}) = g\sum_{\boldsymbol{h}} \boldsymbol{e}_{\boldsymbol{k}}(\boldsymbol{h})\mathrm{e}^{\mathrm{i}(\boldsymbol{k}+\boldsymbol{h})\cdot\boldsymbol{r}}$$
$$= \sum_{\boldsymbol{h}} (g\boldsymbol{e}_{\boldsymbol{k}}(\boldsymbol{h}))\,\mathrm{e}^{\mathrm{i}(\boldsymbol{k}+\boldsymbol{h})\cdot(g^{-1}\boldsymbol{r})}.$$

Since

$$(\boldsymbol{k}+\boldsymbol{h})\cdot(g^{-1}\boldsymbol{r}) = (g\boldsymbol{k}+g\boldsymbol{h})\cdot\boldsymbol{r}$$

and $g\boldsymbol{h}$ is another reciprocal lattice vector $\boldsymbol{h}'$, we find

$$g\boldsymbol{E}_{\boldsymbol{k}}(\boldsymbol{r}) = \sum_{\boldsymbol{h}'} g\left(\boldsymbol{e}_{\boldsymbol{k}}(g^{-1}\boldsymbol{h}')\right)\mathrm{e}^{\mathrm{i}(g\boldsymbol{k}+\boldsymbol{h}')\cdot\boldsymbol{r}}$$
$$\equiv \sum_{\boldsymbol{h}'} \boldsymbol{e}_{g\boldsymbol{k}}(\boldsymbol{h}')\mathrm{e}^{\mathrm{i}(g\boldsymbol{k}+\boldsymbol{h}')\cdot\boldsymbol{r}}.$$

This last equation shows that $g\boldsymbol{E}_{\boldsymbol{k}}(\boldsymbol{r})$ is a Bloch state belonging to the wavevector $g\boldsymbol{k}$, because

$$T(-\boldsymbol{R})\,(g\boldsymbol{E}_{\boldsymbol{k}}(\boldsymbol{r})) = \sum_{\boldsymbol{h}'} \boldsymbol{e}_{g\boldsymbol{k}}(\boldsymbol{h}')\mathrm{e}^{\mathrm{i}(g\boldsymbol{k}+\boldsymbol{h}')\cdot(\boldsymbol{r}+\boldsymbol{R})}$$
$$= \mathrm{e}^{\mathrm{i}g\boldsymbol{k}\cdot\boldsymbol{R}}\,(g\boldsymbol{E}_{\boldsymbol{k}}(\boldsymbol{r}))\,.$$

The point $g\boldsymbol{k}$ of the BZ is called a star of $\boldsymbol{k}$. The PBs at the stars of $\boldsymbol{k}$ have identical energies. This is why we need to calculate the band structure only in the limited area of the first BZ: in the 2D photonic lattice of C_{4v}, for example, we calculate 1/8 of the region of the first BZ.

(3) The set of operators g, which involves the rotational part that leaves $\boldsymbol{k}$ under consideration at rest, forms a subgroup of the space group.
For every wavevector $\boldsymbol{k}$ in the Brilloin zone, we can find such a group $\mathcal{G}(\boldsymbol{k})$, called the group of $\boldsymbol{k}$, which consists of the operators g which leaves $\boldsymbol{k}$ unaltered. At a point of high symmetry, as $\boldsymbol{k}=0$ or $\boldsymbol{k}=(k_x,0,0)$, $\mathcal{G}(\boldsymbol{k})$ consists of several operations. For $\boldsymbol{k}=0$, the full point group of the lattice is involved in $\mathcal{G}(\boldsymbol{k})$ and for $\boldsymbol{k}=(k_x,0,0)$ the operations which leave the k_x axis invariant comprise $\mathcal{G}(\boldsymbol{k})$. In contrast, for $\boldsymbol{k}$ of a general point of the BZ such as $\boldsymbol{k}=(0.1,0.3,-0.1)$ in units of the BZ edges, $\mathcal{G}(\boldsymbol{k})$ is composed solely of the identity operation.

(4) We can specify any band at $\boldsymbol{k}$ by the irreducible representations of $\mathcal{G}(\boldsymbol{k})$. In other words, the band index n specifies one of the irreducible representations of $\mathcal{G}(\boldsymbol{k})$. The dimension of an irreducible representation is equal to the degeneracy of the bands which belong to it. Therefore we can predict the degeneracy of the band structure without making an actual band calculation, if only we know the irreducible representation of that band. The group $\mathcal{G}(\boldsymbol{k})$ and its irreducible representation are usually tabulated in textbooks of group theory.

(5) For a given $\boldsymbol{k}$ in the BZ, we can construct the basis functions of any irreducible representation of $\mathcal{G}(\boldsymbol{k})$ from the assembly of the plane waves of Bloch type. They are called symmetry-adapted basis functions. This can be carried out by using the projection operator for each irreducible representation of $\mathcal{G}(\boldsymbol{k})$. For the details of the projection operators, see, e.g., the book of group theory cited above.

4.2.3 Symmetry-Related Polarizations of Plane Waves

We here examine the symmetry operation on a Bloch electric field $\boldsymbol{E}_{\boldsymbol{k}}^{(n)}(\boldsymbol{r})$ of wavevector $\boldsymbol{k}$ and band index n.

The irreducible representation of state $(n, \boldsymbol{k})$ determines the symmetry adapted basis functions. This means that, without a detailed BS calculation, the amplitudes $\boldsymbol{e}_{\boldsymbol{k}}^{(n)}(\boldsymbol{h})$ of the plane waves $\boldsymbol{h}$ are fixed for the set of $\boldsymbol{h}$ points. The set is made of the plane waves associated with the reciprocal lattice points, connected mutally by the operations in $\mathcal{G}(\boldsymbol{k})$. They are reciprocal lattice points located on the shell of equal length from the origin $\boldsymbol{h} = 0$ of the reciprocal lattice.

To be more specific, let us consider a PC of fcc lattice of cube size a and consider the PBs for $\boldsymbol{k}$ in the (001) direction. The reciprocal lattice is a bcc lattice. The group $\mathcal{G}(\boldsymbol{k})$ for this $\boldsymbol{k}$ contains the point group C_{4v}, which consists of the $\pm\pi/2$ and π rotations around the z axis and the mirror reflections in the planes $x = 0$, $y = 0$ and $x = \pm y$. The point group C_{4v} has five irreducible representations, A_1, A_2, B_1, B_2, and E, of which E is two-dimensional and the rest are one-dimensional. The first shell of the bcc reciprocal lattice points mutually connected by the C_{4v} operations is composed only of $\boldsymbol{h} = 0$, the second shell is composed of four body-center points $(1, 1)(2\pi/a)$, $(1, -1)(2\pi/a), \ldots$, third shell is made of cube-corners $(\pm 2, 0)(2\pi/a)$, $(0, \pm 2)(2\pi/a)$, and so on (with the z components of $\boldsymbol{h}$ dropped).

Figure 4.4 shows the diagram for the $\boldsymbol{e}_{\boldsymbol{k}}^{(n)}(\boldsymbol{h})$ with $\boldsymbol{k}$ in the (001) direction for the four $\boldsymbol{h}$ in the second shell [18]. In the group σ, $\boldsymbol{e}_{\boldsymbol{k}}(\boldsymbol{h})$ is directed parallel to $\boldsymbol{h}$, while in the group π it is directed perpendicularly. In the two-dimensional irreducible representation of E, the diagram for only one of the two basis functions is given. Its partner is obtained by rotation through $\pi/2$ about the z axis. A similar diagram is obtained for the higher shell.

The polarizations of the plane waves shown in Fig. 4.4 also provide the basis functions of the PBs of $(k_x, k_y) = (0, 0)$ in a slab PC of fcc symmetry bounded in the z direction by two parallel (001) surfaces. They are the PBs at the $\boldsymbol{\Gamma}$ point of the 2D BZ. Any band belonging to, say, the A_2 representation, is expressed by the linear combination of the basis functions of $A_2(\pi)$, $A_2(\sigma)$ of first, second, ... shells. A similar diagram for the cubic group O_h is given in [17].

To excite the PBs at Γ point in our slab system, we use incident light of normal incidence with the lateral wavevector $k_x = k_y = 0$. A pair of incident

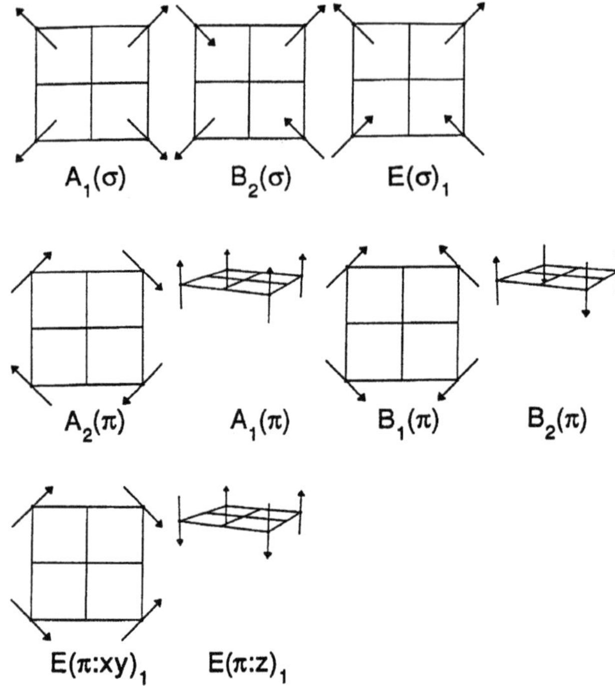

Fig. 4.4. Symmetry-adapted basis vectors for the point group C_{4v}. For the four reciprocal lattice points nearest to the z axis, whose x and y components are (1 1), (1 −1), (−1, 1) and (−1 −1), the basis fuctions for the five irreducible representation of C_{4v} are given

light waves of polarization in the x and y directions are degenerate and they are the partners of the basis functions for the irreducible representation E. As a consequence, the PBs that can respond to them are only those of E symmetry at the Γ point. This means that the PBs active to light of normal incidence must be two-fold degenerate and that all the PBs except those of E symmetry are silent at Γ to incident light of normal incidence. These silent PBs are usually called optically inactive modes. The lack of the coupling of the one-dimensional irreducible representations is connected with the fact that the first shell $\boldsymbol{h} = 0$ produces only the basis functions of the E representation.

When light of normal incidence gives rise to diffracted light associated with, e.g., four reciprocal lattice points of the second shell, the polarizations of the four diffracted light waves, transmitted or reflected, are determined by the polarizations of the symmetry-adapted plane waves of the second shell.

When the frequency of the incident light is short for the diffraction, the second-shell plane waves of the PBs of E symmetry produce evanescent light outside the surfaces. Near the surface, the pile up of their amplitudes gives rise to an enhanced near-field intensity. As in the case of diffraction, we can predict the intensity profile and the polarization of the near-field fairly well from the properties of symmetry-adapted plane-wave basis funtions. An example of the analysis for the near-field intensity in relation to the symmetry of excited PBs is given in Miyazaki and Ohtaka [28].

4.2.4 Application of Group Theory

Orthogonality of Different Bands. The basis functions of different irreducible representations are orthogonal. The eigenvectors of PBs of different $\boldsymbol{k}$ are thus orthogonal, because they belong to different irreducible representations of the translation group. It thus holds that for $\boldsymbol{k} \neq \boldsymbol{k}'$

$$\int_{\mathrm{pc}} \mathrm{d}\boldsymbol{r} \boldsymbol{E}_{\boldsymbol{k}}^{(n)}(\boldsymbol{r})^* \cdot \boldsymbol{E}_{\boldsymbol{k}'}^{(n')}(\boldsymbol{r}) = 0, \tag{4.36}$$

irrespective of whether $n = n'$ or not (the symbol pc stands for the integral over the entire PC volume). The orthogonality of the two states $\boldsymbol{k} \neq \boldsymbol{k}'$ is established not only between the electric fields but also between different fields, between the elecric and magnetic fields, for example.

The orthogonality still holds for a pair of band states of the same $\boldsymbol{k}$, if they belong to different irreducible representations of the space group. Since irreducible representations of the space group are constructed from those of $\mathcal{G}(\boldsymbol{k})$, the orthogonality is established for the different irreducible representations of $\mathcal{G}(\boldsymbol{k})$. This should hold, again, irrespective of the kind of fields in question. In the plane-wave expansion of the $(\mathrm{n}, \boldsymbol{k})$ state

$$\boldsymbol{E}_{\boldsymbol{k}}^{(n)}(\boldsymbol{r}) = \sum_{\boldsymbol{h}} \boldsymbol{e}_{\boldsymbol{k}}^{(n)}(\boldsymbol{h}) \exp[\mathrm{i}(\boldsymbol{k}+\boldsymbol{h}) \cdot \boldsymbol{r}], \tag{4.37}$$

$\boldsymbol{h}$ being a reciprocal lattice vector, the orthogonality between the two electric fields of the same $\boldsymbol{k}$ is defined by using the inner product over the unit cell

$$\int_{\mathrm{uc}} \mathrm{d}\boldsymbol{r} \boldsymbol{E}_{\boldsymbol{k}}^{(n)}(\boldsymbol{r})^* \cdot \boldsymbol{E}_{\boldsymbol{k}}^{(n')}(\boldsymbol{r}) = v_{\mathrm{c}} \sum_{\boldsymbol{h}} \boldsymbol{e}_{\boldsymbol{k}}^{(n)}(\boldsymbol{h})^* \cdot \boldsymbol{e}_{\boldsymbol{k}}^{(n')}(\boldsymbol{h}) = 0 \tag{4.38}$$

with v_{c} the volume of the unit cell. To summarize, the inner product vanishes similarly in the DD scheme, i.e., between the displacement fields of two bands, and ED, EB, EH, HB, etc., when the two PB states belong to different irreducible representations.

If an operator is hermitian, its two eigenvectors which have *different* eigenvalues are orthogonal. In the case of the photonic bands, the orthogonality in this sense does not necessarily hold between the two electric fields. Instead, the orthogonality between bands of *different* eigenvalues is established between the electric and displacement fields. This is obvious from the orthogonality between the eigenvector

$$\boldsymbol{Q}_{\boldsymbol{k}}^{(n)}(\boldsymbol{r}) \equiv \sqrt{\varepsilon(\boldsymbol{r})} \boldsymbol{E}_{\boldsymbol{k}}^{(n)}(\boldsymbol{r}),$$

which will be introduced in Sect. 4.5.1 in relation to the hermitian formalism of PBs. We find the inner product between $\boldsymbol{Q}_{\boldsymbol{k}}^{(n)}$ and $\boldsymbol{Q}_{\boldsymbol{k}}^{(m)}$ turns out to be

$$\begin{aligned}\int_{\mathrm{uc}} \mathrm{d}\boldsymbol{r} \boldsymbol{Q}_{\boldsymbol{k}}^{(n)}(\boldsymbol{r})^* \cdot \boldsymbol{Q}_{\boldsymbol{k}}^{(m)}(\boldsymbol{r}) &= \int_{\mathrm{uc}} \mathrm{d}\boldsymbol{r} \boldsymbol{E}_{\boldsymbol{k}}^{(n)}(\boldsymbol{r})^* \cdot \varepsilon(\boldsymbol{r}) \boldsymbol{E}_{\boldsymbol{k}}^{(m)}(\boldsymbol{r}) \qquad (4.39)\\ &= \int_{\mathrm{uc}} \mathrm{d}\boldsymbol{r} \boldsymbol{E}_{\boldsymbol{k}}^{(n)}(\boldsymbol{r})^* \cdot \boldsymbol{D}_{\boldsymbol{k}}^{(m)}(\boldsymbol{r}).\end{aligned}$$

In conclusion, what establishes between bands of different frequencies is the ED orthogonality, whether or not they are of the same irreducible representation of $\mathcal{G}(\boldsymbol{k})$. When, in particular, they are of different irreducible representations of $\mathcal{G}(\boldsymbol{k})$, the two bands are orthogonal between any pair of fields. In [18], a calculation is given to demonstrate the features described here.

Response of a Photonic Band to External Light. Suppose a PC with its (100) surface on the xy plane has plane-wave light propagatong in the $+z$ direction incident upon it. Due to the continuity of the wavevector components parallel to the entrance surface, the PBs excited inside are those of $\boldsymbol{k} = (00k_z)$. Therefore a complete stop-band of total reflection appears in the reflectivity, whenever the frequency of the light lies in the region where degenerate bands are absent (an example will be given in Fig. 4.5(a) below).

Suppose next that the incident condition changes and the wavevector of an excited Bloch wave shifts to a general point in the BZ ($\boldsymbol{k} = (k_x k_y k_z)$ with three different components). The point group of $\mathcal{G}(\boldsymbol{k})$ is then composed solely of the identity operation with the result that all the photonic bands are nondegenerate. Therefore a slight altering of the wavevector off the symmetry axis of the incident light makes all the PBs active to the incident light.

We hereafter consider the intermediate case between the two: some of the nondegenerate bands are active and others not. This case is realized for $\boldsymbol{k} = (k_x, 0, k_z)$ and corresponds to the incident wavevector lying in the xz plane but inclined off the Δ axis. For $\boldsymbol{k} = (k_x, 0, k_z)$, the point group of $\mathcal{G}(\boldsymbol{k})$ consists of the identity operation and the mirror reflection in the xz plane. This simple point group has two one-dimensional irreducible representaions, A$'$ and A$''$. Thus all the bands for $\boldsymbol{k} = (k_x, 0, k_z)$ are nondegenerate and have polarization $\boldsymbol{E}_{\boldsymbol{k}}(\boldsymbol{h})$ in the xz plane (in the case of A$'$ symmetry) or perpendicular to it (A$''$).

These PBs are excited by incident light of wavevector component $(k_x, 0)$ parallel to the surface. The incident light has choices of p- and s-polarizations. The PBs of A$'$ are excited by the p-polarized light (polarized in the xz plane) and those of A$''$ respond only to the s-polarized light (polarized in the y direction) of 2D wavevector component $(k_x, 0)$.

Figure 4.5 (a) shows the reflectivity of light of normal incidence [18]. We are considering the (100) surface of a PC composed of $N = 32$ layers of arrayed spheres in the fcc lattice. The parameters of the spheres are the same as those used in the PBS shown in Fig. 4.2. The panel (b) shows the

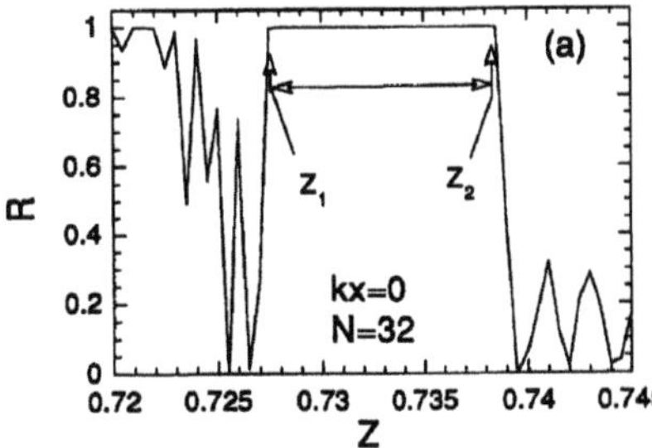

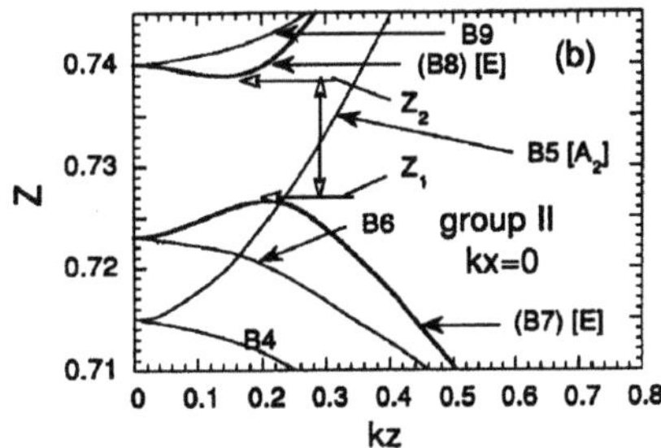

Fig. 4.5. Reflectivity of light of normal incidence from a fcc PC of arrayed spheres. The result is for the PC composed of 32 layers of the (100) planes of spheres. Panel (a) shows normal indidence and panel (b) gives the band structure of the related range of the normalized frequency Z, defined by $\omega a/(2\pi c)$, a being the size of fcc cube. The edge of the broad stop-band is marked by the arrows in both panels

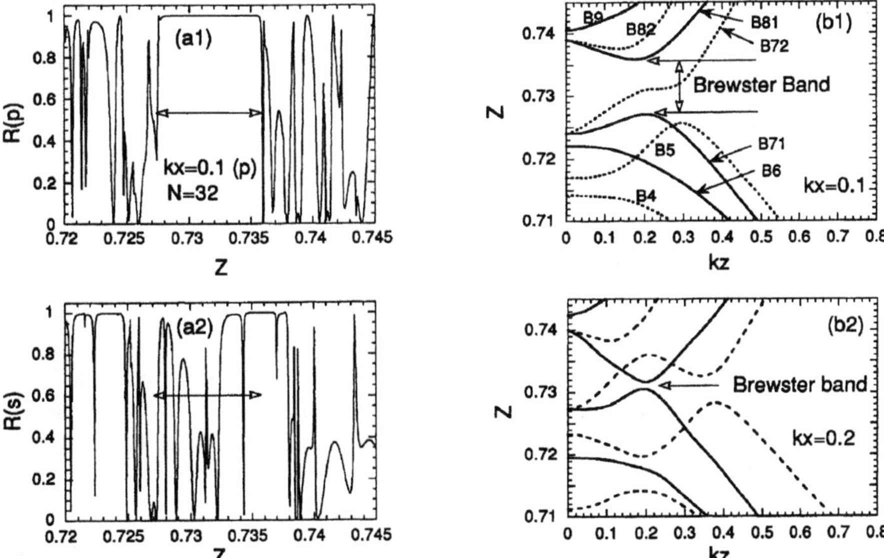

Fig. 4.6. Reflectivities of p- and s-polarized light in oblique incidence: (a1) for the p-polarized incident light with $k_x = 0.1$ and (a2) for the s-polarized light. Panel (b1) shows the band structure for the relevant frequency range. The solid curves show the p-active bands and the dashed ones the s-active bands. Panel (b2) shows the Brewster band with $k_x = 0.2$ just before its disappearance

band structure of the relevent frequency range. We see that a broad stop-band of reflectivity exists in the frequency range $Z_1 \leq Z \leq Z_2$, whose edges are marked by the arrows in the two panels. We note the existence of the nondegenerate band B5 within the stop-band, which leaves no trace in the reflectivity.

The reflectivity of oblique incidence is shown in Fig. 4.6 [18] for the p-polarized light in (a1) and for the s-polarized in (a2). We took $k_x = 0.1$. Panel (b1) shows the relevant band structure of $k_x = 0.1$, in which the p-active bands are shown by solid curves and the s-active bands by the dashed curves. In the frequency range marked as the "Brewster band" in panel (b1), the s-active band of B7-2 exists but no p-active band is found. This feature is very well reflected in (a1) and (a2), in that the p-polarized light suffers from total reflection, while the s-polarized light passes through the PC.

Panel (b2) for $k_x = 0.2$ shows the band structure just before the disappearance of the stop-band for the p polarization. When k_x is increased further, we indeed confirmed that there is no longer a complementary characteristic in the p and s reflectivities.

A special incident angle of light which gives the total reflection of one of the polarizations is known as the Brewster angle in the light reflection and transmission through a homegeneous system. It is to be noted that the argument given above for a PC holds for a range of incident angle or of the incident frequency. In a PC, therefore, elliptically polarized light is linearly polarized after passing through it in a range of incident parameters. This characteristic of PCs may be more appropriately called the existence of the Brewster band of angle or frequency. The feature examined above is an example revealing the possibility of PCs as a useful polarizer of light.

4.3 Leaky Modes of Slab-Type Photonic Crystals

In a slab PC, extended to infinity in the x and y directions and having a finite thickness in the z direction, there arise two types of photonic bands, leaky and nonleaky bands found inside and outside, respectively, of the light cone in the $(\omega, \boldsymbol{k})$ phase space (Sect. 3.4). The calculation of the dispersion relations of nonleaky bands is relatively easy. They are now receiving intensive attention as promising transporters of information in future optical circuits. The properties of nonleaky PBs will be treated later in Sect. 6.1. This section is devoted to the calculation of the dispersion relation and lifetime of leaky PBs, which are essentially affected by the incomplete confimement effect.

Experimentally we have a number of well-established methods of measuring the width of a resonant signal due to the excitation of leaky PBs. In light reflection and transmission, for example, the finesse of the interference fringes has direct information about the leaky modes. Another example is the emission spectrum of a photon from an electron in a PC. When it is observed via the excitation of a leaky PB, the sharpness of the resonant signals reflects directly the radiative lifetime of the leaky PB.

Theoretical treatment of the dispersion relations and lifetimes may be carrid out, if we can somehow make a successful calculation of the density of states (DOS) of these leaky modes. The DOS of the electromagnetic modes of the entire system, a slab PC plus free space surrounding it, should have sharp increases as a function of frequency, which reveal the existence of the leaky modes. Here it is essential to take into account the modes of the exterior region, because the leaky PBs cannot be separated from them. The width in frequency of a DOS peak determines the lifetime of the leaky mode manifested by the peak. If we follow the peak position as a function of $\boldsymbol{k}_{\|}$, we can get the dispersion relation of that PB, where $\boldsymbol{k}_{\|}$ is the wavevector in the lateral plane, which we can use in specifying the normal modes of the entire system. The width of the peak may also vary as a function of $\boldsymbol{k}_{\|}$, which reveals that the radiative lifetime or the degree of the confinement depends on the value of $\boldsymbol{k}_{\|}$ of the PB. In this way we may obtain the band structure of lifetime-broadened PBs in the leaky region.

The exact DOS including the lifetime broadening can be obtained from the tramsmitted and reflected complex amplitudes of a plane-wave light incident on that system. In this section we summarize the essence of the DOS formula.

4.3.1 DOS Formula for a Homogeneous Dielectric Slab

We start by considering a *homogeneous* dielectric slab of no internal structure and examine the leaky modes set up in it. We are examining the change of the DOS due to the presence of the slab, relative to the free-space DOS. The formula of the change of DOS due to a PC is an extension of the one obtained here for the dielectric slab.

Suppose there is a slab of a homegeneous dielectric of arbitrary thickness $2L$ bounded by two plane surfaces, as shown in Fig. 4.7. We take the thickness direction as the z direction and assume an infinite lateral extension in the x and y directions. We choose the origin $z = 0$ at the middle of the slab, letting the two surfaces be located at $z = \pm L$. To examine the effects of the exterior region, we place this dielectric slab between two perfect mirrors positioned

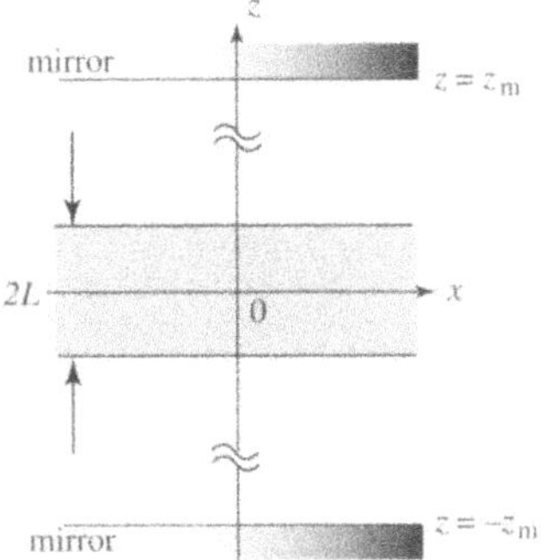

Fig. 4.7. Geometry of the system used in the DOS calculation. Perfect mirrors are placed at $z = \pm z_{\mathrm{m}}$ to give rise to a standing wave. A dielectric slab of thickness $2L$ occupies the region $-L < z < L$. The whole system has mirror symmetry with respect to the plane $z = 0$ and classifies the modes by their parities

at $z = \pm z_{\mathrm{m}}$, where z_{m}, much larger than L, specifies a remote position from the slab. We impose the boundary condition that the components of the electromagnetic fields tangential to the mirror surfaces vanish there. We thereby calculate the normal modes set up in the region $-z_{\mathrm{m}} < z < z_{\mathrm{m}}$ with and without the dielectric slab. The difference of the DOS of the two cases is the change of DOS we are looking for. Since the states of different wavevectors $\boldsymbol{k}_{\parallel} = (k_x, k_y)$ decouple, due to the lateral translational invariance, we confine ourselves to the DOS of the modes having $\boldsymbol{k}_{\parallel} = 0$. To obtain the general result for $\boldsymbol{k}_{\parallel} \neq 0$, we have only to add the $\boldsymbol{k}_{\parallel}$-dependence to the quantities appearing in the final formula.

Therefore, we consider the propagation of a light wave along the $\pm z$ direction, with the boundary condition that the electric field polarized in x and magnetic field polarized in y or vice versa should both vanish at the mirrors. Without the dielelctric slab, the system is just an ordinary Fabry–Perot cavity. The wavelength of the fields is quantized by the boundary condition at $z = \pm z_{\mathrm{m}}$. Apart from the normalization factor, the electric field with polarization in the y direction is

$$E(z) = \sin kz, \qquad \cos kz, \tag{4.40}$$

depending on the parity of the mode with respect to the center $z = 0$ of the cavity. Here

$$k = \frac{\omega}{c}$$

defines the wavenumber. By the boundary condition at the mirrors, the wavenumber is quantized:

$$k_p = \frac{p\pi}{z_{\mathrm{m}}}, \qquad \frac{(p+1/2)\pi}{z_{\mathrm{m}}} \tag{4.41}$$

for the pth odd- and even-parity states, respectively, p being an integer. The frequency of the quantized normal modes is then

$$\omega_p = c\frac{p\pi}{z_{\mathrm{m}}}, \qquad c\frac{(p+1/2)\pi}{z_{\mathrm{m}}}. \tag{4.42}$$

The spacing between the neighboring modes is π/z_{m} in k and $c\pi/z_{\mathrm{m}}$ in ω for both parities.

When a dielectric slab occupies the region $-L < z < L$, the electric fields are modified from (4.40). The standing wave solutions in $-z_{\mathrm{m}} < z < z_{\mathrm{m}}$ are expressed *outside* the slab by

$$\begin{aligned} E(z) &= \frac{z}{|z|}\sin\left(k|z| + \delta_{\mathrm{o}}(k)\right), \\ E(z) &= \cos\left(k|z| + \delta_{\mathrm{e}}(k)\right) \end{aligned} \tag{4.43}$$

for both parities, because they are expressed by a superposition of $\sin kz$ and $\cos kz$, due to the scattering by the dielectric slab. The phases $\delta_{\mathrm{o}}(k)$

and $\delta_e(k)$ of (4.43) are the phase shifts of the respective parities, which give the phase change of the incident light relative to the phase of the incident wave. Obviously, they are determined by the transmission and reflection of the incident light through the dielectric slab. They vary as functions of k. For the details, see Appendix B. The procedure given there for obtaining $\delta_o(k)$ and $\delta_e(k)$ is summarized as follows. First we should obtain the transmitted and reflected amplitudes, $t(k)$ and $r(k)$ for the incident plane wave of wavenumber k. From the two eigenvalues of the 2×2 matrix

$$\begin{pmatrix} t(k) & r(k) \\ r(k) & t(k) \end{pmatrix}, \tag{4.44}$$

which can be expressed as $\exp(2i\delta_e(k))$ and $\exp(2i\delta_o(k))$, we obtain the two phase shifts.

Because of the presence of the phase shifts in (4.43), the quantization rule for k changes from that of the Fabry–Perot cavity given by (4.41):

$$kz_m + \delta_o(k) = p\pi. \tag{4.45}$$

From now on we concentrate on the odd-parity states for the treatment is quite similar for both parities. The spacing δk between two neighboring quantized values of k for p and $p-1$ is determined by

$$\delta k \left(z_m + \frac{\partial}{\partial k}\delta_o(k) \right) = \pi.$$

The quantity $1/\delta k$ is the number of allowed values of k per unit wavelength and, by definition, equal to the DOS (per unit wavelength) of the whole system having a dielectric slab inside (see Appendix A). The change of DOS due to the dielectric slab is the change of $1/\delta k$. Therefore, writing the change of DOS per wavelength as $\Delta\rho(k)$, we find

$$\begin{aligned} \Delta\rho(k) &= \frac{1}{\pi}\left(z_m + \frac{\partial}{\partial k}\delta_o(k) \right) - \frac{z_m}{\pi} \\ &= \frac{1}{\pi}\frac{\partial}{\partial k}\delta_o(k). \end{aligned} \tag{4.46}$$

The second term of the first line is the DOS in the absence of the dielectric slab. Note that z_m disappears in the change of DOS, as it should. Combining the even-parity solution, we find the change of DOS is expressed by

$$\Delta\rho(k) = \frac{1}{\pi}\frac{\partial}{\partial k}\left(\delta_o(k) + \delta_e(k) \right). \tag{4.47}$$

The conclusion is that the k-derivative of the sum of the phase shifts gives the change of DOS. Since the polarization of the electric field can be directed both in x and y, the change of DOS should be twice as large due to this

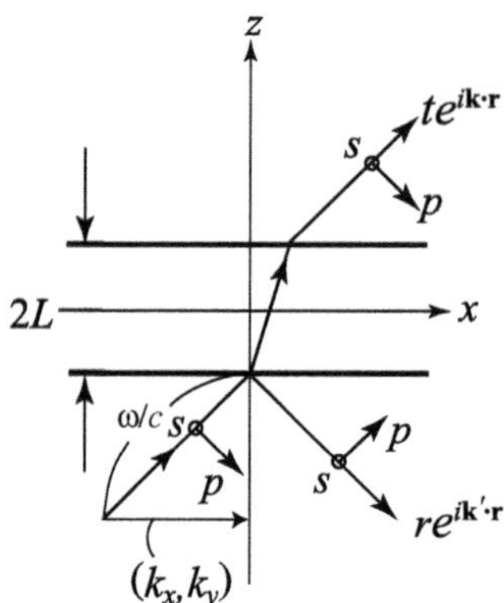

Fig. 4.8. Polarizations s and p of incident plane-wave light

degeneracy. If the derivative $\partial/\partial k$ is replaced by $\partial/\partial\omega$, the above formula gives $\Delta\rho(\omega)$, the change of DOS per frequency:

$$\Delta\rho(\omega) = \frac{1}{\pi}\frac{\partial}{\partial\omega}\left(\delta_{\mathrm{o}}(k) + \delta_{\mathrm{e}}(k)\right). \tag{4.48}$$

So far we have considered the case of $\boldsymbol{k}_{\|} = 0$ using the transmission and reflection data of normal incidence. For the DOS change for $(k_x,\, k_y) \neq 0$, the above formulae still give the DOS change per unit k and ω, respectively. However, k now stands for the wavevector compoment in the thickness direction z and we have to calculate $\delta_{\mathrm{o}}(k)$ and $\delta_{\mathrm{e}}(k)$ for new k defined by

$$k = \sqrt{\left(\frac{\omega}{c}\right)^2 - k_x^2 - k_y^2}$$

from the 2×2 matrix (4.44) for the oblique incidence shown in Fig. 4.8. Since in the oblique case the p-polarized and s-polarized light have different phase shifts, we have different $\delta_{\mathrm{e}}(k)$ and $\delta_{\mathrm{o}}(k)$ for the two polarizations. The sum

$$\delta_{\mathrm{o}}^{\mathrm{p}}(k) + \delta_{\mathrm{e}}^{\mathrm{p}}(k)$$

then gives the DOS change of p-polarized modes and

$$\delta_{\mathrm{o}}^{\mathrm{s}}(k) + \delta_{\mathrm{e}}^{\mathrm{s}}(k)$$

yields that of s-polarized modes. They are associated with the modes active to p- and s-polarized incident light, respectively.

Figure 4.9 is an application of the above theory to obtain the dispersion relations of the leaky modes of a slab of homogeneous dielectric. We consider a free standing dielectric slab of finite thickness $2L$ and dielectric constant 12. The dispersion curves of the s- and p-active leaky modes were drawn by plotting the frequencies of the DOS peaks as a function of k_x. The modes

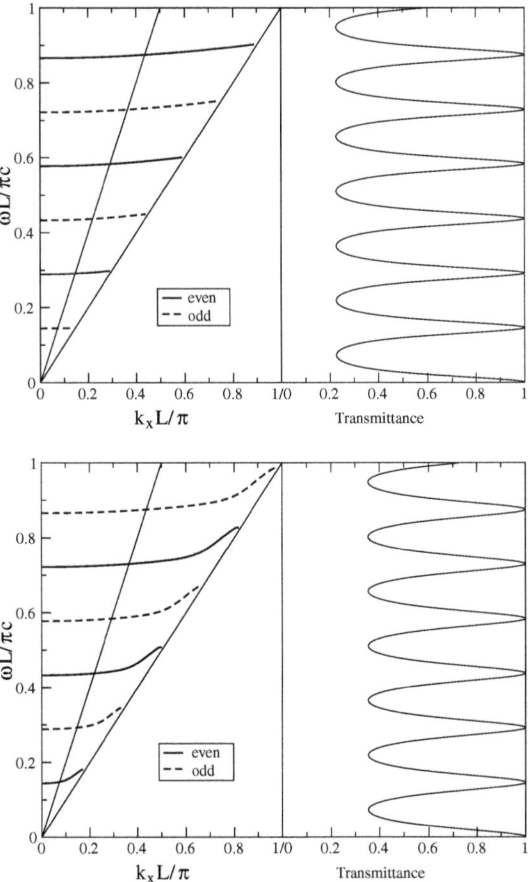

Fig. 4.9. Dispersion relation of leaky modes of $\boldsymbol{k}_{\parallel} = (k_x, 0)$ of s-polarization (above) and p-polarization (below) and the transmittance of incident light of respective polarizations. The result is given for a free-standing homogeneous dielectric slab of thickness $2L$ and dielectric constant 12 in the air. Modes are classified according to the parity with respect to the center of the slab. The incidence angle of the light is 30°. The straight line drawn inside the light cone shows the dispersion relation of the incident light. The dispersion curves are obtained by plotting the peaks of the DOS given by (4.48) as functions of k_x

are further classified into even and odd parities with respect to the center $z = 0$ of the slab. The transmittance of the s- and p-light is given in the right panels with the incidence angle kept fixed to be 30°. Since the incident light has no definite parity, it excites the leaky modes of both parities. The fringes seen in the transmittance result from the interference between the entrance and exit surfaces. But it can be interpreted in terms of the excitation of the slab modes which have radiative lifetimes due to the leak. Therefore, the finesse of an interference fringe, or the width of each of the transmittance peaks, provides a measure of the lifetime of the excited leaky mode. This interpretation using the excited normal modes is important when we study the optical properties of a slab of PCs, because the interference fringes thereof are often very complicated because of the crossing, splitting and beat among many coexisting PBs (see, e.g., the transmittance of a metallic PC given in Sect. 8.4).

4.3.2 DOS Formula for a Spherical Scatterer

The formula of the change of DOS given above, which is expressed using the phase shifts of incoming wave, reminds us of the formula of electrons by the presence of an impurity atom in a metal matrix. The change of electronic DOS at the wavenumber k of electrons is determined by the phase shifts $\delta_\ell(k)$ of the partial wave ℓ, acquired by electrons by the impurity scattering. The formula reads

$$\Delta\rho(k) = \frac{2}{\pi}\frac{\partial}{\partial k}\sum_{\ell=0}(2\ell+1)\delta_\ell(k), \tag{4.49}$$

where the factor 2 takes account of the spin degeneracy of electrons and the factor $(2\ell+1)$ incorporates the degeneracy of the quantum number ℓ of angular momentum [29]. In the same way, the change of the DOS of electromagnetic modes due to the presence of a dielectric sphere is expressed by [19]

$$\Delta\rho(k) = \frac{1}{\pi}\frac{\partial}{\partial k}\sum_{\ell=1}(2\ell+1)\left(\delta_\ell^{\mathrm{M}}(k)+\delta_\ell^{\mathrm{N}}(k)\right) \tag{4.50}$$

where $\delta_\ell^\beta(k)$ (β = M and N) are the phase shifts of the two partial waves M and N of whispering gallery modes (WGM) (Mie resonance modes). The increase of the photonic DOS due to the appearance of WGM in a sphere, presented in Fig. 3.9, was obtained from this formula. Note that $\Delta\rho(\omega)$, the change of the DOS per unit frequency is obtained from (4.49) and (4.50), with $\frac{\partial}{\partial k}$ replaced simply by $\frac{\partial}{\partial\omega}$. A Mie resonance of light scattering occurs every time incident light excites a WGM. The existence of WGM of a sphere is clearly demonstrated by the peaks in $\Delta\rho(\omega)$. The lifetime of a WGM may be estimated by the frequency width of the peak of $\Delta\rho(\omega)$, which is just the factor determining the Q value of the resonant optical response (Sect. 3.5).

4.3.3 DOS Formula for a Slab Photonic Crystal

We are now in a position to consider the change of DOS by the presence of a slab PC. Conceptually, the method of deriving the DOS formula is similar to the one used in the homogeneous dielectric slab given in Sect. 4.3.1. We confine the slab PC between the pair of perfect mirrors placed symmetrically

[30]. In Appendix B the derivation of the DOS change is gi ven for the simple case where an s and p incident waves are reflected and transmitted as s and p waves, respectively. The scattering by the PC slab is summarized by the two phase shift $\delta_{\mathrm{o}}^{\mathrm{s}}(k)$ and $\delta_{\mathrm{e}}^{\mathrm{s}}(k)$ and $\delta_{\mathrm{o}}^{\mathrm{p}}(k)$ and $\delta_{\mathrm{e}}^{\mathrm{p}}(k)$ for s and p lights, respectively. The DOS change is given by the $k-$ and $\omega-$ derivative of the sum of the four phase shifts. The formula is thus identical to the one for a homegenous dielectric slab treated above.

Generally in the light transmission and reflection by a slab PC, the mixing between s and p polarizations takes place and diffracted plane waves are produced by incident light. The treatment for these cases is a much more involved. We mention here the DOS formula for these general cases for a slab PC of a finite thickness. Because of the conservation of the lateral wave vector component $\boldsymbol{k}_{\|}$ of the incident light, the DOS change is given as a function of $\boldsymbol{k}_{\|}$.

Since the amplitudes of incident, transmitted and reflected light are all transverse vectors outside the photonic slab, any of the plane-wave electric fields before and after the scattering by the slab PC has two degrees of freedom. Consequently, in describing the process of an incident plane wave being converted to another plane wave in the slab, we need a 2×2 matrix to take into account the mixing of the polarizations. Considering the simultaneous incidence in the $+z$ and $-z$ directions of the incident lights to form the standing waves between the mirrors, we need to express the scatterig by the PC slab in terms of a 4×4 scattering marix (S matrix) of the form

$$\begin{pmatrix} \hat{\boldsymbol{t}}^{(++)} & \hat{\boldsymbol{r}}^{(-+)} \\ \hat{\boldsymbol{r}}^{(+-)} & \hat{\boldsymbol{t}}^{(++)} \end{pmatrix} . \tag{4.51}$$

Here the superscript of $\boldsymbol{t}^{(++)}$ shows that it is a transmitted amplitude of the wave in the $+z$ direction converted from the light incident in the $+z$ direction and $r^{(-+)}$ describes the amplitude of the $-z$-directed reflected wave obtained from the incident light in the $+z$ direction. Others are similarly defined. The symbol $\hat{}$ on t^{++}, etc., shows that it is a 2×2 tensor. Thus the S matrix defined by (4.51) is a 4×4 matrix. Since this marix is unitary, it has eigenvalues expressed by $\mathrm{e}^{2\mathrm{i}\delta_j}$ with $j = 1, \cdots, 4$. We have therefore four phase shifts in the case of no diffraction. In conclusion, four phase shifts are obtained and the change of the DOS is given by the k-derivative (ω-derivative for the DOS change per unit frequency) of the sum of the four.

When diffracted plane waves are produced, we have to take into account the boundary conditions at the mirror surfaces for them, too. If there are diffractions related to N 2D reciprocal lattice vectors (the case of N open channels), an incident light wave comes out of the PC in the form of N transmitted and N reflected plane waves. Because each light wave has two components, the dimension of the S matrix is $4N \times 4N$. In this general case, therefore, we obtain $4N$ phase shifts from $4N$ eigenvalues of the S matrix, which are by unitarity expressed as

$$e^{2i\delta_j(k)} \ (j = 1, 2, \cdots, 4N).$$

They are all functions of k or $\omega = ck$. The conclusion for the general case is that it is the sum of these $4N$ phase shifts that gives the change of DOS. The situation treated above of the frequency region of no diffraction is just the case of $N = 1$. In summary, all the $4N$ phase shifts of the general case are obtained as functions of the lateral wavevector component $\boldsymbol{k}_{\|}$ of incident light and the increment of the DOS of the modes of $\boldsymbol{k}_{\|}$ is given by [31]

$$\Delta\rho_{\boldsymbol{k}_{\|}}(k) = \frac{1}{\pi}\sum_{j=1}^{4N} \frac{\partial}{\partial k}\delta_j(k). \tag{4.52}$$

The change of DOS per unit frequency is obtained by changing $\partial/\partial k$ to $\partial/\partial\omega$.

4.3.4 Application of DOS Formula to a PC Slab

Here we examine the fine structure of the transmittance in connection with the excitations of leaky PBs.

Figure 4.10 shows the PBS of the arrayed spheres mutually in contact in the square lattice [31]. The calculation is made for the dielectric spheres of Si_3N_4 of diameter $\frac{1}{8}$ inch to obtain the PBS in the gigaherts (GHz) range (the dielectric constant is 2.96^2 in the frequency range concerned). We stack the 2D sheets of spheres. The left figure is the PBS for the wavevector $\boldsymbol{k} = (0.1, 0.2, k_z)$ as a function of k_z. The portion magnified for the range between 45 and 47 GHz is shown in the right figure. We examine the correlation between the transmittance and DOS in this frequency range.

Figure 4.11 shows the calculated transmittance $T(\omega)$ of s-polarized light (left) and the cange of DOS $\Delta\rho(\omega)$ (right) for the lateral wavevector equal to $\boldsymbol{k}_{\|} = (0.1,\ 0.2)$ [31]. The result is given for a slab of stacked layers $N = 8$. The change of DOS is given separately for the DOS for two different parities with respect to the center of the slab. The marks given on the upper and lower horizontal lines are, respectively, the position of the peaks and dips of $T(\omega)$ shown in the left figure. We learn that the interference fringes of $T(\omega)$ are not at all sinusoidal. In view of the band structure shown in Fig. 4.10, this is due partly to the fact that there are simultaneously a number of excited bands: interference among the several propagating light waves complicates the line shape. We see a wide range of inhibition of transmission (less than a few percent) in the middle of the transmittance, which is in good agreement with the band structure calculation given in Fig. 4.10. This shows the thickness of $N = 8$ is already enough for the development of the band gap, which is to appear in a system of $N = \infty$. Since there is no mirror reflection involved in $\mathcal{G}(\boldsymbol{k})$ (for the group of $\boldsymbol{k}$, see Sect. 4.2.2) for $\boldsymbol{k} = (0.1, 0.2, k_z)$ used in the calculation, the polarization of an excited PB is generally a mixture of p- and s-polarizations. Consequently, all the PBs are active in the transmittance

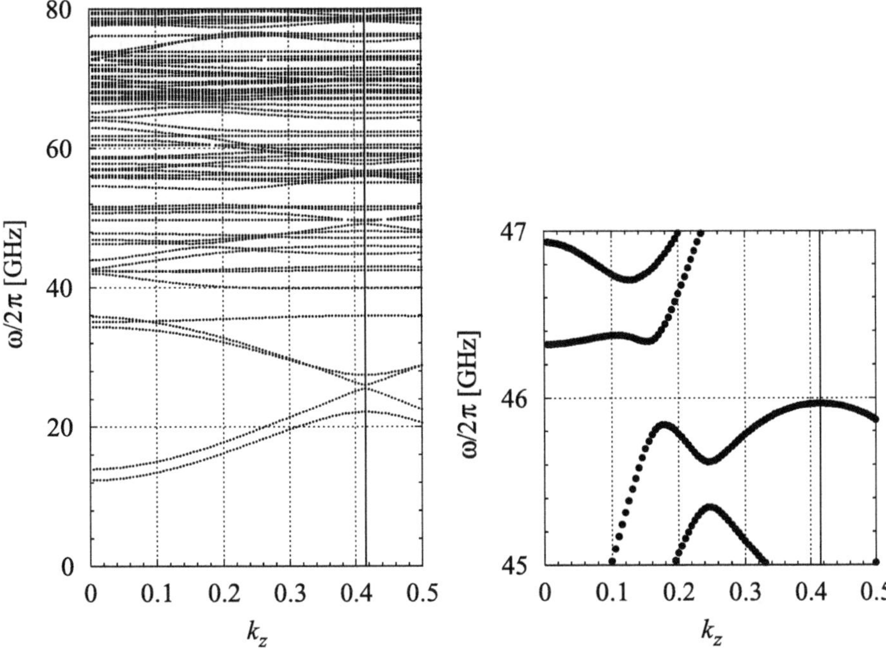

Fig. 4.10. Band structure of cubic lattice of Si_3N_4 spheres of diameter $\frac{1}{8}$ inch. See the text for the other parameters used in the calculation. A space of thickness 0.204 times the lattice constant a (a is equal to the diameter), is made between the stacked layers for experimental reasons. The BZ edge in the stacking direction z is thus 0.415 in units of $2\pi/a$, which is shown by the vertical solid line. In the right figure a portion of the band structure is magnified in a narrow frequency range $45\,\mathrm{GHz} < \omega/2\pi < 47\,\mathrm{GHz}$

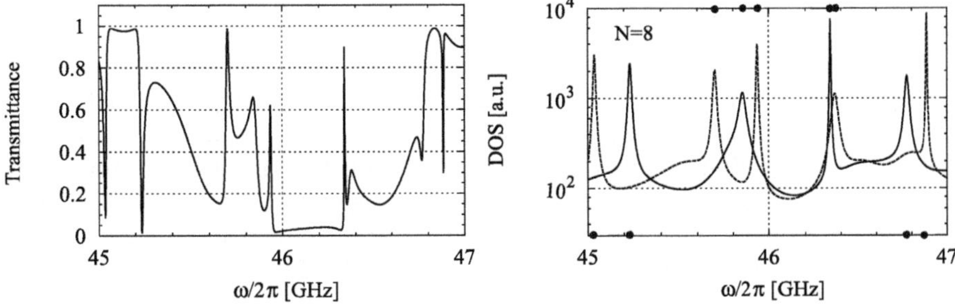

Fig. 4.11. Transmittance of s-polarized light (left) and change of DOS (right) of the slab PC of $N = 8$. The incident s-polarized light of $\boldsymbol{k}_{\parallel} = (0.1, 0.2)$ is used in the calculation of the transmittance. In the DOS the parities of the modes with respect to the middle of the slab are classified by the solid (even parity) and dotted curves (odd parity)

$T(\omega)$ of s light shown in the left panel. It is interesting that a DOS peak sometimes causes a dip and other times a peak in the transmittance. The peaks of the DOS marked in the upper (lower) horizontal line produce peaks (dips) in the transmittance, as seen by comparison between the left and right figures. Whether a peak or dip appears presumably depends on the relative phase between the residues of the nearby poles. Also the width of the DOS peak differs from peak to peak. In understanding the optical response of a PC, therefore, we need information about band structure, change of DOS, parity and polarization of PBs together.

4.4 Layer-Doubling Method

So far we have presented a number of numerical results related to the transmission of light through a slab of PC. Here we describe the layer-doubling method, which has been so successfully applied to the calculation of the reflectance and transmittance of many PCs of finite thickness.

4.4.1 Procedures of Layer Doubling in Light Propagation

We are concerned with slabs of PCs bounded by two plane surface layers, which are entrance and exit surfaces of external light. The wavevectors and reciprocal lattice vectors involved here are mainly two-dimensional due to the lateral translational invariance of the sytem. To avoid notational complexities, we use the following simplified notation by dropping the subscript $_\parallel$ to stand for 2D vectors:

$$\boldsymbol{h}_\parallel \rightarrow \boldsymbol{h}, \quad \boldsymbol{k}_\parallel + \boldsymbol{h}_\parallel \rightarrow \boldsymbol{k}_{\boldsymbol{h}}. \tag{4.53}$$

The difficulty in the treatment of light scattering due to the finiteness of the thickness may be overcome by an algorithm called the layer-doubling method developed in the treatment of low-energy electron diffraction [22, 32]. The underlying idea is to treat the system of finite thickness as a collection of parallel layers each having a complete 2D periodicity and to treat the scattering from the entire system as a repetition of the single-layer scatterings. The layer doubling means the procedure of constructing the scattering data of the slab of thickness $2L$ from the scattering data of the slab of thickness L, the scattering data of $4L$ from that of $2L$ obtained in the previous run and so on. This process is illustrated in Fig. 4.12. In this way we can obtain the complex reflected and transmitted amplitudes of a thick system including diffracted amplitudes.

For the layer-doubling method to be reliable enough to reproduce the PBS obtained by band calculation in the limit of infinite thickness, there are two requirements to be met; first, we should start from accurate scattering data for a single layer, which incorporates exactly the multiple light scattering within it. The second point is that the method is free from divergence when the doubling procedure is repeated to cover thick slab systems. Both can indeed be satisfied by the layer-doubling method.

We begin with the scattering data of a monolayer of PC. We consider light scattering from it by assuming an infinite lateral extension in the xy plane with 2D periodicity. An example is a free-standing 2D periodic array of substances in free space, as shown in Fig. 4.13. Let an incident light wave have unit amplitude and propagate in the $+z$ direction with lateral component of wavevector $\boldsymbol{k}_{\|}$. After multiple scattering *within* this monolayer, the reflected or transmitted light emerges having a lateral wavevector component $\boldsymbol{k_h}$ and z component given by

$$\pm \Gamma_{\boldsymbol{h}} = \pm \left[\left(\frac{\omega}{c} \right)^2 - (\boldsymbol{k_h})^2 \right]^{\frac{1}{2}} . \tag{4.54}$$

The $\boldsymbol{h}$ wave represents both evanescent light and plane-wave light depending on the sign of the argument of the square root of (4.54). The scattering data of a 2D monolayer consists of the reflected amplitudes $\boldsymbol{r}^{(-+)}(\boldsymbol{h}'i'; \boldsymbol{h}i)$ and the transmitted amplitudes $\boldsymbol{t}^{(++)}(\boldsymbol{h}'i'; \boldsymbol{h}i)$ with $i, i' = x, y, z$. The argument $\boldsymbol{h}i$ of $\boldsymbol{r}^{(-+)}(\boldsymbol{h}'i'; \boldsymbol{h}i)$ specifies the incident light in the channel $\boldsymbol{h}$, open or closed, with 2D wavevector $\boldsymbol{k_h}$ and polarization in the i direction. The argument $\boldsymbol{h}'i'$ refers to the i' component of the reflected wave in the channel $\boldsymbol{h}'$ of the wavevector $\boldsymbol{k_{h'}}$. The superscript $(-+)$ of $\boldsymbol{r}^{(-+)}(\boldsymbol{h}'i'; \boldsymbol{h}i)$, etc. is the same as used in Sect. 4.3.3 and refers to the process for the incident wave incoming in

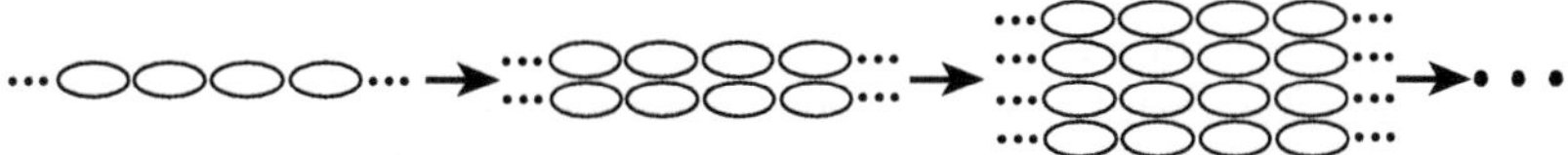

Fig. 4.12. Layer-doubling procedure in increasing the thickness of a slab PC

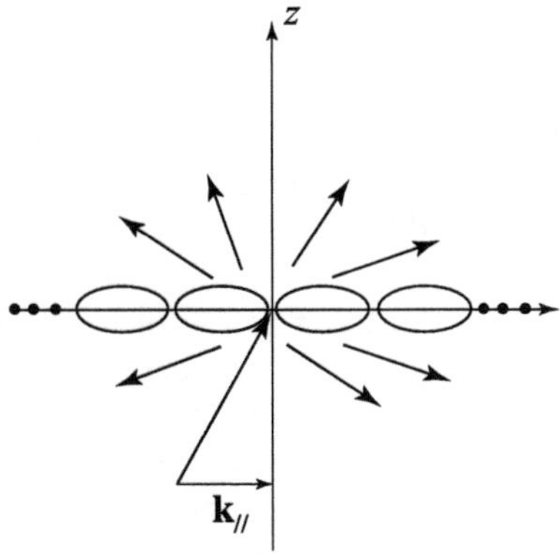

Fig. 4.13. Scattering of an incident light wave from a monolayer PC. Monolayer scattering data consist of reflected and transmitted amplitudes of various diffraction channels

the $+z$ direction to be reflected back in the $-z$ direction. In the transmission, we have $\boldsymbol{t}^{(++)}$ and $\boldsymbol{t}^{(--)}$. Though we are considering the case of the incident light coming in the $+z$ direction with wavevector $\boldsymbol{k} = (\boldsymbol{k}_{\|}, \Gamma_0)$, we need to prepare $\boldsymbol{r}^{(+-)}$ and $\boldsymbol{t}^{(--)}$ for those light waves incident in the $-z$ direction with various wavevectors related to $\boldsymbol{k}_{\boldsymbol{h}}$. This is because each of the stacked layers is exposed to many incoming light waves from both below and above the layer with diffracted wavevector components $\boldsymbol{k}_{\boldsymbol{h}}$. These four scattering amplitudes for the monolayer are treated as matrices, $\boldsymbol{r}^{(+-)}, \boldsymbol{t}^{(++)}$, etc., whose rows and columns are labeled by the indices $\boldsymbol{h}'i'$ and $\boldsymbol{h}i$, respectively.

The layer-doubling method [22, 32, 33] consists of calculating the new matrices $\boldsymbol{r}^{(+-)}, \boldsymbol{t}^{(++)}$, etc. for the system of two layers using the scattering data of the single layer and repeating this process to obtain the scattering matrices for 4, 8, ..., 2^n, ... layers. The first run to treat the scattering from the double layers is given here.

The matrix $\boldsymbol{t}_{21}^{(++)}$, which expresses the transmission through a pair of layers 1 and 2, is made out of

$$\begin{aligned}\boldsymbol{t}_{21}^{(++)} &= \boldsymbol{t}_2^{(++)}[\boldsymbol{I} + \boldsymbol{r}_1^{(+-)}\boldsymbol{r}_2^{(-+)} + (\boldsymbol{r}_1^{(+-)}\boldsymbol{r}_2^{(-+)})^2 + \cdots]\boldsymbol{t}_1^{(++)} \\ &= \boldsymbol{t}_2^{(++)}[\boldsymbol{I} - \boldsymbol{r}_1^{(+-)}\boldsymbol{r}_2^{(-+)}]^{-1}\boldsymbol{t}_1^{(++)} .\end{aligned} \tag{4.55}$$

Here $\boldsymbol{t}_1^{(++)}$ and $\boldsymbol{t}_2^{(++)}$ ($\boldsymbol{r}_1^{(+-)}$ and $\boldsymbol{r}_2^{(-+)}$) are the transmission (reflection) matrices of layer 1 and 2, respectively, which incorporate the phase difference caused by the light propagation between the two layers [22, 32, 33]. The unit matrix within the bracket of the first line corresponds to the process of a photon passing through the two layers after transmitting through layers 1 and 2 both in the $+z$ direction and the second and higher-order terms describe the reflection between the two layers once, twice, $\cdots$, before passing through them. These scattering processes are shown in Fig. 4.14, by dropping the superscripts (++), etc., for simplicity. By collecting the geometrical series of repeated processes into an inverse matrix, we can avoid the divergence of the algorithm often encountered in a naive iterative treatment for repeated scatterings. If we are to obtain the transmission coefficient of light of wavevector $\boldsymbol{k}$ through the system of two layers, we have only to use the $(\boldsymbol{0}i', \boldsymbol{0}i)$ element of the matrix $\boldsymbol{t}_{12}^{(++)}$. The other matrices for the two-layer system are obtained similarly. Repeating the doubling n times, we obtain the scattering data for the 2^n layers.

The coding of the layer doubling procedure is straightforward. Therefore, the feasibility of this method wholly depends upon whether the scattering data of a monolayer of 2D system is calculated. For the case of arrayed spheres or cylinders, the vector KKR method is the strongest and most convenient way. By exact scattering data we mean the exact treatment not only of the hopping process within the layer but also the coupling of the excited modes with the plane-wave light outside the layer. The coupling to the outside region leads to a finite lifetime of the leaky PBs due to the leakage. Examples of the detailed treament are found in [14, 16, 33].

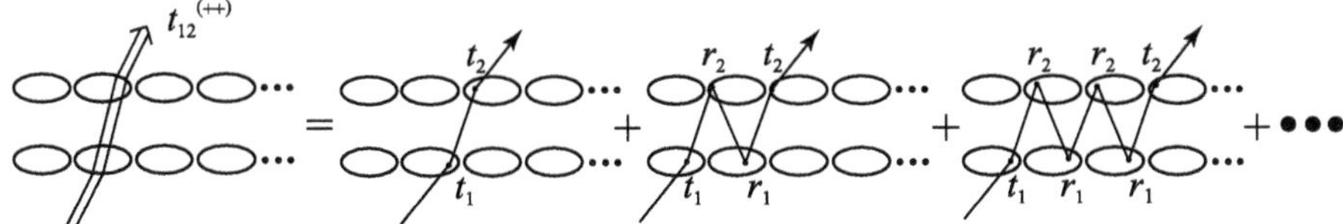

Fig. 4.14. Calculation of transmitted amplitudes of double layers using the single-layer scattering data

4.4.2 Applications of Layer-Doubling Method

Formation of Band Gaps versus PC Thickness Now we present an example of the layer-doubling calculation. We examine the formation of the band gaps as the number N of the stacking layers increases.

Figure 4.15 shows the reflectance of the fcc array of spheres with radius r given by $r/a = 0.3$ relative to the cube size a of the lattice [33]. The surface layer is the (111) plane. The band structure of Figure 4.15(a) is obtained by the band calculation based on the vector KKR method for the infinite system. The result is given as a function of wavevector component k in the [111] direction using the dimensionless frequency $Z = \omega a/2\pi c$. The dielectric constant of the spheres is taken to be $\epsilon_< = 3.2^2$. Figure 4.15(b) through (e) show the calculated reflectance of the plane-wave light of normal incidence (hence propagating in the [111] direction) on the stacked (111) planes of spheres. The quantity $R(n)$ is the reflectance obtained by the layer-doubling procedure applied $n-1$ times, i.e., the reflectance of the slab with $N = 2^{n-1}$. We see a precise agreement of the location of the band gaps as N increases. It is to be noted that for the higher band gaps, denoted as g_2 and g_3, to be formed, a thickness of $N = 4$ is almost sufficient in such a high $\varepsilon_<$ case as used here. This feature obviously reflects a higher confinement of WGMs of spheres which have higher ℓ (see Sect. 3.5). Also the density of fringes of $R(n)$ changes with N in accordance with the feature discussed in Sect. 3.4.

Enhancement of Near-Field by an Excited Photonic Band The second example of the application is the calculation of the local field near the surface of a PC [28, 33]. The precise magnitudes of the field intensity near the surface of the PC is determined by considering the intensities of the evanesent light together with the reflected or transmitted plane-wave lights. These evanescent light waves arise because of the presence of the closed diffraction channels (see Sect. 3.3.3), characterised by

$$\left(\frac{\omega}{c}\right)^2 < |\boldsymbol{k_h}|^2, \tag{4.56}$$

for a PB of lateral wavevector $\boldsymbol{k}_\parallel$. The pile up of the evenescent light waves from various $\boldsymbol{h}$ causes an enhanced near field near the surfaces of the slab PC. Obviously the enhancement will depend on the PB mode excited inside.

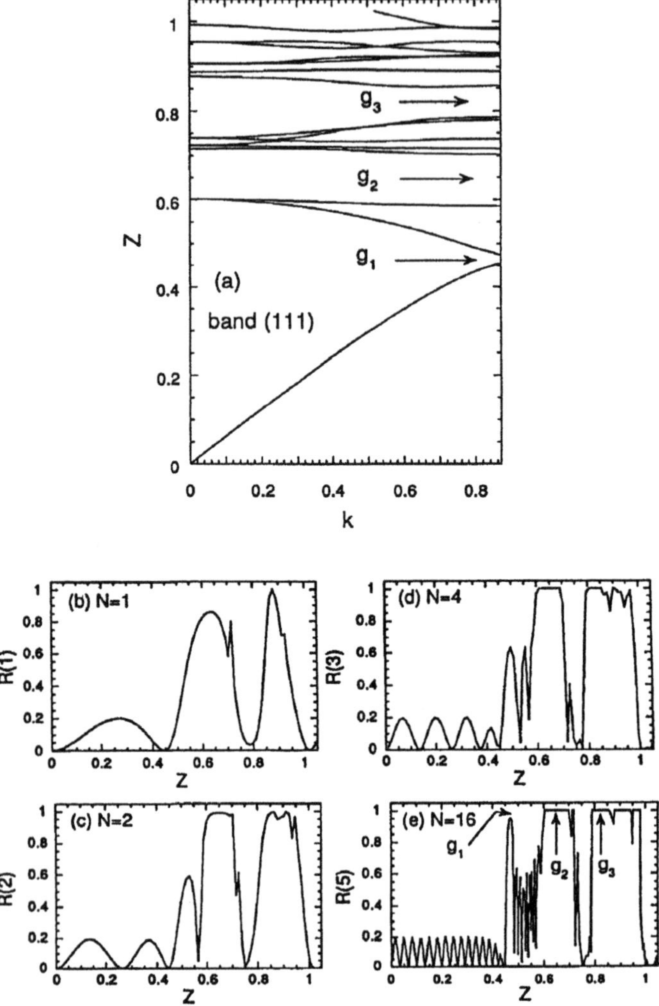

Fig. 4.15. Band-gap formation of a PC of arrayed spheres with increase of the thickness of the slab. $R(n)$ of (b) through (e) shows the reflectance of $N = 2^{n-1}$ layers with $Z = \omega a/2\pi c$. As the number of layers N increases, the band gaps seen in panel (a) of an infinite PC manifest themselves in the reflectance

The layer doubling method formulated above is powerful in the treatment of the near field, because it may give all the scattered light, whether evanescent or not. Figure 4.16 shows the local field intensity as a function of frequency of incident light of normal incidence with unit amplitude [33]. The PC slab is $N = 16$ stacked layers of the (001) plane of spheres. The intensity is calculated at the top of the spheres of the surface layer. The PC is a fcc crystal of spheres of $\varepsilon = 3.2^2$ and $r/a = 0.3$ as used in Figs. 4.2 and 4.15.

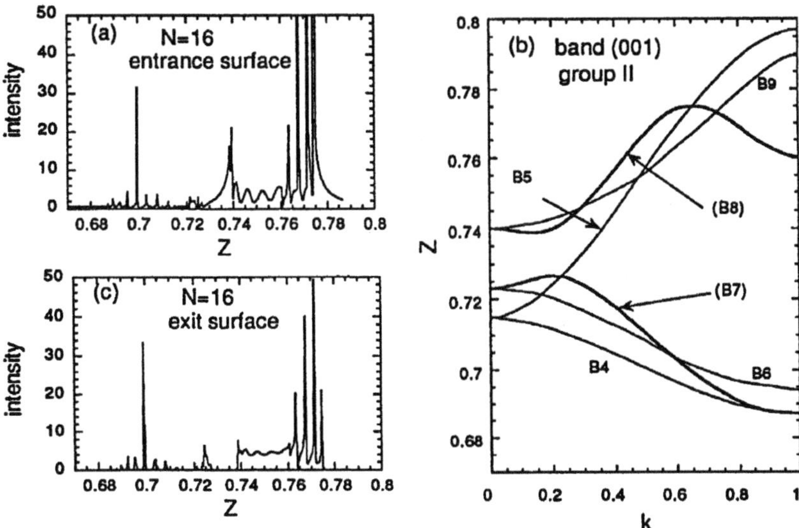

Fig. 4.16. Intensity of the local field and related band structure of a fcc PC. Z is the normalized frequency $\omega a/(2\pi c)$ and k is the wavenumber normalized by π/a. Panels (a) and (c) show the intensity of the local fields excited by the light of unit intensity, (a) at the top position of a sphere located in the entrance surface and (c) that of the exit surface. Panel (b) is the band structure for $\boldsymbol{k}$ in the [001] direction for the same range of frequency. See Fig. 4.2 for the numbering of the bands and group naming of the bands. Only the bands shown by thick solid curves with band index within braces are optically active

In the band structure shown in Fig. 4.16(b), only the bands B7 and B8 are optically active (for the origin of active and inactive bands, see Sect. 4.2.3). By comparing Fig. 4.16(a) and (c) with (b), it is seen that the local fields are enhanced strongly when a PB mode of small group velocity or very large band bending is excited by the incident light (see, e.g., the frequency regions $Z \simeq 0.74$ and 0.77). It is remarkable that the intensity becomes as large as several tens as compared to the incident intensity.

We have given two applications for PCs of arrayed dielectric spheres but the layer-doubling method is quite general covering any kind of stacked PCs, if the scattering data of the monolayer system is available.

4.5 Origin of Band Gap Modes

By introducing a defect into an otherwise perfect PC, we may create defect modes within a band gap. A defect mode, when it is lying in the frequency region of a band gap, is localized spatially to the defect site, because no PB of equal frequency is available for the mode to get delocalized by an energy-conserving coupling process. In this section we describe the basic properties of the photonic modes which appear in the band gap region of a PC having an impurity unit cell. We call here a band gap mode any normal mode existing in a band gap region in an imperfect PC.

4.5.1 Completeness of Photonic Band Solutions

To understand the properties of the band-gap modes of PCs in a quantitatively reliable manner like those of the donor or acceptor states of electrons, the best way is to expand their fields in terms of the normal modes of the host PC of perfect periodicity. Therefore it is worth devoting some of our efforts to discussing the completeness of the PB solutions [34].

Maxwell's equations to determine the electric fields reduce to a Hamiltonian form familiar in the Schrödinger equation by changing $\varepsilon(\boldsymbol{r})$ of Maxwell's equations to a product form $\sqrt{\varepsilon(\boldsymbol{r})}\sqrt{\varepsilon(\boldsymbol{r})}$. We are considering here only the case of the positive-definite dielectric constant $\varepsilon(\boldsymbol{r})$. Otherwise, the hermitian property of the PBs as well as their well-defined nature as normal modes is lost. Of the two $\sqrt{\varepsilon(\boldsymbol{r})}$, one is combined with $\boldsymbol{E}(\boldsymbol{r})$ to lead to $\boldsymbol{Q}(\boldsymbol{r})$ defined below and the other is removed by dividing Maxwell's equations by $\sqrt{\varepsilon(\boldsymbol{r})}$. Maxwell's equation for the electric field, given by (3.12), then becomes

$$\left(\frac{\omega^2}{c^2} - H\right)\boldsymbol{Q}(\boldsymbol{r}) = 0, \tag{4.57}$$

with the solution $\boldsymbol{Q}(\boldsymbol{r})$ defined by

$$\boldsymbol{Q}(\boldsymbol{r}) = \sqrt{\varepsilon(\boldsymbol{r})}\boldsymbol{E}(\boldsymbol{r}), \tag{4.58}$$

and the "Hamiltonian" given by

$$H = \frac{1}{\sqrt{\varepsilon(\boldsymbol{r})}}\boldsymbol{\nabla}\times\left\{\boldsymbol{\nabla}\times\frac{1}{\sqrt{\varepsilon(\boldsymbol{r})}}\right\}. \tag{4.59}$$

The point of this transformation is that H turns out to be a hermitian operator. That H is indeed hermitian may be seen at once if we note that two commutable momentum operators, $p_x p_y$, for example, are sandwiched in H by the same functions of $\boldsymbol{r}$ ($p_x = \frac{1}{\mathrm{i}}\frac{\partial}{\partial x}$, etc.). The solution $\boldsymbol{Q}_{n\boldsymbol{k}}$ of the PB state (n, $\boldsymbol{k}$) and wavevector $\boldsymbol{k}$ then satisfies the ordinary eigenvalue equation

$$H\boldsymbol{Q}_{n\boldsymbol{k}}^{(T)}(\boldsymbol{r}) = (\omega_{n\boldsymbol{k}}^2/c^2)\boldsymbol{Q}_{n\boldsymbol{k}}^{(T)}(\boldsymbol{r}), \tag{4.60}$$

for the operator H with eigenvalues $\omega_{n\boldsymbol{k}}^2/c^2$, where the superscript T is added to show that the solution satisfies

$$\boldsymbol{\nabla}\cdot\left\{\sqrt{\varepsilon(\boldsymbol{r})}\boldsymbol{Q}_{n\boldsymbol{k}}^{(T)}(\boldsymbol{r})\right\} = 0, \tag{4.61}$$

because the quantity in the curly bracket is $\boldsymbol{D}_{n\boldsymbol{k}}(\boldsymbol{r})$, the displacement field of PB. Therefore, $\boldsymbol{Q}^{(T)}_{n\boldsymbol{k}}(\boldsymbol{r})$ is a solution which tends to a transverse electromagnetic field in the free space of $\varepsilon(\boldsymbol{r}) = 1$. Although H is hermitian, the set of eigenvectors $(n,\, \boldsymbol{k})$ of $H$is not by itself a complete set, because H has static solutions of zero eigenvalue in addition, which have nothing to do with PB solutions. Since any longitudinal vector field is irrotational, the field

$$\boldsymbol{Q}(\boldsymbol{r}) = \sqrt{\varepsilon(\boldsymbol{r})}\boldsymbol{f}(\boldsymbol{r})$$

is a solution of (4.60) with zero eigenvalue, if $\boldsymbol{f}(\boldsymbol{r})$ is irrotational ($\boldsymbol{\nabla} \times \boldsymbol{f}(\boldsymbol{r}) = 0$). An example of such solutions of $\boldsymbol{Q}(\boldsymbol{r})$, which is compatible with the Bloch theorem, is

$$\boldsymbol{Q}_{\boldsymbol{kh}}(\boldsymbol{r}) = C\sqrt{\varepsilon(\boldsymbol{r})}\frac{\boldsymbol{k}+\boldsymbol{h}}{|\boldsymbol{k}+\boldsymbol{h}|}\exp\{\mathrm{i}(\boldsymbol{k}+\boldsymbol{h})\cdot\boldsymbol{r}\} \tag{4.62}$$

for an arbitrary reciprocal lattice vector $\boldsymbol{h}$, C being a normalization constant. We call these plane waves as longitudinal solutions.

From the hermiticity of H, the completeness is then recovered for the entire set of solutions, made up of the transversal and longitudinal solutions:

$$\sum_{n\boldsymbol{k}} \boldsymbol{Q}^{(T)}_{n\boldsymbol{k}}(\boldsymbol{r})\boldsymbol{Q}^{(T)}_{n\boldsymbol{k}}(\boldsymbol{r}) + \sum_{n\boldsymbol{k}} \boldsymbol{Q}^{(L)}_{n\boldsymbol{k}}(\boldsymbol{r})\boldsymbol{Q}^{(L)}_{n\boldsymbol{k}}(\boldsymbol{r}) = V_{\mathrm{pc}}\boldsymbol{I}\delta(\boldsymbol{r}-\boldsymbol{r}'), \tag{4.63}$$

where the left-hand side stands for the second rank tensor with $\boldsymbol{I}$ on the right-hand side being the 3×3 unit tensor and V_{pc} the volume of the PC. The labeling in terms of the index n for the longitudinal solutions will be explained below.

4.5.2 Formal Treatment of a Single Defect

With these preparations we now return to a PC having a single defect in an otherwise perfect PC. An example is a photonic crystal of arrayed spheres or cylinders with one of the spheres or cylinders replaced by a distinct one. In this case, $\varepsilon(\boldsymbol{r})$ of the whole system may be expressed by

$$\varepsilon(\boldsymbol{r}) = \varepsilon_0(\boldsymbol{r}) + \delta\varepsilon(\boldsymbol{r}), \tag{4.64}$$

where $\delta\varepsilon(\boldsymbol{r})$ stands for the difference of the dielectric constant of the defect region from $\varepsilon_0(\boldsymbol{r})$, the dielectric function of the perfect PC. This form covers general classes of PCs involving defects, in which the composition of one or a few of the unit cells is different from all the others. Therefore, the approach in what follows applies to a wide class of defects in PCs.

The eigenvalue equation to solve is then

$$\left(H_0(\boldsymbol{r}) - \frac{\omega^2}{c^2}\right)\boldsymbol{Q}(\boldsymbol{r}) + H_1(\boldsymbol{r})\boldsymbol{Q}(\boldsymbol{r}) = 0 \tag{4.65}$$

where

$$H_0(\boldsymbol{r}) = \frac{1}{\sqrt{\varepsilon_0(\boldsymbol{r})}} \boldsymbol{\nabla} \times \left\{ \boldsymbol{\nabla} \times \frac{1}{\sqrt{\varepsilon_0(\boldsymbol{r})}} \right\} \tag{4.66}$$

is the Hamiltonian for the perfect PC and

$$H_1(\boldsymbol{r}) = -\frac{\omega^2}{c^2} \frac{\delta\varepsilon(\boldsymbol{r})}{\varepsilon_0(\boldsymbol{r})} \tag{4.67}$$

is treated as a perturbation to H_0 from the defect. Except for the ω dependence of the perturbing Hamiltonian H_1, the main effect of the defect is thus analogous to that of the electronic problem.

We expand the solution $\boldsymbol{Q}(\boldsymbol{r})$ of (4.65) in terms of the complete set involving PB solutions of the perfect system:

$$\boldsymbol{Q}(\boldsymbol{r}) = \sum_{n\boldsymbol{k}} c_{n\boldsymbol{k}}^{(T)} \boldsymbol{Q}_{n\boldsymbol{k}}^{(T)} + \sum_{n\boldsymbol{k}} c_{n\boldsymbol{k}}^{(L)} \boldsymbol{Q}_{n\boldsymbol{k}}^{(L)}, \tag{4.68}$$

with unknowns $c_{n\boldsymbol{k}}^{(T)}$ and $c_{n\boldsymbol{k}}^{(L)}$ to be determined to satisfy (4.65). The index n for the longitudinal basis functions is introduced because the wavefunctions $\boldsymbol{Q}_{\boldsymbol{kh}}(\boldsymbol{r})$ defined in (4.62) are not mutually orthogonal because of the presence of the factor $\sqrt{\varepsilon_0(\boldsymbol{r})}$ in its definition. The orthonormality is meant here using the inner product defined by the integral over the volume V_{pc}

$$(\boldsymbol{Q}, \boldsymbol{Q}') = \int_{\mathrm{pc}} \mathrm{d}\boldsymbol{r} \boldsymbol{Q}^*(\boldsymbol{r}) \cdot \boldsymbol{Q}(\boldsymbol{r}). \tag{4.69}$$

The longitudinal solutions $\{\boldsymbol{Q}_{\boldsymbol{kh}}(\boldsymbol{r})\}$ are made orthonormal by, say, the Schmidt method to introduce the orthonormalized set $\boldsymbol{Q}_{n\boldsymbol{k}}^{(L)}$, with the same symbol n adopted, for brevity, to specify them.

The unknown constants of (4.68) are determined so that (4.68) satisfies (4.65). Inserting (4.68) into (4.65) and using the orthonormality (α, β = T, L)

$$(\boldsymbol{Q}_{n\boldsymbol{k}}^{(\alpha)}, \boldsymbol{Q}_{n'\boldsymbol{k}'}^{(\beta)}) = V_{\mathrm{pc}} \delta_{\alpha\beta} \delta_{nn'} \delta_{\boldsymbol{k}\boldsymbol{k}'}, \tag{4.70}$$

we find

$$\left(\frac{\omega_{n\boldsymbol{k}}^2}{c^2} - \frac{\omega^2}{c^2} \right) c_{n\boldsymbol{k}}^{(T)} + \frac{\omega^2}{c^2} \sum_{n'\boldsymbol{k}'} h(n\boldsymbol{k}, n'\boldsymbol{k}') c_{n'\boldsymbol{k}'}^{(T)} = 0 \tag{4.71}$$

with

$$h(n\boldsymbol{k}, n'\boldsymbol{k}') = -\int_{\mathrm{pc}} \mathrm{d}\boldsymbol{r} \boldsymbol{Q}_{n\boldsymbol{k}}^{(T)*} \cdot \left(\frac{\delta\varepsilon(\boldsymbol{r})}{\varepsilon_0(\boldsymbol{r})} \boldsymbol{Q}_{n'\boldsymbol{k}'}^{(T)} \right). \tag{4.72}$$

The quantity $h(n\boldsymbol{k}, n'\boldsymbol{k}')$ is the matrix element of the perturbation $H_1(\boldsymbol{r})$ between the states (n, $\boldsymbol{k}$, T) and (n', $\boldsymbol{k}'$, T). Here we have retained only the coupling between the PB states. The mixing among the PB modes caused by the defect is induced by this term.

To proceed further, we introduce a simplification to $h(n\boldsymbol{k}, n'\boldsymbol{k}')$. We assume the following form of the mixing

$$h(n\boldsymbol{k}, n'\boldsymbol{k}') = h, \tag{4.73}$$

h being a constant independent of the indices $(n\boldsymbol{k})$ and $(n'\boldsymbol{k}')$. Let us, for brevity, take h to be a real constant. Equation (4.71) then turns out to be

$$c_{n\boldsymbol{k}}^{(T)} + \frac{\omega^2}{c^2} \frac{h}{\omega_{n\boldsymbol{k}}^2/c^2 - \omega^2/c^2} \sum_{n'\boldsymbol{k}'} c_{n'\boldsymbol{k}'}^{(T)} = 0. \tag{4.74}$$

From this we can derive the equation for the unknown

$$\sum_{n'\boldsymbol{k}'} c_{n'\boldsymbol{k}'}^{(T)}$$

by summing (4.74) over $(n,\ \boldsymbol{k})$, which finally leads to the secular equation of our problem

$$\frac{1}{h}\frac{c^2}{\omega^2} + \sum_{n\boldsymbol{k}} \frac{1}{\omega_{n\boldsymbol{k}}^2/c^2 - \omega^2/c^2} = 0. \tag{4.75}$$

Except for the frequency dependence of the first term, this is a typical form which is well known as the Koster–Slater model for a donor or acceptor impurity state of semiconductors [35]. The second term has a series of poles at the frequencies of the PB states of the host perfect PC. For ω within the continuous spectrum of the band states $(n, \boldsymbol{k})$, the second term of this equation is complex, having imaginary part related to the DOS of the band states of the perfect photonic crystal; for $\omega > 0$ it is found that

$$\sum_{n\boldsymbol{k}} \frac{1}{\omega_{n\boldsymbol{k}}^2/c^2 - \omega^2/c^2} = \sum_{n\boldsymbol{k}} \left[\frac{\mathcal{P}}{\omega_{n\boldsymbol{k}}^2/c^2 - \omega^2/c^2} + \mathrm{i}\pi\delta(\omega_{n\boldsymbol{k}}^2/c^2 - \omega^2/c^2) \right], \tag{4.76}$$

where $\mathcal{P}$ stands for the principal part when the sum over $\boldsymbol{k}$ is converted to an integral. Due to the second purely imaginary term, which is nonzero unless ω is in the band gap region, a real solution of (4.75) does not exist within the continuous spectrum of host PBs. The defect-induced real eigenvalue can thus be found only within the frequency region of band gaps where the second term vanishes. This is the origin of band gap modes. This simple model explains why the continuous spectrum of PBs of host perfect PC is important in considering the appearance of defect modes.

A sketch of the second term of (4.75) as well as minus the first term is given in Fig. 4.17. It shows a graphical solution of (4.75) when the values of $\boldsymbol{k}$ in the sum are treated as discrete. The intersections of the two curves give the normal modes of the system of the PC containing a defect. We see that the states within the PB continuum are shifted slightly from the orginal ones of the perfect PC.

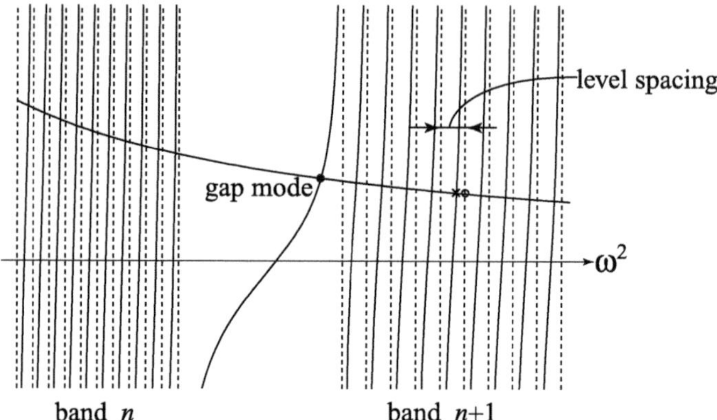

Fig. 4.17. Graphical solutions of (4.75). The second term of (4.75) consisting of many divergences at band energies is crossed with the curve $-1/(h\omega^2)$, which is a monotonically decreasing funtion of ω^2 for the case $h < 0$. A gap mode appears in the band gap between the bands n and $n+1$. Within the band an unperturbed band state marked by an open circle is shifted by the effect of the defect to the state marked by the cross

Within a band gap, however, the situation is quite different and the impurity effect manifests itself as appearance of a state as a well-defined normal mode. By measuring the optical signal in the band gap region, we can catch a band gap mode. Since the signal of the gap modes is proportional to the density of the defects, we need to introduce a substantial density of defects for the gap mode to be observed experimentally.

If we remove one whole layer from a PC or replace one whole layer by a new layer different from the host layers, the resultant system still has 2D translational symmetry in the plane of layers (taken as the xy plane). For such a defect, the band gap modes are specified by the 2D wavevector (k_x, k_y) and their frequencies have dependences on the wavevector [36]. Such modes are localized in the defect layer but propagate freely in the xy plane. This type of defect is still covered by the above model. An example of this type of 2D defect mode is given by Fig. 2.18.

4.5.3 Practical Treatment of Defect Modes

Usually the band structure of a photonic crystal needs to be calculated very precisely to make comparison with experimental data. The frequency of gap modes too needs to be obtained with comparable precision. For this purpose, we may no longer rely on a simplified treatment as used above. Also, the calculation of the whole spectrum of $\omega_{n\boldsymbol{k}}$ and the eigenvector $\boldsymbol{Q}_{n\boldsymbol{k}}^{(T)}$ of the host PC is not practical.

We describe two methods of calculation used practically for the treatment of the band gap mode, by taking a 2D PC of arrayed cylinders which contains a single impurity cylinder as an example. The first is the supercell method.

Supercell Method. The supercell method is a band structure calculation by assuming that defects are present periodically. We assume, for example, that one impurity cylinder exists in every L^2 original unit cells, i.e., we treat a PC with a defect as if it were a periodic system which has L^2 cylinders in a new unit cell. Then we can apply various methods of PBS calculation of a perfect system. If we take L sufficiently large for the coupling between the impurity cylinders of neighboring unit supercells to be neglected, the PBS of this PC should give a precise gap-mode frequency. Since the unit cell of the supercell method is large, the BZ is so much reduced. The $\boldsymbol{k}$ dependence of the gap mode remains, if the size L is not sufficent, because the $\boldsymbol{k}$ dependence arises from the hopping via the impurity sites of the fictitious periodic PC [37, 38].

Figure 4.18 is an example of the use of the supercell method, based on the vector KKR formulation for the calculation of a gap mode in the hexagonal lattice of dielectric cylinders of radius r [38]. In the host hexagonal PC of lattice constant a, the dielectric constant of the cylinders is $\varepsilon = 13$ with radius given by $r/a = 0.2$. The impurity cylinder is assumed to have a greater radius of $r_{\mathrm{i}}/a = 0.3$ and the same dielectric constant as the host cylinders. Figure 4.18(a) shows the unit cell of lattice constant $3a$, i.e., $L = 3$. The impurity cylinder is assumed to be located at the center position. We see the gap mode arise at the normalized frequency $Z(= \omega a/2\pi c) = 0.35$ in Fig. 4.18(b), which shows the band structure of $L = 3$. Figure 4.18(c) shows that as L becomes larger, the dispersion over the BZ becomes smaller.

For the same system, the transmittance calculation was also given, which shows the appearance of a sharp peak in the transmittance spectrum just at the gap mode frequency $Z = 0.35$. Defect-induced transmission peaks have been confirmed in many ways both theoretically and experimantally.

Response of an Imbedded Test Dipole In the second method of the treatment of the band gap modes, we introduce a fictitious point or line dipole which is oscillating with frequency ω in a PC with a defect and calculate the radiated energy from it as a function of ω. The introduced dipole brings about an inhomogeneous term in Maxwell's equations and its effect and hence the electric field irradiated from it can be calculated formally by using the Green function for a PC with a defect (see Sect. 4.6). The calculated radiation spectrum should have a resonant structure at $\omega = \omega_g$, if there exists a band gap mode with frequency ω_g:

$$U(\omega) = \frac{p_0^2 W}{(\omega - \omega_g)^2 + \gamma^2}, \tag{4.77}$$

where $U(\omega)$ is the stationary energy outflow of the radiation from the test dipole with a dipole moment of strength p_0, W is a constant which has

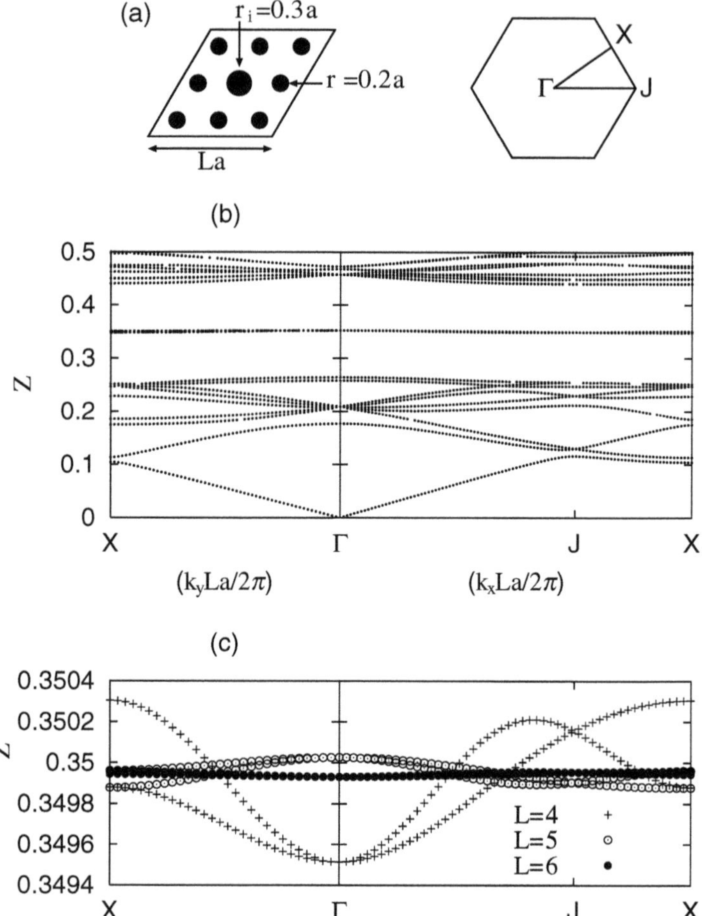

Fig. 4.18. Calculation of the gap mode by the supercell method. Panel (a) shows the supercell of triangular lattice of $L = 3$ and the 2D BZ. (b) shows the appearance of the gap mode at Z=0.35. The calculation is for the super-unit-cell of $L = 3$. (c) shows a gradual vanishing of the dispersion of the gap mode over the BZ as L becomes larger. The gap mode is two-fold degenerate at Γ

information about the eigenvector of the gap mode and γ is the damping constant, if any, of the band gap mode.

As an actual procedure of catching the gap mode we calculate first the radiated electromagnetic energy from the dipole in the time domain and then search for the resonant peak in the Fourier spectrum which would have the typical Lorentzian peak given above. Usually the periodic boundary condition is added in the numerical calculation and so the translational symmetry of a supercell is imposed in this method, too. The FDTD calculation has been successfully applyed in obtaining the time-domain response.

This method may be applied to examine band gap modes in complicated PCs, defect modes in a slab photonic crystal, for example. By the leakage of the energy out to the exterior of the host system, the host PBs of the slab are already lifetime-broadened. The defect mode is inevitably influenced by the radiative damping of host PBs. In such a system the lifetime of the defect mode can be obtained from the line shape of $U(\omega)$. A number of examples of the analysis are found in the book by Sakoda [36].

4.6 Inhomogeneous Maxwell Equations

The optical responses of PCs treated so far are all covered by the homeogeneous Maxwell equations. This section treats a point dipole and a nonlinear substance introduced externally as a perturbation to an otherwise perfect PC. They are both very important examples which can be treated by inhomegeneous Maxwell equations.

4.6.1 Green's Function and Inhomogeneous Maxwell Equations

In the sense that the perturbation of a perfect system plays a key role, the problem of light emission from a dipole or nonlinear substance imbedded in PCs is in essence similar to that of the gap modes treated in Sect. 4.5.2.

Let us write the dipole moment introduced externally to a PC as $\boldsymbol{P}_{\text{ex}}(\boldsymbol{r}, t)$, which is supposed to be a dipole moment of a light-emitting atom or a nonlinear substance imbedded in it. It can be localized at a specified point or region of the PC or can be distributed more or less homogeneously throughout it. The effect of $\boldsymbol{P}_{\text{ex}}(\boldsymbol{r}, t)$ appears through the change of the charge and current densities in Maxwell's equations:

$$\begin{aligned}
&\boldsymbol{\nabla} \times \boldsymbol{E}(\boldsymbol{r}) = \mathrm{i}\omega \boldsymbol{B}(\boldsymbol{r}), \\
&\boldsymbol{\nabla} \cdot (\varepsilon_0 \varepsilon(\boldsymbol{r}) \boldsymbol{E}(\boldsymbol{r})) = -\boldsymbol{\nabla} \cdot \boldsymbol{P}_{\text{ex}}(\boldsymbol{r}), \\
&\boldsymbol{\nabla} \times \boldsymbol{B}(\boldsymbol{r}) = -\mathrm{i}\omega \mu_0 \boldsymbol{P}_{\text{ex}}(\boldsymbol{r}) - \frac{\mathrm{i}\omega}{c^2} \varepsilon(\boldsymbol{r}) \boldsymbol{E}(\boldsymbol{r}), \\
&\boldsymbol{\nabla} \cdot \boldsymbol{B}(\boldsymbol{r}) = 0.
\end{aligned} \tag{4.78}$$

Here we assumed the harmonic time dependence

$$\boldsymbol{P}_{\text{ex}}(\boldsymbol{r}, t) = \boldsymbol{P}_{\text{ex}}(\boldsymbol{r}) \mathrm{e}^{-\mathrm{i}\omega t}$$

and put

$$\boldsymbol{E}(\boldsymbol{r}) = \boldsymbol{E}(\boldsymbol{r}) \mathrm{e}^{-\mathrm{i}\omega t},$$

etc. Except the presence of the terms involving $\boldsymbol{P}_{\text{ex}}(\boldsymbol{r})$, equation (4.78) is identical to (3.5) and (3.15) for a host PC of periodic dielectric constant $\varepsilon(\boldsymbol{r})$.

Hereafter we shall treat $\varepsilon(\boldsymbol{r})$ as a real positive constant, *independent* of ω, when making Fourier inverse transforms back to the time domain. This assumption is made in order to avoid the complexities in the required contour integral over the complex ω plane. In this approximation, Maxwell's equations given above are equivalent to the ones in the time domain:

$$\begin{aligned} &\boldsymbol{\nabla} \times \boldsymbol{E}(\boldsymbol{r},t) = -\frac{\partial}{\partial t}\boldsymbol{B}(\boldsymbol{r},t), \\ &\boldsymbol{\nabla} \cdot (\varepsilon_0 \varepsilon(\boldsymbol{r})\boldsymbol{E}(\boldsymbol{r},t)) = -\boldsymbol{\nabla} \cdot \boldsymbol{P}_{\mathrm{ex}}(\boldsymbol{r},t), \\ &\boldsymbol{\nabla} \times \boldsymbol{B}(\boldsymbol{r},t) = \mu_0 \frac{\partial}{\partial t}\boldsymbol{P}_{\mathrm{ex}}(\boldsymbol{r},t) + \frac{1}{c^2}\frac{\partial}{\partial t}\varepsilon(\boldsymbol{r})\boldsymbol{E}(\boldsymbol{r},t), \\ &\boldsymbol{\nabla} \cdot \boldsymbol{B}(\boldsymbol{r},t) = 0. \end{aligned} \tag{4.79}$$

They are our starting equations. Eliminating $\boldsymbol{B}(\boldsymbol{r},t)$ from the first and third of (4.79), we find

$$-\left(\frac{1}{c^2}\frac{\partial^2}{\partial t^2} + H\right)\boldsymbol{Q}(\boldsymbol{r},t) = \frac{\mu_0}{\sqrt{\varepsilon(\boldsymbol{r})}}\frac{\partial^2}{\partial t^2}\boldsymbol{P}_{\mathrm{ex}}(\boldsymbol{r},t), \tag{4.80}$$

where the operator H and $\boldsymbol{Q}(\boldsymbol{r},t)$ are defined by (see Sect. 4.5.1)

$$H\boldsymbol{Q}(\boldsymbol{r},t) = \frac{1}{\sqrt{\varepsilon(\boldsymbol{r})}}\boldsymbol{\nabla} \times \left\{\boldsymbol{\nabla} \times \frac{1}{\sqrt{\varepsilon(\boldsymbol{r})}}\boldsymbol{Q}(\boldsymbol{r},t)\right\} \tag{4.81}$$

with

$$\boldsymbol{Q}(\boldsymbol{r},t) = \sqrt{\varepsilon(\boldsymbol{r})}\boldsymbol{E}(\boldsymbol{r},t). \tag{4.82}$$

Equation (4.80) is an inhomogeneous equation. Following the familiar procedure, we introduce the tensor Green's function $\boldsymbol{G}(\boldsymbol{r},\boldsymbol{r}',t-t')$ of the host PC by the following equation :

$$-\left(\frac{1}{c^2}\frac{\partial^2}{\partial t^2} + H\right)\boldsymbol{G}(\boldsymbol{r},\boldsymbol{r}',t-t') = \boldsymbol{I}\delta(\boldsymbol{r}-\boldsymbol{r}')\delta(t-t'), \tag{4.83}$$

$\boldsymbol{I}$ being the unit tensor, with the retarded boundary condition

$$\boldsymbol{G}(\boldsymbol{r},\boldsymbol{r}',t-t') = 0 \qquad \text{for} \quad t-t' < 0. \tag{4.84}$$

The vanishing of the Green's function for negative $t-t'$ is the causality requirement, which dictates that the system should not respond to the input signal before it is switched on. Note that $\boldsymbol{G}$ incorporates all the periodic scattering of electromagnetic fields which result in the appearance of the PBS.

The electric field emitted by $\boldsymbol{P}_{\mathrm{ex}}(\boldsymbol{r},t)$, i.e., the solution of (4.80), is then given by [34]

$$\begin{aligned} \boldsymbol{E}(\boldsymbol{r},t) &\equiv \frac{1}{\sqrt{\varepsilon(\boldsymbol{r})}}\boldsymbol{Q}(\boldsymbol{r},t) \\ &= \frac{1}{\sqrt{\varepsilon(\boldsymbol{r})}}\int_{\text{pc}} \mathrm{d}\boldsymbol{r}' \int_{-\infty}^{\infty} \mathrm{d}t' \boldsymbol{G} \frac{\mu_0}{\sqrt{\varepsilon(\boldsymbol{r}')}} \frac{\partial^2}{\partial t'^2} \boldsymbol{P}_{\text{ex}}(\boldsymbol{r}',t'), \end{aligned} \tag{4.85}$$

where $\boldsymbol{G}$ stands for $\boldsymbol{G}(\boldsymbol{r},\boldsymbol{r}',t-t')$. Its Fourier transform $\boldsymbol{G}(\boldsymbol{r},\boldsymbol{r}',\omega)$ is now given by

$$\begin{aligned} \boldsymbol{G}(\boldsymbol{r},\boldsymbol{r}',\omega) &\equiv \int_{-\infty}^{\infty} \mathrm{d}t \boldsymbol{G}(\boldsymbol{r},\boldsymbol{r}',t) \mathrm{e}^{\mathrm{i}\omega t} \\ &= \frac{c^2}{V_{\text{pc}}} \sum_{\boldsymbol{k}n} \left[\frac{\boldsymbol{Q}_{n\boldsymbol{k}}^{(T)}(\boldsymbol{r}) : \boldsymbol{Q}_{n\boldsymbol{k}}^{(T)}(\boldsymbol{r})^*}{(\omega+\mathrm{i}\delta)^2 - \left(\omega_{n\boldsymbol{k}}^{(T)}\right)^2} + \frac{\boldsymbol{Q}_{n\boldsymbol{k}}^{(L)}(\boldsymbol{r}) : \boldsymbol{Q}_{n\boldsymbol{k}}^{(L)}(\boldsymbol{r})^*}{(\omega+\mathrm{i}\delta)^2} \right]. \end{aligned} \tag{4.86}$$

The small imaginary part $\mathrm{i}\delta$ $(\delta > 0)$ in the denominators is added to ω, as usual, so that the Green's function satisfies the causality condition (4.84), when transformed back to t space. The vectors $\boldsymbol{Q}_{n\boldsymbol{k}}^{(T)}(\boldsymbol{r})$ and $\boldsymbol{Q}_{n\boldsymbol{k}}^{(L)}(\boldsymbol{r})$ were defined in Sect. 4.5.1. They satisfy the orthonormality and closure relations as explaned there. The symbol $\boldsymbol{Q} : \boldsymbol{Q}$ is introduced to stand for a second-rank tensor. That the expression (4.86) indeed satisfies the Fourier tranform of (4.83) is seen if we operate with $(\omega^2/c^2 - H)$ on $\boldsymbol{G}(\boldsymbol{r},\boldsymbol{r}',\omega)$ and make use of the eigenvalue equation (4.60) and closure relation (4.63).

We obtain $\boldsymbol{G}(\boldsymbol{r},\boldsymbol{r}',t)$ from $\boldsymbol{G}(\boldsymbol{r},\boldsymbol{r}',\omega)$ by the inverse Fourier transform. From the residues at the poles at $\omega = \omega_{n\boldsymbol{k}}^{(T)} - \mathrm{i}\delta$ and $\omega = -\mathrm{i}\delta$, we find

$$\begin{aligned} \boldsymbol{G}(\boldsymbol{r},\boldsymbol{r}',t-t') = &-\frac{c^2}{V_{\text{pc}}} \sum_{\boldsymbol{k},n} [\frac{\sin\omega_{n\boldsymbol{k}}^{(T)}(t-t')}{\omega_{n\boldsymbol{k}}^{(T)}} \boldsymbol{Q}_{n\boldsymbol{k}}^{(T)}(\boldsymbol{r}) : \boldsymbol{Q}_{n\boldsymbol{k}}^{(T)}(\boldsymbol{r})^* \\ &+(t-t')\boldsymbol{Q}_{n\boldsymbol{k}}^{(L)}(\boldsymbol{r}) : \boldsymbol{Q}_{n\boldsymbol{k}}^{(L)}(\boldsymbol{r})^*] \quad (\text{for } t \geq t') \end{aligned} \tag{4.87}$$

and $\boldsymbol{G}(\boldsymbol{r},\boldsymbol{r}',t-t') = 0$ for $t < t'$. Using this expression in (4.85), we obtain the electric field emitted by $\boldsymbol{P}_{\text{ex}}$. Since

$$\begin{aligned} \boldsymbol{G}(\boldsymbol{r},\boldsymbol{r}',t-t')|_{t'=t-} &= 0, \\ \left.\frac{\partial}{\partial t'}\boldsymbol{G}(\boldsymbol{r},\boldsymbol{r}',t-t')\right|_{t'=t-} &= c^2\boldsymbol{I} \end{aligned} \tag{4.88}$$

from (4.87) and the closure relation (4.63), integration twice by parts gives

$$\begin{aligned} &\boldsymbol{E}(\boldsymbol{r},t) + \frac{\boldsymbol{P}_{\text{ex}}(\boldsymbol{r},t)}{\varepsilon_0\varepsilon(\boldsymbol{r})} \\ &= \frac{1}{V_{\text{pc}}} \sum_{n\boldsymbol{k}} \frac{\boldsymbol{Q}_{n\boldsymbol{k}}^{(T)}(\boldsymbol{r})}{\varepsilon_0\sqrt{\varepsilon(\boldsymbol{r})}} \int \mathrm{d}\boldsymbol{r}' \int_{-\infty}^{t} \mathrm{d}t' \frac{\boldsymbol{Q}_{n\boldsymbol{k}}^{(T)}(\boldsymbol{r}')^* \cdot \boldsymbol{P}_{\text{ex}}(\boldsymbol{r}',t')}{\sqrt{\varepsilon(\boldsymbol{r}')}} \\ &\times\omega_{n\boldsymbol{k}}^{(T)} \sin\omega_{n\boldsymbol{k}}^{(T)}(t-t'), \end{aligned} \tag{4.89}$$

where we have assumed that $\boldsymbol{P}_{\text{ex}}(\boldsymbol{r},t)$ was introduced adiabatically from $t=-\infty$, i.e., $\boldsymbol{P}_{\text{ex}}(\boldsymbol{r},-\infty)=0$. We see that the role of the longitudinal field $\boldsymbol{Q}_{n\boldsymbol{k}}^{(L)}(\boldsymbol{r})$ is only to lead to the dipolar term on the left hand side and the propagating radiation field comes only from $\boldsymbol{Q}_{n\boldsymbol{k}}^{(T)}(\boldsymbol{r})$. However, when we multiply both sides of (4.89) by $\varepsilon_0\varepsilon(\boldsymbol{r})$, we see that the displacement field $\boldsymbol{D}(\boldsymbol{r},t)$, involving the polarization $\boldsymbol{P}_{\text{ex}}(\boldsymbol{r},t)$ in addition to those from $\varepsilon(\boldsymbol{r})$, satisfies $\boldsymbol{\nabla}\cdot\boldsymbol{D}(\boldsymbol{r},t)=0$, as it should. To check this statement, note that

$$\boldsymbol{\nabla}\cdot\sqrt{\varepsilon(\boldsymbol{r})}\boldsymbol{Q}_{n\boldsymbol{k}}^{(T)}(\boldsymbol{r})=0$$

by the definition (4.82). Therefore, the longitudinal field $\boldsymbol{Q}_{n\boldsymbol{k}}^{(L)}(\boldsymbol{r})$ is essential for the consistency of the theory.

4.6.2 Applications of the Derived Formula

Emission of Light from a Point Dipole. For the case of an oscillating dipole moment at the position $\boldsymbol{r}_0$, $\boldsymbol{P}_{\text{ex}}(\boldsymbol{r},t)$ has the form

$$\boldsymbol{P}_{\text{ex}}(\boldsymbol{r},t)=\boldsymbol{p}_0\delta(\boldsymbol{r}-\boldsymbol{r}_0)\mathrm{e}^{-\mathrm{i}\omega t}. \tag{4.90}$$

Then the integral over t' is carried out in (4.89) and leads to

$$\begin{aligned}\boldsymbol{E}(\boldsymbol{r},t)+\frac{\boldsymbol{P}_{\text{ex}}(\boldsymbol{r},t)}{\varepsilon_0\varepsilon(\boldsymbol{r})} &= \frac{2\pi}{V_{\text{pc}}}\mathrm{e}^{-\mathrm{i}\omega t}\sum_{n\boldsymbol{k}}\omega_{n\boldsymbol{k}}^{(T)}\boldsymbol{E}_{n\boldsymbol{k}}^{(T)}(\boldsymbol{r})\left(\boldsymbol{E}_{n\boldsymbol{k}}^{(T)}(\boldsymbol{r}_0)^*\cdot\boldsymbol{p}_0\right)\\ &\quad\times\left(\frac{1}{\omega+\omega_{n\boldsymbol{k}}^{(T)}+\mathrm{i}\delta}-\frac{1}{\omega-\omega_{n\boldsymbol{k}}^{(T)}+\mathrm{i}\delta}\right).\end{aligned} \tag{4.91}$$

The energy balance then states that for the dipole to keep oscillating as expressed by (4.90) the work done by $\boldsymbol{E}(\boldsymbol{r},t)$ on $\boldsymbol{P}_{\text{ex}}(\boldsymbol{r},t)$ should be equal to the energy irradiated from $\boldsymbol{P}_{\text{ex}}(\boldsymbol{r},t)$. Finally the time average of the irradiated energy is given by

$$\begin{aligned}U &= \int_{S1}\mathrm{d}\boldsymbol{s}\cdot\langle\boldsymbol{S}(\boldsymbol{r},t)\rangle\\ &= \int_{V1}\mathrm{d}\boldsymbol{r}\boldsymbol{\nabla}\cdot\langle\boldsymbol{S}(\boldsymbol{r},t)\rangle\\ &= \frac{\pi^2\omega^2}{V_{\text{pc}}}\sum_{n\boldsymbol{k}}\left|\boldsymbol{p}_0\cdot\boldsymbol{E}_{n\boldsymbol{k}}^{(T)}(\boldsymbol{r}_0)\right|^2\delta(\omega-\omega_{n\boldsymbol{k}}^{(T)}),\end{aligned} \tag{4.92}$$

where we have introduced the Poynting vector

$$\boldsymbol{S}(\boldsymbol{r},t)=\boldsymbol{E}(\boldsymbol{r},t)\times\boldsymbol{H}(\boldsymbol{r},t)$$

and its time average $\langle\boldsymbol{S}(\boldsymbol{r},t)\rangle$ on the surface S_1 of an arbitrary area V_1 encircling $\boldsymbol{r}_0$.

According to (4.92), the irradiated energy is proportional to ω^2: one factor ω comes from the current density, i.e., from $(\partial/\partial t)\boldsymbol{P}_{\text{ex}}(t)$, and the Poynting vector is proportional to the square of the radiated electric field. The δ function of (4.92) stands for the energy conservation. If the frequency ω of $\boldsymbol{P}_{\text{ex}}$ lies in the band gap, we find $U = 0$, because there is no $(n, \boldsymbol{k})$ that meets the energy conservation condition. When one value of the band index n participates in energy conservation, a particular value for $\boldsymbol{k}$ is selected by the δ function. In this way the emission spectrum sharply reflects the dispersion relation of the PB. It is interesting that the above formula is just the formula obtained by the Fermi golden rule of the quantum mechanical treatment of photon emission due to dipole radiation. The relevant quantities determing the emission intensity are not only the dispersion relations of PBs but also their eigenvectors. Thus, strictly speaking, the calculation of the emission spectrum requires more than the calculation of the DOS of PBs. Such an elaborate calculation is carried out by, for example, Lousse et al. [39].

Raman Scattering Involving a Photonic Crystal. Raman scattering is similarly treated in the present theory. In this case we have only to redefine $\boldsymbol{P}_{\text{ex}}(\boldsymbol{r}, t)$ to be a dipole moment induced by the electric field of the incident light. In (4.92), we can put

$$\boldsymbol{p}_0 = \alpha \boldsymbol{E}_{\text{loc}} \tag{4.93}$$

where α is the polarizability of, say, the molecule under investigation and $\boldsymbol{E}_{\text{loc}}$ is the local field at the site of the target molecule. By expanding α in terms of the coordinates of phonon, magnon, etc., we may incorporate the Stokes and anti-Stokes shift by changing the ω of $\boldsymbol{P}_{\text{ex}}(\boldsymbol{r}, t)$ into $\omega \pm \omega_0$ in the above theory. One interesting difference from the emission case is that $\boldsymbol{p}_0$ is determined by $\boldsymbol{E}_{\text{loc}}$, which in itself is affected by the PBS and has a strong ω dependence. This is the same situation as the enhancement of the near-field intensity of PCs (see Sect. 4.4.2). The positive effect of a PB excitation on the enhancement of the Raman signals was thoretically examined in 1982 [40].

Second and Higher Harmonic Generation. Second, third and higher harmonic generations may be covered by a similar treatment [41]. In the case of second harmonic generation, the key quantity in the expression of the emitted electric field given by (4.89) is given by

$$\begin{aligned} \boldsymbol{E}_{n\boldsymbol{k}}^{(T)*} \cdot \boldsymbol{P}_{\text{ex}}(\boldsymbol{r}, t) \sin \omega_{n\boldsymbol{k}}^{(T)}(t - t') &= \sum_{\boldsymbol{k}1n1} \sum_{\boldsymbol{k}2n2} \chi^{(2)} : \boldsymbol{E}_{n\boldsymbol{k}}^{(T)*} \boldsymbol{E}_{\boldsymbol{k}1n1}^{(T)} \boldsymbol{E}_{\boldsymbol{k}2n2}^{(T)} \\ &\times \sin \omega_{n\boldsymbol{k}}^{(T)}(t - t') \exp\left(-i(\omega_{\boldsymbol{k}1n1}^{(T)} + \omega_{\boldsymbol{k}2n2}^{(T)})t'\right) . \end{aligned} \tag{4.94}$$

Here we have introduced the third-rank tensor $\chi^{(2)}$. Comparing this form with the treatment given above for light emission from a point dipole, we have only to replace the frequency ω there by $\omega_{\boldsymbol{k}1n1}^{(T)} + \omega_{\boldsymbol{k}2n2}^{(T)}$. Accordingly, in place of $\delta(\omega - \omega_{n\boldsymbol{k}})$ of the emission case there appear

$$\delta(\omega_{\boldsymbol{k}1n1} + \omega_{\boldsymbol{k}2n2} - \omega_{n\boldsymbol{k}})$$

for the sum frequency generation. It is then obvious that the density of photons and eigenvectors at the final photon frequency $\omega_{n\boldsymbol{k}}$ plays a vital role in determining the intensity of the sum frequency generation. See the detailed analysis carried out in [41], which shows that the intensity is enhanced due to the PB effect leading to the enhanced density of states of photons. If we take into account the enhancement in the nonlinear $\boldsymbol{P}_{\rm ex}$ due to a local field [28, 33] and to the nonlinear thickness-dependence in the phase matching [42], the figure of merit of the PB effect is huge because nonlinear optical processes involve $\boldsymbol{E}^2_{\rm loc}$ or $\boldsymbol{E}^3_{\rm loc}$ in inducing $\boldsymbol{P}_{\rm ex}$.

4.7 Optics of Photonic Crystals

We here discuss the optics of PCs. To understand the origin of the similarity between the PC optics and crystal optics seen in anisotropic crystals of a homogeneous substance, it is necessary to handle the optics of both systems group-theoretically. Such an analysis was given in [43]. This section is based on this work.

Optics involving photonic crystals is similar in many ways to that of crystal optics of anisotropic materials. Much before the active study of PCs began, the analogy was noticed in layered systems of 1D periodicity, which presented the phenonemon of double refraction [44, 45]. Here we show that PCs may be viewed as large-scale anisotropic crystals which embody the crystal optics much more strikingly. Detailed descriptions of the characteristic features of crystal optics are found in many books on optics or electromagnetism [44, 46, 47].

We summarize several characteristic features expected in the optics of PCs. All of them have their counterparts in a homogeneous anisotropic crystal. For the numerical demonstration given here, we use a PC of dielectric spheres of refractive index 3.2 arrayed in a fcc lattice. The radius r of the spheres is taken to be $r = 0.3a$, a being the size of the fcc cube and the numerical results are shown using the dimensionless frequency Z and wavevector $\boldsymbol{k}$, scaled by $2\pi c/a$ and $2\pi/a$, respectively.

(1) Difference of the directions and magnitudes of phase and group velocities. A constant-frequency surface (CFS) of frequency ω is obtained by sweeping $\boldsymbol{k}$ in the entire BZ to look for the band states of ω. There are in general several bands participating simultaneously in constructing an equi-frequency surface. The number of relevant bands depends on the band population. Figure 4.19 shows an example for $Z = 0.75$ [43].
    We observe at once that the surface consists of several sheets and they are not at all spherical in shape. The latter fact shows at once that the direction of group velocity, which is normal to the CFS, does not coincide with that of $\boldsymbol{k}$. It also shows that if we change the direction of $\boldsymbol{k}$ only slightly, a drastic change can take place in the direction of the group

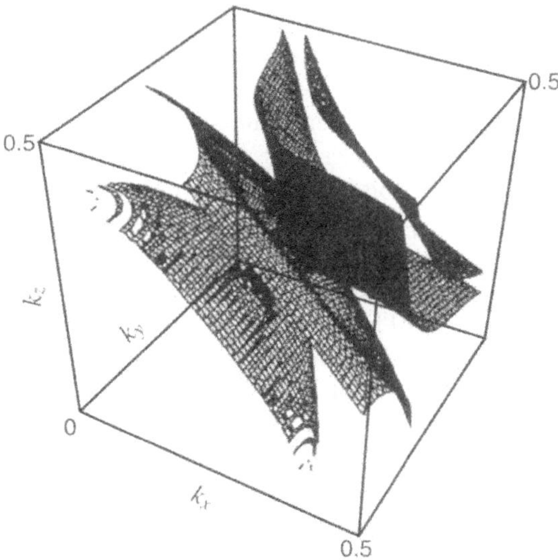

Fig. 4.19. Example of CFS in the cubic range of $0 < k_x, k_y, k_z < 0.5$. The frequency is $Z = 0.75$ for a fcc array of dielectric spheres of refractive index 3.2 and r/a (ratio of radius to cube size) 0.3. The CFS consists of several sheets

velocity. This is the origins of superprism effect [48] and superlensing effect of PCs [49, 50].

(2) Difference in direction between the electric field and displacement field. We know that the fields involved in PB states are neither transverse nor longitudinal in a strict sense. The calculation of the angle between $\boldsymbol{E}$ and $\boldsymbol{D}$ for several PB states was given in [17], which showed that the angles are generally of the order of 0.1 radians in our PC for relatively low-lying bands in the range $Z < 1.0$.
(3) Existence of several CFSs for a given ω in the $\boldsymbol{k}$ space and the double or more than double refraction as a result of it. We have seen in Fig. 4.19 an example of multiple CFSs of frequency $Z = 0.75$. If all of them are excited simultaneously by incident light of frequency $Z = 0.75$, the refraction will be double, triple, ..., according to the number of excited bands [51]. We have given Fig. 3.5, which presents clearly the multi-refraction phenomenon for a PC of arrayed spheres.
(4) Existence of the singular points in the constant frequency surface and the possibility of conical reflection as a result of it.

In what follows we discuss the item (4) in some detail. When a PB state $(n, \boldsymbol{k})$ is exited by an incident light beam, the propagation direction inside the PC is given by $\boldsymbol{\nabla}_{\boldsymbol{k}}\omega^{(n)}(\boldsymbol{k})$, which is normal to the CFS of the band n. By singular points in the title of the item (4) we mean the points of the $\boldsymbol{k}$ space, where the vector normal to the CFS is not determined uniquely. Such a case happens at the intersections between two CFSs in the $\boldsymbol{k}$ space, because the vectors normal to these CFSs are generally directed differently. If two CFSs belong to the bands of *different* irreducible representations, the intersection can indeed occur between them.

In an anisotropic crystal such a crossing actually happens when the CFSs of extraordinary and ordinary waves cross each other. The singular light scattering involving the crossing point has received much attention in the history of crystal optics (see, e.g., [44], which treats conical refraction). Generally the CFSs of equal frequency repel each other and the CFS crossing rarely happens. When $\boldsymbol{k}$ is in a direction of good symmetry, however, the crossing of two CFSs can take place. This happens because the group of $\boldsymbol{k}$, $\mathcal{G}(\boldsymbol{k})$, of such a $\boldsymbol{k}$ point is composed of more than two operations and the extraordinary and ordinary waves thereof belong to different irreducible representations of $\mathcal{G}(\boldsymbol{k})$. For a biaxial crystal of three principal dielectric constants, $\varepsilon_x = 1.0$, $\varepsilon_y = 1.5^2$ and $\varepsilon_z = 2.0^2$, Fig. 4.20 shows the CFSs of the extraordinary and ordinary modes. On the plane $k_y = 0$, we indeed see one crossing. Their CFSs repel each other when $k_y \neq 0$. Consequently, around the singular point of Fig. 4.20, two CFSs form the two cones sharing their vertices in the plane $k_y = 0$. In terms of the group-theoretical language, $\mathcal{G}(\boldsymbol{k})$ for $\boldsymbol{k}$ with $k_y = 0$ is $\mathrm{C_{1h}}$ and two CFSs belong to different irreducible representations of A' and A'' to allow the crossing, while, when $k_y \neq 0$, only an identity operation remains in $\mathcal{G}(\boldsymbol{k})$ and two CFSs belong to the same irreducible representation (identity representation) to repel each other. We can show that there are only four such points in the plane $k_y = 0$, the one shown in Fig.4.20 and those obtained from it by symmetry.

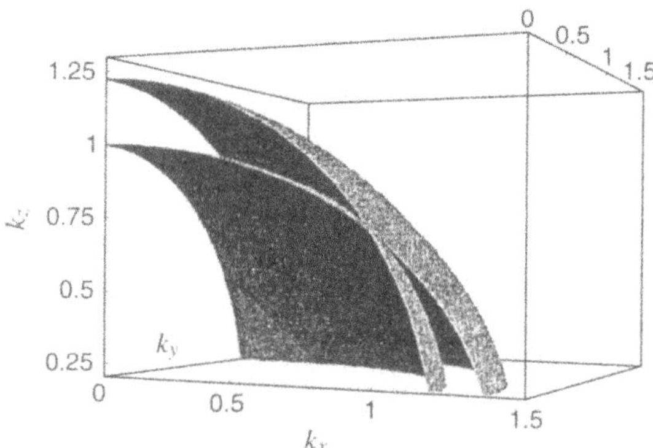

Fig. 4.20. Crossing of two CFSs of a frequency ω in a biaxial anisotropic crystal. The front face is the section with $k_y = 0$. The result is given for $\varepsilon_x = 1.0$, $\varepsilon_y = 1.5^2$ and $\varepsilon_z = 2.0^2$. The two sheets repel each other except in the plane $k_y = 0$. The crossing point in the plane $k_y = 0$ is the singular point. When ω varies, the same situation takes place because of the conformal change of the CFSs and the singular point moves on the straight line passing though the origin $\boldsymbol{k} = 0$

Now we turn to PCs. In the PC of our example, a fcc array of dielectric spheres, $\mathcal{G}(\boldsymbol{k})$ for a wavevector $\boldsymbol{k}$ on the plane $k_y = 0$ and that for a $\boldsymbol{k}$ in the region $k_y \neq 0$ are both identical to the anisotropic case. Since we have multiples of PB dispersion surfaces in the Brillouin zone, we shall see a lot of crossings between CFSs of PBs on the plane $k_y = 0$ and these crossed CFSs should repel each other once $\boldsymbol{k}$ is away from the plane $k_y = 0$. Namely we have many pairs of cones of CFS whose vertices meet at a point in the plane $k_y = 0$. The same thing happens in the plane $k_x = 0$ and $k_z = 0$, i.e., at the stars of $\boldsymbol{k}$ or even on the other symmetric planes such as the plane $k_x = k_y$ and their equivalents. With change of ω, each of the singular points moves along a complex curve on the plane $k_y = 0$, for example, in contrast to the crystal optics of anisotropic material in which the singular point is known to be on the straight line passing $\boldsymbol{k} = 0$ (see the caption of Fig. 4.20).

Since the appearance of the singular points results from the lowering of the symmetry of $\mathcal{G}(\boldsymbol{k})$ as $\boldsymbol{k}$ deviates from a symmetry plane, the same feature arises in principle in electrons, phonons, magnons, etc.. To the best of the author's knowledge, however, no attention has so far been paid except for the anisotropic crystal optics known since old days.

Figure 4.21 shows the CFSs of $Z = 0.96$ of the same PC treated in Fig. 4.19, for the $\boldsymbol{k}$ space of $0 < k_x, k_y, k_z < 1.0$ in (a) and $0 < k_y < 0.15$ in (b) [43]. If we make up the other side of the plane $k_y = 0$ by mirror reflection, several singular surfaces come in, each composed of two cones connected at a point.

Conical refraction inside an anisotropic crystal originates from the presence of an infinite number of directions of the energy flow of light associ-

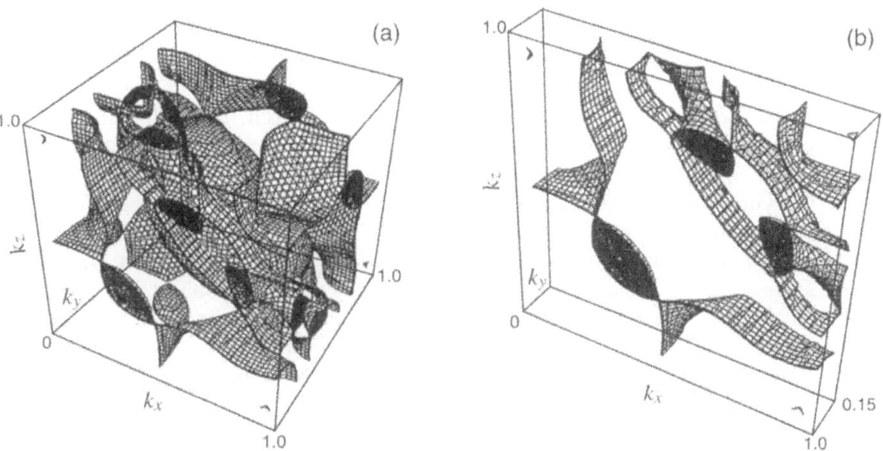

Fig. 4.21. Constant-frequency surface at $Z = 0.96$. Singular points are seen in the plane $k_y = 0$. Panel (a) shows the region of the cube $0 < k_x,\ k_y,\ k_z < 1.0$ and (b) shows the part of (a) within the narrow region $0 < k_y < 0.15$.

ated with singular points in the $\boldsymbol{k}$ space [44]. The phenomenon therefore prerequires the points in the $\boldsymbol{k}$ space, where $\nabla_{\boldsymbol{k}}\omega^{(n)}(\boldsymbol{k})$ is not determined uniquely. The common vertices of connected two cones discussed above are typical examples of such points.

Let $\boldsymbol{k}^0$ be one of such singular points and suppose a plane slab of PC exposed to light. If we cover the entrance surface with a diaphragm with a small aperture, which is arranged so that a narrow beam of the wavevector $\boldsymbol{k} = \boldsymbol{k}^0$ is allowed inside, the energy propagation inside the PC occurs along the surface of a cone (a surface weaved by the vectors $\nabla_{\boldsymbol{k}}\omega^{(n)}(\boldsymbol{k})$ with $\boldsymbol{k}$ swept near, but not exactly equal to, the point $\boldsymbol{k}^0$). The transmitted light goes out of the system from the intersection of that cone with the exit surface. The parallel rays thus emerge lying on the surface of the cylinder (not necessarily circular), with a common wavevector determined by the inside wavevector $\boldsymbol{k}_0$. This is the internal conical refraction known in crystal optics. When a number of diffraction channels are open, the leaving rays are lying not on a single cylinder but on several cylindrical surfaces, which are related mutually by reciprocal lattice vectors. Likewise, the external conical refraction will also be expected in the case of PCs.

To summarize, we discuss the possibility of superprism effect, multi-birefringence effect and the existence of singular points in the equi-frequency surfaces, the last of which can lead to internal and external conical refractions involving PBs. To investigate these phenomena experimentally, a search for a PC with a large and lossless dielectric constant and with composition as perfect as possible is essential.

References

1. C. M. Soukoulis (ed.): *Photonic Band Gaps and Localization*, (Plenum, New York 1993)
2. C. M. Bowden, J. P. Dowling and H. O. Everitt (ed.): *Development and Applications of Materials exhibiting Photonic Band Gaps*, J. Opt. Soc. Am. B **10** (1993) pp. 279–413
3. G. Kurizki and J. W. Haus (ed.): *Photonic Band Structures*, J. Mod. Opt. **41** (1994) pp.171–404
4. J. D. Joannopoulos, R. D. Meade, J. N. Winn: *Photonic Crystals: Molding the Flow of Light*, (Princeton University Press, Princeton 1995)
5. C. M. Soukoulis (ed.): *Photonic Band Gap Materials*, (Kluwer, Dordrecht 1996)
6. A. Scherer, T. Doll, E. Yablonovitch, H. O. Everitt and J. A. Higgins (eds.): *Special Section on Electromagnetic Crystal Structures, Design, Synthesis and Applications*, J. Lightwave Technology **17** (1999) pp.1928–2411
7. S. G. Johnson and J. D. Joannopoulos: *Photonic Crystals: The Road from Theory to Practice*, (Kluwer, Dordrecht 2002)
8. K. Sakoda: *Optical Properties of Photonic Crystals* (Springer, Berlin Heidelberg New York 2001)
9. De la Rue (ed.): *PECS 2000*, Opt. Quantum Electron. **34** No.1/3 (2002) pp.1-309

10. S. Noda and T. Baba: *Roadmap on Photonic Crystals*, (Kluwer, Dordrecht, 2003)
11. K. S. Yee: IEEE Trans. Antennas Propag. **14**, 302 (1966)
12. A. Taflove and S. C. Hagness: *Computational Electrodynamics: The Finite-Difference Time-Domain Method*, 2nd ed. (Artech House, Boston 2000)
13. J. Korringa: Physica, **13**, 392 (1947);
W. Kohn and N. Rostoker: Phys. Rev. **94**, 1111 (1954)
14. K. Ohtaka: J. Phys. C **13**, 667 (1980)
15. K. Ohtaka: Phys. Rev. B **19**, 5057 (1979)
16. K. Ohtaka, T. Ueta and K. Amemiya: Phys. Rev. B **57**, 2550 (1998)
17. K. Ohtaka and Y. Tanabe: J. Phys. Soc. Jpn. **65**, 2265 (1996)
18. K. Ohtaka and Y. Tanabe: J. Phys. Soc. Jpn. **65**, 2670 (1996)
19. K. Ohtaka and M. Inoue: Phys. Rev. B **25**, 677 (1982);
M. Inoue, K. Ohtaka and S. Yanagawa: Phys. Rev. B **25**, 689 (1982)
20. N. Stefanou, V. Karathanos and A. Modinos: J. Phys. Condens. Matter **4**, 7389 (1992);
X. Wang, X.-G. Zhang, Q. Yu and B. N. Harmon: Phys. Rev. B **47**, 4161 (1993);
A. Moroz: J. Phys. Condens. Matter **6**, 171 (1994);
N. A. Nicorovici, R. C. McPhedron and L. C. Botten: Phys. Rev. E **52**, 1135 (1995)
21. A. Moroz: Opt. Lett. **26**, 1119 (2001);
K. Amemiya and K. Ohtaka: J. Phys. Soc. Jpn. **72**, 1244 (2003)
22. J. B. Pendry: *Low Energy Electron Diffraction* (Academic, London 1974)
23. J. B. Pendry and A. MacKinnon: Phys. Rev. Lett. **69**, 2772 (1992)
24. N. Stefanou, V. Karathanos and A. Modinos: J. Phys. Condens. Matter **4**, 7389 (1992)
25. W. M. Robertson, J. Arjavalingam, R. D. Meade, K. D. Brommer, A. M. Rappe and J. D. Joannopoulos: Phys. Rev. Lett. **68**, 2023 (1993)
26. K. Sakoda: *Optical Properties of Photonic Crystals* (Springer Berlin Heidelberg New York 2001) pp 43–80
27. T. Inui, Y. Tanabe and Y. Onodera: *Group Theory and Its Applications in Physics* (Springer, Berlin Heidelberg New York 1990) pp 234–290
28. H. Miyazaki and K. Ohtaka: Phys. Rev. B **58**, 6920 (1998)
29. J. M. Ziman *Theory of Solids* (Cambridge University Press, Cambridge, 1999) pp 157–163
30. K. Ohtaka, et al., Phys. Rev. B **61**, 5267 (2000)
31. K. Ohtaka: J. Lightwave Tech. **17**, 2161 (1999)
32. S. Y. Tong: Progr. Surf. Sci. **7**, 1 (1975)
33. K. Ohtaka and Y. Tanabe: J. Phys. Soc. Jpn. **65**, 2276 (1996)
34. K. Sakoda and K. Ohtaka: Phys. Rev. B **54**, 5732 (1996)
35. J. Callaway: *Quantum Theory of the Solid State, Part B* (Academic Press, New York London 1974) pp 371-464
36. K. Sakoda: *Optical Properties of Photonic Crystals* (Springer Berlin Heidelberg New York 2001) pp 125–150
37. X.-P. Feng and Y. Arakawa: Jpn. J. Appl. Phys. **36**, L120 (1997)
38. K. Amemiya and K. Ohtaka: J. Phys. Soc. Jpn. **72**, 1244 (2003)
39. V. Lousse, J.-P. Vigneron, X. Bouju, J.-M. Vigoureux: Phys. Rev. B **64**, 201104 (2001)

40. M. Inoue and K. Ohtaka: Phys. Rev. B **26**, 3487 (1982)
41. K. Sakoda and K. Ohtaka: Phys. Rev. B **54**, 5742 (1996)
42. Y. Dumeige, I. Sagnes, P. Monnier, P. Vidakovic, I. Abram, C. Mériadec and A. Levenson: Phys. Rev. Lett. **22**, 043901-1 (2002)
43. K. Ohtaka, T. Ueta and Y. Tanabe: J. Phys. Soc. Jpn. **65**, 3068 (1996)
44. M. Born and E. Wolf: *Principles of Optics* (Pergamon Press, Oxford London 1975) p 705
45. L. M. Brekhovskikh: *Waves in Layered Media* (Academic, London, 1960) p 79
46. L. D. Landau, E. M. Lifshitz abd L. P. Pitaevskii: *Electrodynamics of Continuous Media* (Pergamon Press, Oxford, London, 1984) pp 268
47. A. Yariv: *Quantum Electronics* 3rd ed. (John Wiley and Sons, New york, 1989) pp 87–105
48. H. Kosaka, T. Kawashima, A. Tomita, M. Notomi, T. Tamamura, T. Sato and S. Kawakami: Phys. Rev. B **58**, R10096 (1998)
49. M. Notomi: Phys. Rev. B **62**, 10696 (2000)
50. C. Luo, S. G. Johnson and J. D. Joannopoulos: Phys. Rev. B **68**, 045115 (2003) and the references therein
51. M. C. Netti, A. Harris, J. J. Baumberg, D. M. Whittaker, M. B. D. Charlton, M. E. Zoorob and G. J. Parker: Phys. Rev. Lett. **81**, 3927 (2001)

5 Two-Dimensional Photonic Crystals

K. Inoue

Fabrication of two-dimensional (2D) photonic crystal (PC) samples is relatively easy compared to that of 3D PCs. Moreover, calculation of the band structure is also far easier. Although a full band gap does not exist in a strict sense, 2D PCs are important from the viewpoint of both academic and application uses. As already described in Chap. 1, there are many cases where a 2D PC suffices for observing novel phenomena. The term 2D PC is usually distinguished from a 2D PC slab, which is quasi-three-dimensional in a sense. Namely, for the former, the dimension (size) h in the direction perpendicular to the propagation direction is such that $h \gg a$ or λ (λ is the wavelength of light), so that an analysis based on plane waves is justified, while the reverse is true for the latter, that is, for the former, the eigenmodes with $\boldsymbol{k}$ parallel to the 2D plane are expressed as a combination of a set of plane waves. 2D PCs are classified into two types; one is of air cylinder type, and the other, of dielectric pillar type. We describe here three examples of 2D PCs that have mainly been fabricated and studied in Japan, together with the fabrication method.

In this chapter we also briefly describe PC fibers, since these are closely related to 2D PCs; the propagation direction of light, however, is perpendicular to the 2D plane.

5.1 2D Photonic Crystal of Arrayed Fiber Type

This kind of 2D PC sample at near-infrared wavelengths was fabricated in the early stage of PC research history, i.e., in 1993 [1]. The sample was fabricated by means of the technology for producing a microchannel plate. First, an ordered hexagonal array of optical fibers made of PbO glass was formed. In order to decrease the lattice constant, the array was pulled and elongated parallel to the fiber axis. Then, the lattice constant a of, e.g.,1.17 μm and the core diameter R of 0.90 μm were attained. The distance between two opposite faces of the hexagon was 64 μm. Second, about 400 hexagonal arrays thus fabricated were placed together to form a quasi-circular bundle of hexagons; the diameter of the bundle was about 1.5 mm and the length along the fiber axis was around 1.0 mm. The triangular lattice structure was supported outside by Pb-O glass similar to the cladding material of the original optical

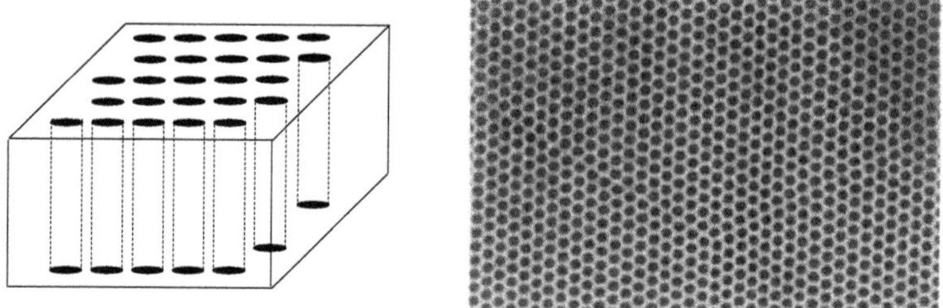

Fig. 5.1. *Left:* a schematic drawing of a 2D PC sample of arrayed air rods, and *right:* an optical microscope top-view picture (x 1,000) of a sample

fiber. Finally, the core glass was dissolved with use of HCl and we obtained a regular triangular lattice with $\varepsilon_1 = 1.0$ (air) and $\varepsilon_2 = 2.6$ (PbO glass).

Direct inspection of the sample with use of an optical microscope reveals that the performance is satisfying enough over the entire region, as shown in Fig. 5.1. A SEM picture (top view) of a sample (not shown) also reveals that the regularity as well as the uniformity of the air cylinders (holes) are perfect except for the boundary array of holes being in contact with the support glass. We measured a and R by using an optical microscope and an electron microscope, respectively: a and R of the two kinds of lattices are 1.17 ± 0.03 μm and 1.02 ± 0.06 μm, and 0.90 ± 0.08 μm and 0.69 ± 0.08 μm, respectively [2].

Independently, a was also determined from the reflected angles of the Bragg spots (see Fig. 5.2) emerging when the specimen was irradiated at normal incidence by a parallel light beam [3]. For this, two polished specimens were prepared with the flat surfaces of the lattice with the normal parallel to either the $\Gamma - X(M)$ or $\Gamma - J(K)$ directions. Depending on the wavelength,

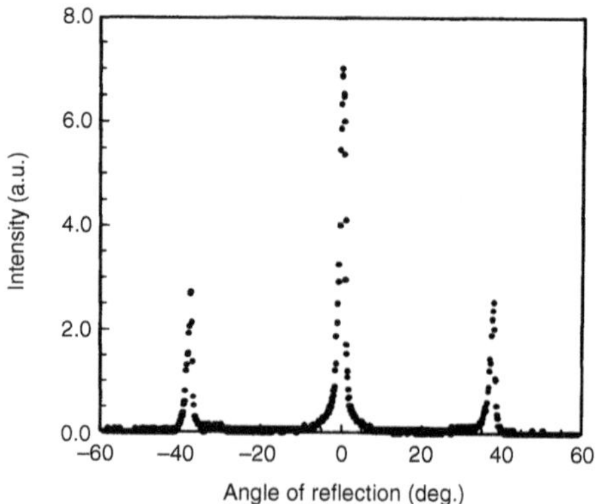

Fig. 5.2. Bragg reflection spectrum of a 2D PC sample of arrayed air rods, which reveals a width of 1.4° for the zeroth-order spectral line; quoted from [3]

from one to three Bragg-reflected spots were observed on one side of the zeroth-order spot [3]. For lattice A, for example, the first- and second-order spots were observed at 26.1° and 60.9°, respectively, with a 880-nm light beam incident parallel to the $\Gamma - J(K)$ directions. From the diffraction angles, a can be evaluated as 1.02 and 1.04 μm, respectively, confirming the value of 1.02 μm measured by an optical microscope. Next, we are concerned with to what extent the present lattice is homogeneous. Judging from the observed fluctuations of a and R as described above, the homogeneity is rather good. The fluctuation of a can also be estimated from the spectral width of the zeroth-order Bragg reflection line, shown in Fig. 5.2. It is as narrow as 1.4°, indicating again that the present sample is of good quality.

Figure 5.3 shows the observed spectra for H-polarization in the $\Gamma - X(M)$ and $\Gamma - J(K)$ directions, respectively. The respective spectra clearly show the existence of a common nontransmission frequency region for H-polarized light. This result is found to be consistent with the corresponding calculated band structure (BS) shown in Fig. 5.4, because the BS indicates that a common gap exists between the first and second bands for H-polarization. For E-polarization, however, the degeneracy of those two bands at the $K(J)$ point prevents the gap from opening [1, 2].

A feature of this kind of sample is that the size is large, and the aspect ratio of the hole, i.e., the length (depth) along the holes against the hole diameter is very large as well; as for the former the size is typically $1.0 \times 1.3\ \text{mm}^2$ with the height of 1 mm, and as for the latter, the ratio is around 1000.

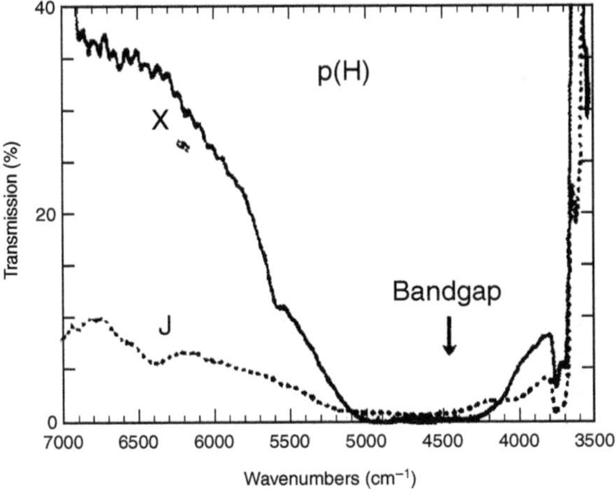

Fig. 5.3. Transmission spectra observed for H-polarization in the $\Gamma - X(M)$ and $\Gamma - J(K)$ directions in a 2D sample of an array of air rods, revealing the existence of a common 2D band gap for H-polarized light; a and R of the sample used are 1.02 μm and 0.69 μm, respectively

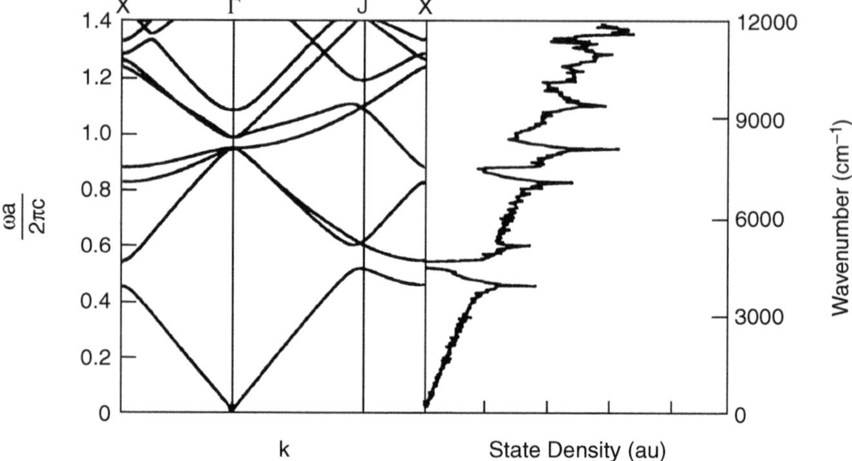

Fig. 5.4. The calculated PBS (left) for H(left)-polarized light and the density of states (right) for the sample used in Fig. 5.3. A common band gap is seen to open for H-polalized light: in the PBS for E-polarized light (not shown) such a gap does not open; quoted from [2]

5.2 2D Photonic Crystals Fabricated Based on Anodic Porous Alumina

2D PCs with a band gap in the visible wavelength region can be fabricated using a different method, i.e., using highly ordered anodic porous alumina, which H. Masuda invented [4]. Anodic porous alumina, which is formed by anodization of Al in acidic solution, consists of an ordered triangular array of holes with high aspect ratio in the alumina matrix. The configuration of the geometrical structure of anodic porous alumina is strongly dependent on the anodizing conditions, and the combined process of a pretexuring treatment of Al and an appropriate anodizing condition generates the ideally arranged configuration of the hole array in the alumina matrix.

PBG structures based on anodic alumina are convenient to prepare and it is easy to control the geometry. In addition, anodic porous alumina is suitable for the fabrication of the optical waveguide, because it is easy to form a composite slab structure such as an alumina/Al structure based on the anodizing process of Al plates.

Fabrication of anodic porous alumina with an ideally arranged configuration was carried out using the process schematically shown in Fig. 5.5. It was such that the development of the holes was initiated by the pretexured array of concavities on the Al surface and then the defect-free configuration was generated on the millimeter scale. For the pretexuring of Al, H. Masuda and his coworkers used the nano-indentation process using a SiC mold which has a hexagonally arranged array of convexities formed by EB lithography [5].

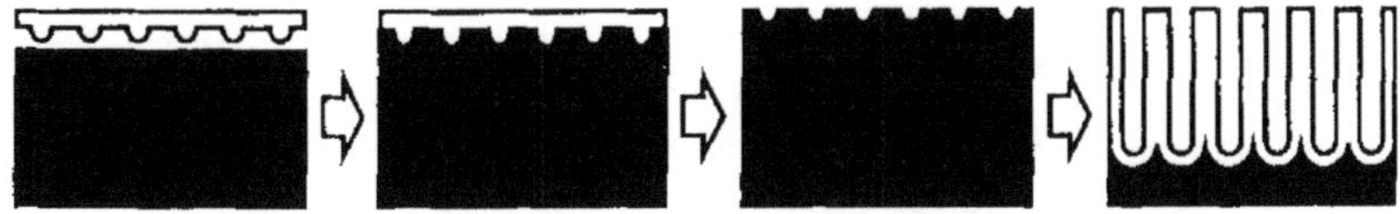

Fig. 5.5. Schematic illustration showing the fabrication process of the anodization method

The shallow concavities formed by the nano-indentation process on the Al surface introduced the development of the holes and guided the growth of the holes with a high aspect ratio and ideal configuration as well.

The master mold was placed on an Al sheet of 99.99% purity, which was polished electrochemically in a mixture solution of perchloric acid and ethanol, and pressed with an oil press under a pressure of 2800 kg/cm^2. This process generated an array of concavities on the surface of the Al. The anodization of Al was carried out in a 0.04 mol oxalic acid solution under a constant voltage at 17°C. The anodic porous alumina samples had a 2D triangular lattice with a lattice constant typically from 200 to 250 nm. The depth of the air cylinders with an ideally ordered configuration was approximately 20 µm [5]. The filling factor of air was adjusted by post-eching treatment using 5 wt% phosphoric acid solution at 30°C. Transmission spectra (not shown) observed in such a sample reveal that a common gap exists for H-polarization as is the case with the sample already described in Sect. 5.1; the refractive index n of 1.67 for porous alumina is similar to 1.65 for PbO [5].

A 2D PC of an air-hole array with an extremely high aspect ratio of over 200 can also be fabricated by adopting the most appropriate anodizing condition. An example of such a sample is shown in Fig. 5.6 [6]. Typical transmission spectra observed for the $\Gamma - M$ or $\Gamma - K$ directions and for H- and E-polarization are also shown in Fig. 5.6. The observed energy position and width of the gaps in the transmission spectra are almost in good agreement with the calculated band structure. For H-polarization a distinct dip of the transmission due to the SB is observed for both the $\Gamma - J(K)$ and $\Gamma - X(M)$ directions, but the overlap of the dips is narrow in this case. However, it has been confirmed that increasing the air-filling factor to 0.3 makes the widths of both dips larger, resulting therby in a relatively wide band gap for H-polarization [6].

The ordered air-cylinder array in anodic porous alumina can be transferred to other materials using a replicating process [7, 8]. Thereby, the replicated structures such as TiO_2 with a higher refractive index than alumina can be fabricated, which should be useful as a PC with a wider band gap in the visible wavelength region. In Fig. 5.7 is shown a schematic illustration of the replicating method. In Fig. 5.8 is shown a SEM photograph of a sample of $LiNbO_3$ thus fabricated [9]. It is noted that even metallic PCs such as one made of Pt can also be fabricated by using this method [6].

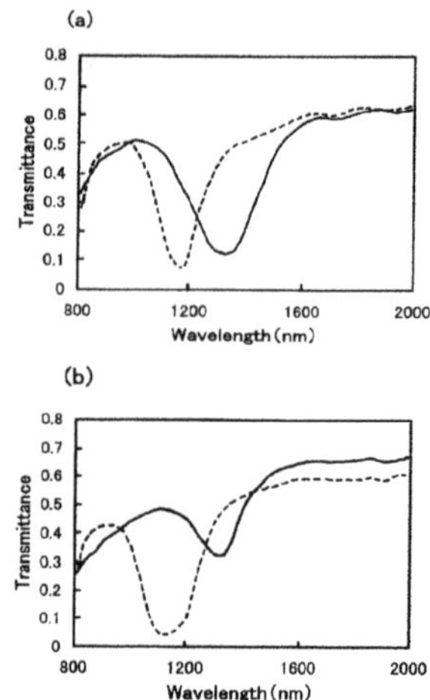

Fig. 5.6. *Left:* SEM monographs of a 2D PC sample of alumina with a high aspect ratio of 200, and *right:* the observed transmission spectra: (a) H- and (b) E-polarized light for the $\Gamma - J(K)$ (dotted line) and $\Gamma - X(M)$ (solid line) directions. The lattice constant and air filling factor were 500 nm and 0.1, respectively; quoted from [6]

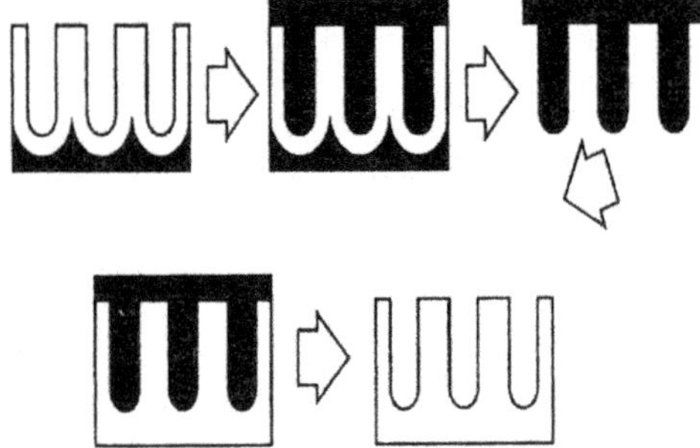

Fig. 5.7. A schematic drawing of the two-step inverse process for fabricating 2D PCs of different material

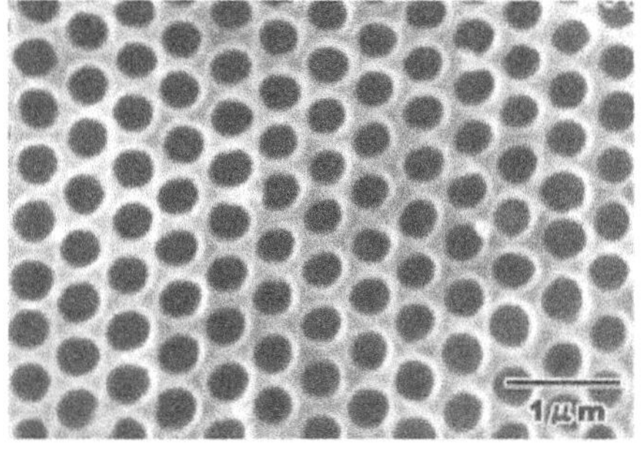

Fig. 5.8. SEM photograph (top view) of a sample of $LiNbO_3$; by courtesy of Professor H. Masuda

5.3 Other Methods and 2D Photonic Crystals with a 2D Band Gap

There is another unique method of fabricating a 2D PC, which was developed by H. Misawa and his collaborators [10–13]. The method is based on solidization of a special chemical material resulting from two-photon absorption due to illumination of a femtosecond pulsed laser, as shown in Fig. 5.9. By using three or four laser beams denoted by 1 to 4, a 2D pattern due to their interference is drawn on the focal plane; as is well-known, parallel interference lines can be drawn by using two beams.

The experimental setup for this method is shown in Fig. 5.10, together with a SEM picture (top view) of a sample thus fabricated. By using a diffractive beam splitter the beam configuration corresponding to either the three- or four-laser-beam case shown in Fig. 5.9 can be easily realized, which makes this method very simple.

The portions of the material illuminated strongly according to the 2D pattern are solidized due to multiple-photon absorption (typically, two-photon absorption). By moving the material perpendicularly to the light propagation direction, a 2D array of solidized pillars is formed. Then, by dissolving away only those pillars, an array of air-hole PCs can be obtained. The two-step inverse process already described in the preceding section can also be applied for fabricating a variety of 2D PC samples composed of different material on the basis of this prototype crystal.

In Fig. 5.11 are presented SEM pictures of two kinds of samples: the left and right pictures correspond to the samples fabricated by using the three- and four-beam methods shown in Fig. 5.9, respectively. It is seen that the cross-section of one rod for the left sample is elliptic, whereas that for the right one is circular, as it should be. It is remarked that this method has also

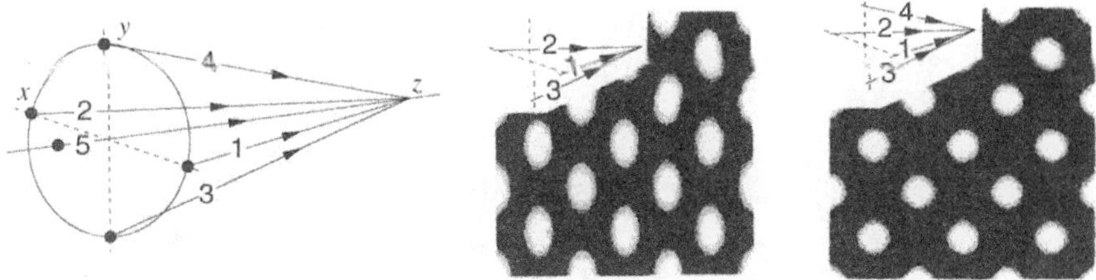

Fig. 5.9. Schematic showing how to fabricate a 2D PC with a combination of solidization of a special chemical material and multiple-laser-beam illumination. Beams 1 and 2, and 3 and 4 are on the xz plane, and yz plane, respectively, with the beam 5 being the center beam on the z axis. As shown schematically, (left) by using three beams 1, 2, and 3, and (right) four beams 1, 2, 3, and 4, a 2D square lattice is formed with the cross-section of the rods being elliptic and circular, respectively; quoted from [13] by courtesy of Professor H. Misawa

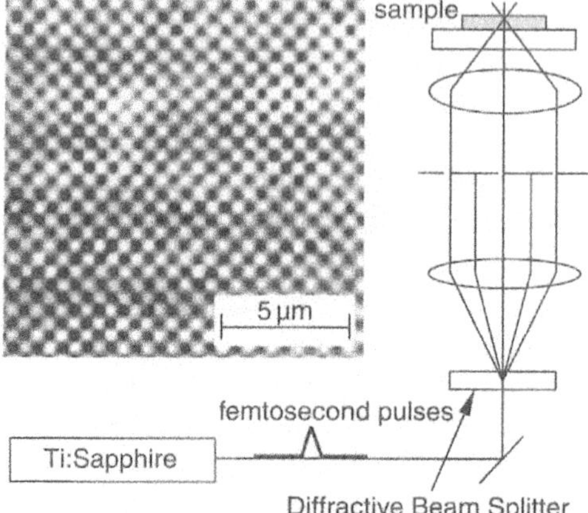

Fig. 5.10. Schematic illustration of experimental setup for fabricating a 2D PC with combined use of solidization and illumination of multiple laser beams. In this case three laser beams are employed with combined use of a diffractive beam splitter and a beam stopper. The inset shows a SEM picture (top view) of a sample fabricated with this method; quoted from [13] by courtesy of Professor H. Misawa

already been verified to be very useful in fabrication of 3D PCs. However, we do not pursue this here.

We also mention that it is possible to fabricate a 2D PC by the selective dry etching method for a semiconductor crystalline plate such as a Si plate, for example, so long as the hole diameter is not too small, i.e., say, 1 μm or larger than that. This method is based on the fact that the speed of etching depends strongly on the crystalline direction, and thereby, air holes can be perforated preferentially in a special direction [14, 15].

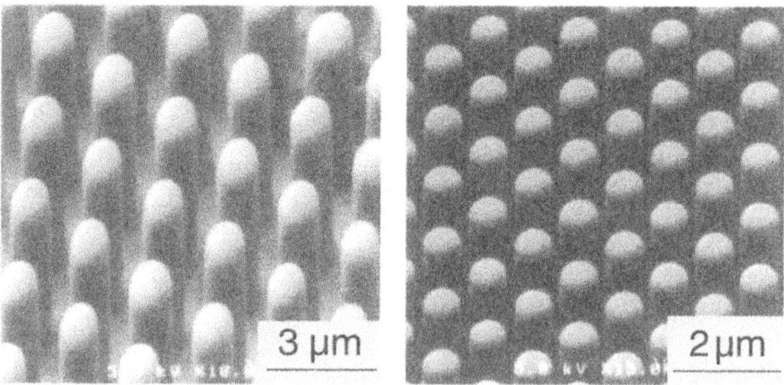

Fig. 5.11. Comparison between two 2D PC samples with different cross-sections of dielectric rods that were fabricated by using three- and four-laser-beam methods, respectively. *Left:* SEM picture of a sample with elliptic cross-section and *right:* that with circular cross-section; quoted from [13] by courtesy of Professor H. Misawa

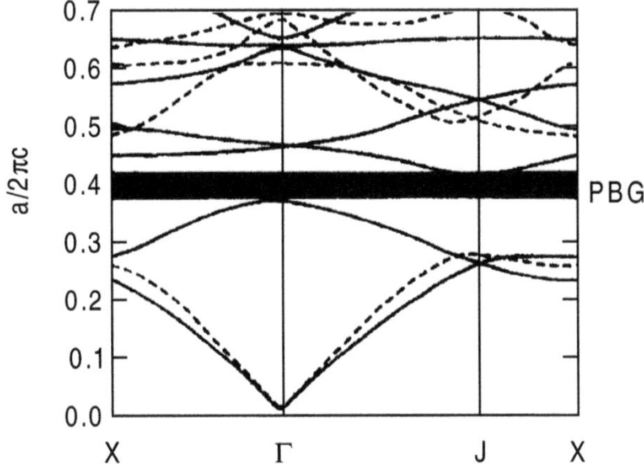

Fig. 5.12. A photonic band structure of a 2D PC of triangular lattice of air-rod type with ε being as large as 11.4, showing a 2D band gap (essentially the same as Fig. 2.14)

The 2D PCs fabricated with the use of the methods described in Sects. 5.1 and 5.2 are the triangular lattice. As already described, for this lattice a BG, i.e., a common gap for both E- and H-polarizations, does not open between the first and second bands, because of the degeneracy at the $K(J)$ point in the BZ. However, it should be noted that for even this lattice structure, a wide BG opens at the higher-energy region. Such an example of an air-rod lattice is shown in Fig. 5.12, which is calculated for the dielectric constant of 11.4.

5.4 Photonic Crystal Fibers

In the above we have described the PBS in parallel to the 2D plane, i.e., on the plane. Let us consider the photonic band or the dispersion relation of energy versus out-of-plane wavevector, i.e., wavevector in the off-axis direction. Since there is no definite (discrete) periodicity in the direction normal to the 2D plane, the band gap in the out-of-plane direction is expected to become smaller as the angle made with the original 2D plane is larger. Suppose light propagates in either a dielectric core or an air-hole region at the center of the 2D plane of an air-rod lattice with long enough dimension perpendicular to the plane; in the former case the central region is left unperforated, as shown in Fig. 5.13, while in the latter the size of the air hole is made larger than the others;(see Fig. 1.3 in Chap. 2, where a schematic illustration of this type of PC fiber is presented.) This is regarded as a kind of line defect introduced in a 2D PC, as explained in Chap. 2. In the former case, it is evident that under an appropriate condition there are guided modes for which light can propagate along the direction perpendicular to the 2D plane. This is because

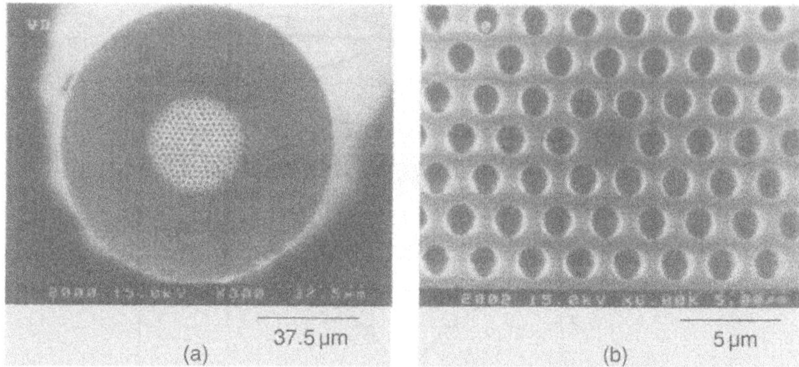

Fig. 5.13. SEM picture of a sample of the index-guiding photonic-crystal fiber; quoted from [19] by courtesy of Professor M. Nakazawa

the refractive index at the central core region is higher than the averaged refractive index of the surrounding area, therefore causing light to be confined in the central core [16]; this is called an index-guiding PC fiber, or a holey fiber. The situation is similar to, or just like, that for a conventional optical fiber, but this holey fiber is a novel type of fiber.

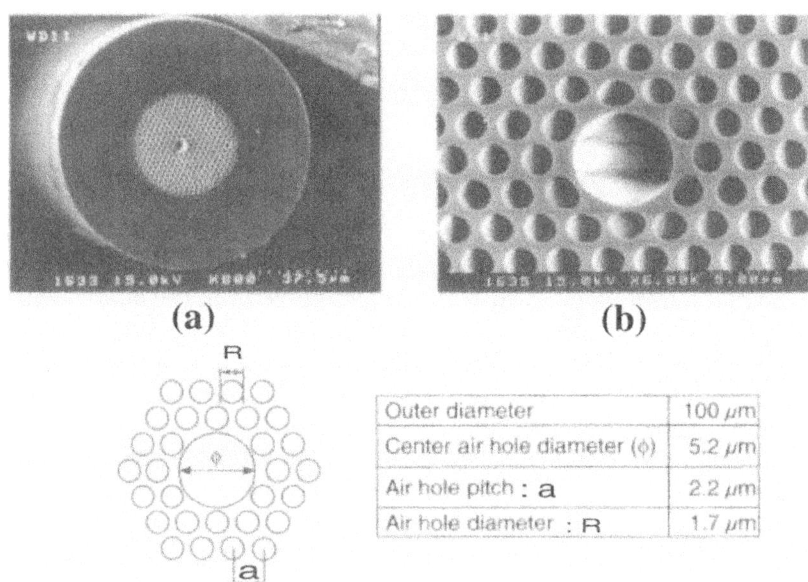

Fig. 5.14. SEM picture of a sample of the photonic band-gap fiber; quoted from [19] by courtesy of Professor M. Nakazawa

On the other hand, the situation is quite different in the latter case (a large air hole at the center). The refractive index at the central air region is of course smaller than that, in the sense of the averaged value, of the surrounding area. However, PBGs may still open in the off-axis directions, depending on the condition. Importantly, it is possible that guided modes exist in this case. Namely, light can be confined in the central region, due to the PBG, instead of the refractive index. Therefore, light can propagate through the air-hole region. This new type of fiber is called a photonic band-gap fiber, first invented by European research groups.[17, 18] The fabrication methods for these kinds of fibers are similar to some extent to the capillary type of arrayed fiber PC described in Sect. 5.1, but are still different. In PC fibers, optical fibers with air holes are arrayed at the first stage, and are then elongated. This fiber is found to exhibit superior characteristics, and therefore is very useful; for more details, see [17, 18].

References

1. K. Inoue, M. Wada, K. Sakoda, A. Yamanaka, M. Hayashi, and J. W. Haus, Jpn. J. Appl. Phys. **33**, L1463 (1994)
2. K. Inoue, M. Wada, K. Sakoda, M. Hayashi, T. Fukushima, and A. Yamanaka, Phys. Rev. B**53**, 1010 (1996)
3. K. Sakoda, M. Sasada, T. Fukushima, A. Yamanaka, N. Kawai, and K. Inoue, J. Opt. Soc. Am. B, **16**, 361 (1999)
4. H. Masuda, M. Ohya, H. Asoh, M. Nakano, M. Nohtomi, and T. Tamamura, Jpn. J. Appl. Phys., **38**, L1403 (1999)
5. H. Masuda, H. Yamada, M. Satoh, H. Asoh, M. Nakano, and T. Tamamura, Appl. Phys. Lett. **71**, 2770 (1997)
6. H. Masuda, M. Ohya, K. Nishino, H. Asoh, M. Nakano, M. Nohtomi, A. Yakoo, and T. Tanimura, Jpn. J. Appl. Phys., **39**, L1039 (2000)
7. H. Masuda and K. Fukuda, Science, **268**, 1466 (1995)
8. P. Hoyer and H. Masuda, J. Mater. Sci. Lett. **15**, 1228 (1996)
9. H. Masuda, unpublished data
10. H. B. Sum, S. Matsuo, and H. Misawa, Appl. Phys. Lett. **74**, 786 (2001)
11. K. Kondo, S. Matsuo, S. Juodkazis, and H. Misawa, Appl. Phys. Lett. **79**, 725 (2001)
12. T. Kondo, S. Matsuo, S. Joudkazis, V. Mizeikis, and H. Misawa, Appl. Phys. Lett. **82**, 2758 (2003)
13. H. Misawa, Rev. Laser Eng. **30**, 435 (2002) (in Japanese)
14. S. Rowson, A. Chelnokov, and J.-M. Louritioz, J. Lightwave Technol. **17**, 1989 (1999)
15. A. Birner, R. B. Wehrspohn, U. M. Gosele, and K. Busch, Adv. Mater. **13**, 377 (2001)
16. J. C. Knight, T. A. Birks, P. St. J. Russell, and D. M. Atkin, Opt. Lett. **21**, 1547 (1996)
17. T. A. Birks, P. J. Roberts, P. St. J. Russell, D. M. Atkin, and T. J. Sheperd, Electron. Lett. **31**, 1941 (1995)

18. J. K. Knight, J. Broeng, T. A. Birks, and P. St. J. Russell, Science, **282**, 1476 (1998)
19. M. Nakazawa, Rev. Laser Eng. **30**, 426 (2002) (in Japanese)

6 Two-Dimensional Photonic Crystal Slabs

K. Inoue, K. Asakawa and K. Ohtaka

In this chapter we treat PC slabs such that the thickness is on the order of the wavelength of light. In this case we cannot treat the problem in terms of only plane waves with $\boldsymbol{k}$ parallel to the slab plane, so the situation is quite different from the 2D PC case. Essentially we need to treat the case in a 3D way, but in a way different also from the 3D case. In Part I we adopt the dielectric-waveguide-based 2D PC (a periodically perforated dielectic waveguide) and an array of dielectric pillars, and consider mainly their guided modes. In Part II, we discuss the radiation modes by adopting a monolayer of 2D arrayed dielectric spheres.

Part I. Semiconductor Waveguide Type

In this type of PC it is important to understand in what cases the guided mode, i.e., light wave capable of propagating within the slab, exists. It depends on the refractive indices of the core layer and the cladding layers above and below the slab. In this sense the situation is similar to the standard dielectric waveguide, but in other respects it is quite different from and more complicated than the dielectric waveguide. This difference is, of cource, due to the periodicity of the PC slab. The condition for the guided mode to exist is much more severe than that for the homogeneous slab.

6.1 The Guided Mode and the Photonic Band Gap

It is very helpful to understand the PC slab case on the basis of the homogeneous dielectric slab. So, first we briefly summarize the guided modes for the latter below. Let us consider as an example such a dielectric waveguide as shown in the left of Fig. 6.1, where the refractive indices of the core and the cladding layers are n_2 and n_1, respectively, and the thickness of the former is d. The guided modes (the dispersion relations) in the case of $n_1 = 1$ and $n_2 = 3.52$ (GaAs) are presented in Fig. 6.1(right), where the frequency ω (the vertical scale) and the wavevector $k_{//}$ (the horizontal scale) are given in units of $\pi c/d$ and π/d, respectively. As is well known, the guided modes are allowed to exist between the two straight light lines of $\omega = ck$ and $\omega =$

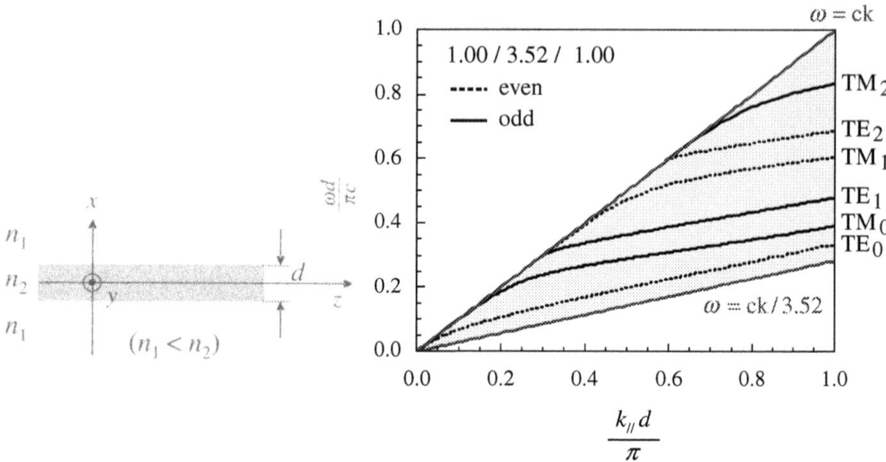

Fig. 6.1. A schematic drawing of the dielectric waveguide (left) and the guided modes (right). The latter dispersion relations are shown with the frequency ω and the wavevector $k_{//}$ in units of $\pi c/d$ and π/d, respectively. For the notation, see the text

$(c/n_2)k$ and are classified into TE- and TM-modes, as shown in Fig. 6.1. The TE-(TM-)mode is such that the electric (magnetic) field is parallel to the slab plane. Concretely, the electric field and the magnetic field for the TE-mode are expressed as $(0, E_y, 0)$ and $(H_x, 0, H_z)$, respectively, while those for the TM mode, are $(E_x, 0, E_z)$ and $(0, H_y, 0)$, respectively. For the TE(TM)-mode, the respective bands are classified by the integer $m\,(0, 1, 2, \ldots)$: m is the number of the node of the electric field in the yz plane.

Now, let us consider the 2D PC slab case of a square lattice of air holes. Perforating periodically air holes of small diameter in the homogeneous waveguide causes the respective bands shown in Fig. 6.1 to be folded into the reduced BZ, so that the stop bands (SBs) manifest themselves at both the zone center and the zone edges, as schematically shown in Fig. 6.2. Furthermore, mode mixing occurs between the TE- and TM-modes. Importantly, the condition under which the guided modes exist in the PC slab is very severe as compared to the homogeneous slab case. Any guided modes are no longer TE(TM)-modes in any case, but the modes are such that the electric (magnetic) field is purely transverse only in the mirror plane, and for this reason, these are called TE(TM)-like modes; in this case the mirror plane is the plane parallel to and in the middle of the core layer. In this connection, for the case where the upper and lower cladding layers differ from each other, the mirror plane does not exist, so that even classification in terms of TE(TM)-like modes is no longer valid. In this case, the situation is very complicated; for an example, see the next section. In this respect, a membrane suspended in air, called an air-bridge type, is the simplest among three types

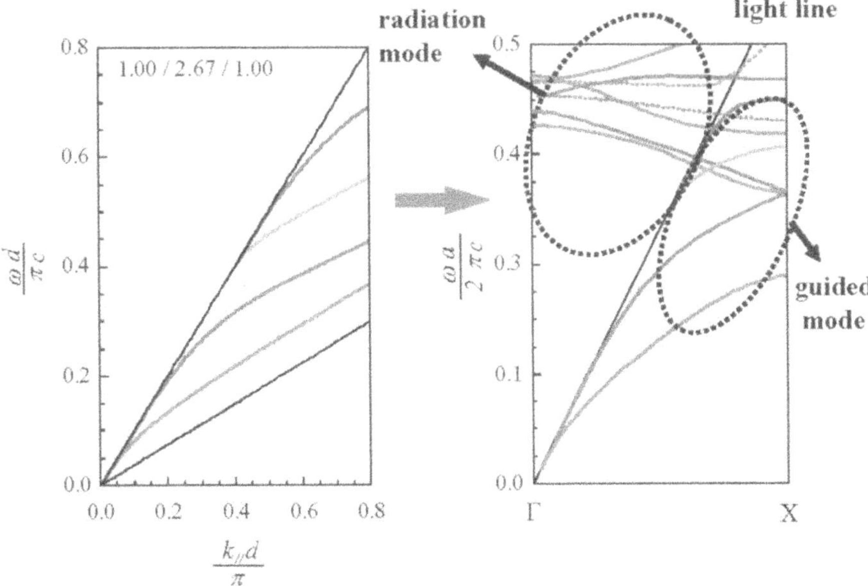

Fig. 6.2. These schematics show how the guided modes in a homogeneous dielectric waveguide (left) are converted by perforating the air holes into the guided and radiation (leaky) modes in a 2D PC slab (right). For the refractive index of the core layer in the former an averaged value of 2.67 for the latter is adopted for comparison. The units on the left are the same with those in Fig. 6.1, while on the right, ω and $k_{//}$ are given by $2\pi c/a$ and π/a, respectively

of PC slabs that are currently studied experimentally; for these three types, see also the next section.

Here, let us consider the photonic band gap (PBG) in the PC slab. Now that there are extended (continuous) modes in air above the air light line, more exactly, within the light cone, any modes within the light cone, of the PC slab can be coupled to the mode in air without fail; we call those modes leaky or radiation modes. Bearing this in mind, it is impossible to expect to have a 2D PBG in the PC slab. Nevertheless, we use the terminology of PBG also in this case, which is defined as a PBG for the guided modes; that is, an energy region over which the guided modes are missing in all 2D directions [1]. It is noted that this concept of the PBG is very useful, as will be shown later in Chaps. 11 and 12.

6.2 Three Types of PC Slabs

It is relatively easy to fabricate a PC slab sample as compared to 3D PC samples. This is because we can utilize state-of-the-art semiconductor microfabrication techniques [2]. Hereafter we focus on the 2D PC of an array of air

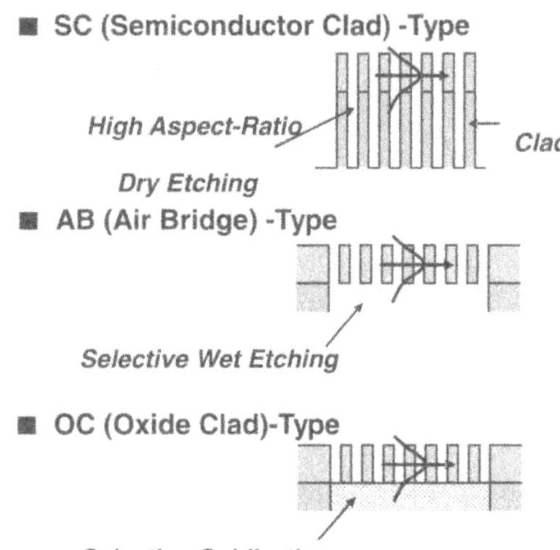

Fig. 6.3. A schematic drawing of three types of 2D PC slabs of air-rod type

holes in a dielectric; from the practical point of view, this kind of slabs are more attractive than the other slab systems of periodically-arrayed dielectric pillars. Currently, three types of PC slabs have been a target: (1) air-bridge type (AB), (2) oxide- or low-refractive-index-cladding type (OC), and (3) conventional semiconductor cladding type (SC), or hetero-structure type [3]. A schematic drawing of those structures is shown in Fig. 6.3. Comparison of the mode property among them is presented in Table 6.1.

(1) AB type: This type is a dielectric membrane suspended in air. Confinement of light in the vertical direction is strongest in this case among the three types; in other words, the area or the region outside the air light cone is the widest, so that the PBG may become broad, or robust. This feature is very attractive. That sample is rather fragile is the only disadvantage as compared to the other two types. In Fig. 6.4 is shown

Table 6.1. Guided modes and their symmetry

type	AB. air bridge	OC. subst.clad	SC. heterostr.type*
modes	guided	guided	quasi-guided
type	TM(TE)-like	mixing	TM(TE)-like
symmetry	even or odd	none	even or odd
(mirror plane)	(yes)	(no)	(yes)
gap	yes (broad)	yes (narrower)	none (in a strict sense)

*made of a conventional semiconductor waveguide by perforating

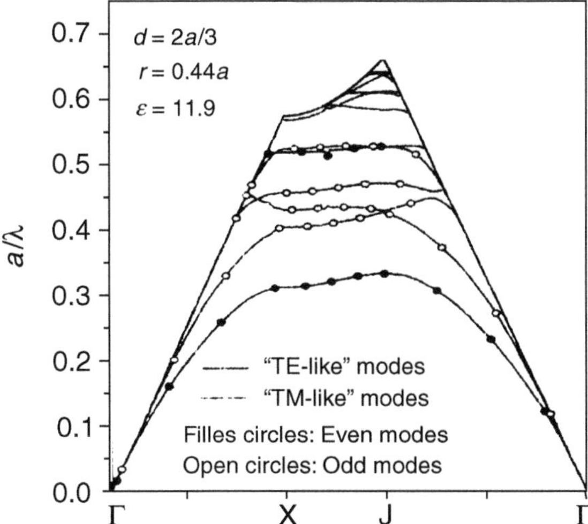

Fig. 6.4. The guided modes in the 2D PC slab of air-rod type outside the light cone; d, a, and r denote the slab thickness, the lattice constant, and the hole radius, respectively; quoted from [6]

an example of the band structure of the triangular lattice of air holes. There, the respective TE-like (TM-like) modes are classified by symmetry into even (open circle) and odd (solid one) modes with respect to the electric field about the mirror plane in the middle of the slab. In Fig. 6.4 the higher modes are also included. It is important to note that there exists a large PBG for the TE-like guided modes (outside the air light cone) with even symmetry; in contrast, there is no PBG for the TM-like eigenmodes. This large band gap is utilized to design a PC line-defect waveguide, which will be described later in detail (see Chaps. 11 and 12). In the case of a square lattice of air holes, there also exists a PBG again for the TE-like modes, but the band gap width is much smaller than that for the triangular lattice. Theoretically, we can also consider a pillar structure (a slab of arrayed dielectric rods) [1][4]. In this case there is a relatively large PBG for the square lattice structure in the TM-like guided modes. In any case it is important to note that in order to attain a large PBG the thickness of the slab (the core layer in this case) is appropriately thin, otherwise the gap disappears; the gap width becomes smaller as the thickness is smaller or larger than the best thickness, which is on the order of the wavelength of the guided light wave, finally ending up disappearing.

(2) OC type: The concept to introduce this type is very simple, i.e., to make a slab sample much more robust for practical use, as compared to the above-mentioned AB type. However, this can be accomplished at the sacrifice of a smaller (narrower) area (region) outside the light cone, which

is given by the refractive index (n) of the upper and lower cladding material. Consequently, in order to have a PBG as large as possible, it is necessary to use a cladding material with a relatively small n. Combinations of Si/SiO_2(SOI) or $SiO_2/Si/SiO_2$, and $AlGaAs/Al_xO_y$ (Al_xO_y is amorphous) are two typical examples for the air-hole slab. By Si/SiO_2, or $AlGaAs/Al_xO_y$ we mean a slab sample without the upper cladding layer, i.e., with a layer of air; as for Si/SiO_2, a starting slab material (wafer) of SOI (silicon on insulator) for this purpose is commercially available. One often prepares such a sample, because it is technically difficult to perforate air holes vertically all the way through a sample with a large aspect ratio. It is remarked that mixing between the TE- and TM-like modes occurs in this case, reflecting essentially the lack of a mirror plane in the middle of the core layer, as already described in the preceding section. Although this mixing is basically troublesome, simulation reveals that it is not serious enough to cause vanishing of a PBG that should otherwise exist.

(3) SC type: From the viewpoint of fabrication, this type of sample is most attractive, as far as an air-hole slab sample is concerned. Strictly speaking, however, there are no guided modes at all as seen in Table 6.1. In other words, all modes are radiative (leaky) or resonant modes that couple to the extended modes in air. Nevertheless this type of sample is considered to be useful in devices, depending on the case [5]. It is obvious that the radiation loss depends on the length of the slab sample (the propagation length of the light wave), so if the loss per unit length is not significant, it may be used in case of a short sample.
In this respect, it is remarked that, as will be described in the next section, it is crucially important to perforate a slab deeply enough into the lower cladding layer, and the difference of the refractive index between the core and cladding layers must be small enough, in contrast to the OC case. The SC samples have been fabricated until now without the upper cladding layer in most cases, unlike the case shown in Table 6.1; the reason is that perforating holes with a high aspect ratio is difficult. In this connection, it is also noted that the radiation loss or the inverse of the Q-value of the mode is known to depend markedly on the position in the same band in the BZ, as well as the band itself [7]. For example, the loss is negligibly small at the Γ-point, although this is an exceptional case.

6.3 Fabrication of Samples

A typical fabrication method is described in detail below, based on modern semiconductor microfabrication technology. Here, we adopt mainly the case for $Al_xGa_{1-x}As$-based samples [8]; the methods for other materials such as GaInAsP and Si that are also currently used, are basically similar to this case. This material has the advantage over the above two materials in that the optically transparent range is broad, covering the light wavelengths used presently for telecommunication; in the case of $x = 0$ with x being the Al content, it is transparent for wavelengths longer than 850 nm, but this critical wavelength is shorter with an increase of x.

The fabrication procedure will be described concretely below in order of the SC, AB and OC types, since it is easier to understand. The samples below are, unless otherwise noted, equipped or sandwiched with a pair of conventional ridge waveguides of stripe shape of 2.4 to 0.9 μm width and typically 300 μm length, which are used for optical access from and exit to outside [9].

For the SC type of PC slab, a periodic air-rod structure was fabricated in molecular-beam-epitaxy (MBE) grown 0.5-μm-thick $Al_{0.10}Ga_{0.90}As$ waveguide (core layer) on a 2.0-μm-thick $Al_{0.35}Ga_{0.65}As$ cladding layer on a GaAs substrate, using electron-beam lithography and Cl_2-reactive-beam-etching (RIBE). More concretely, patterns with a periodic array were first defined in polymethylmethacrylate (PMMA) using electron-beam lithography, and then were directly transferred into AlGaAs using ion-beam etching. The definition of 2.4 to 0.9-μm-wide stripe waveguides was also formed at the same time in a single dry-etching step using a RIBE system. All of the waveguide structures were designed to support only the fundamental TE_0 and TM_0 modes with respect to the vertical confinement of uniform layers in the relevant wavelength region.

We need to achieve high-aspect-ratio, smooth, and vertically shaped nanoholes, so we must control the semiconductor etch profile. An ultra-high-vacuum electron cyclotron-resonance reactive ion-beam etching (RIBE) system was used for this purpose. The etch profile could be controlled by adopting the most appropriate substrate temperature with ion energy fixed at 500 eV (see Sect. 12.4 for more details).

In Fig. 6.5 is shown SEM pictures of an example of a SC sample thus fabricated. The cross-sectional image (right) of the sample shows that the depth of the dry-etched holes approaches 1 μm, which is close to a limiting value attainable by the recent technology.

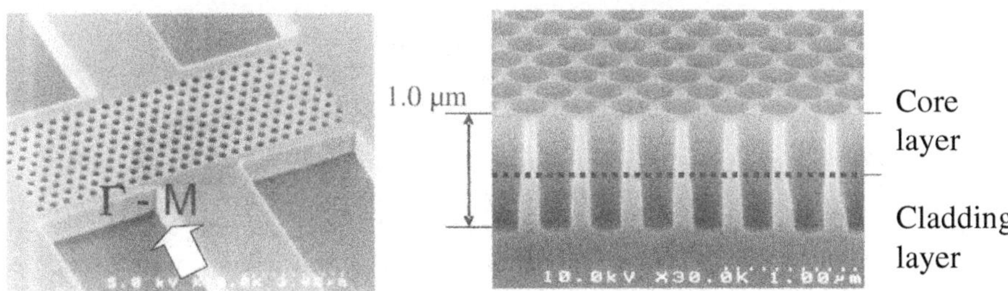

Fig. 6.5. SEM picture of a SC sample

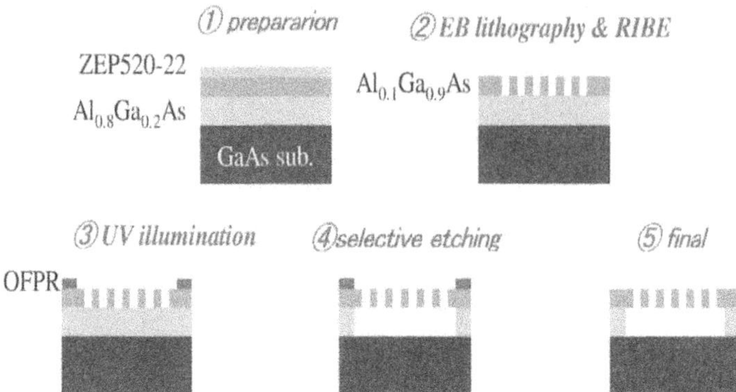

Fig. 6.6. Schematic illustration of fabrication process for a 2D PC slab sample of air-bridge type

For fabrication of the AB type of slab sample, the Al fraction of the 2-μm-thick cladding layer below the $Al_{0.10}Ga_{0.90}As$ core layer was increased to approximately 80%. At the same time, the core-layer thickness was also reduced to 250–270 nm in order to maintain the vertical single-mode behavior in the corresponding uniform waveguide structure. Thanks to high contrast in the Al content between the two layers, only the lower cladding layer can be removed by selective wet etching using a buffered HF solution. A schematic illustration of the fabrication process is shown in Fig. 6.6.

For the PC samples here, the selective etching was performed through subsequent treatment with phosphoric acid and a highly diluted buffered HF solution. The etching rate had to be sufficiently low to prevent damage to

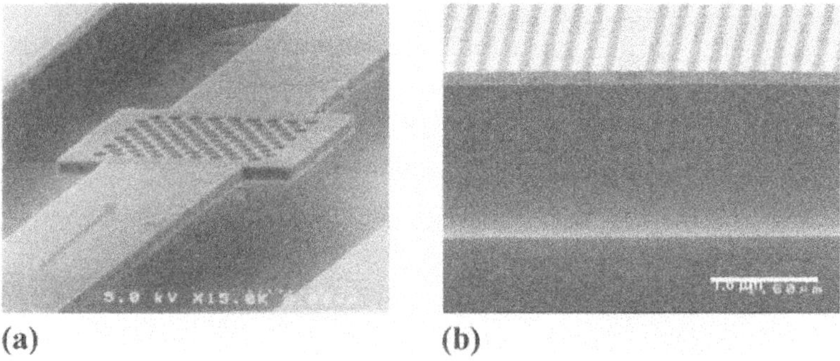

Fig. 6.7. SEM image of the suspended PC membrane: (**a**) a sample with ten rows of air-holes along the $\Gamma - M$ direction, which is sandwiched with a pair of stripe free-hanging waveguides for access of input and output signals, and (**b**) an all-PC sample (a line-defect waveguide) without the stripe waveguides, where single row of air-holes are missing (imperforated) along the $\Gamma - K$ direction; the sample length is 600 μm

the fragile membrane. For the Al compositions that were selected for the core and the underlying sacrificial layer, where the latter also serves as the cladding layer in the stripe waveguides, selectivity with respect to the lateral etching rate exceeds 100. This value is sufficiently high to protect the core layer from morphological degradation. In Fig. 6.7(a) is shown a SEM image of the suspended $Al_{0.1}Ga_{0.9}As$ PC membrane between a pair of stripe free-hanging waveguides [9]. The sacrificial layer was cleanly dissolved during the selective wet-etching step. In Fig. 6.7(b) is shown an example of an AB sample without a pair of stripe waveguides, i.e., the so-called all-PC sample of a line defect waveguide with one row of air holes missing along the $\Gamma - K$ direction. See also the SEM images for a variety of air-bridge type PC waveguides in Chaps. 11 and 12.

For fabrication of the OC type of slab sample, we need to find a solid substrate that has as low a refractive index as possible, as already described. Moreover, this should preferably be achieved using a combination of materials that allows the use of conventional manufacturing technology for PC sample fabrication. In the case of AlGaAs, Al_xO_y compounds are very promising candidates as the cladding layer material in PC waveguide applications; in contrast, a SOI-based PC slab of this type has an advantage over AlGaAs-based one in being easily fabricated, since one has only to remove the SiO_2 layer. Al_xO_y can be directly converted from AlGaAs alloys with the use of a sufficiently high Al content, by using steam oxidation. The resulting Al_xO_y compound has a low refractive index of $n \sim 1.6$, which is approximately half of that for an ordinary $Al_{0.35}Ga_{0.65}As$ cladding layer ($n \sim 3.30$). Similar to

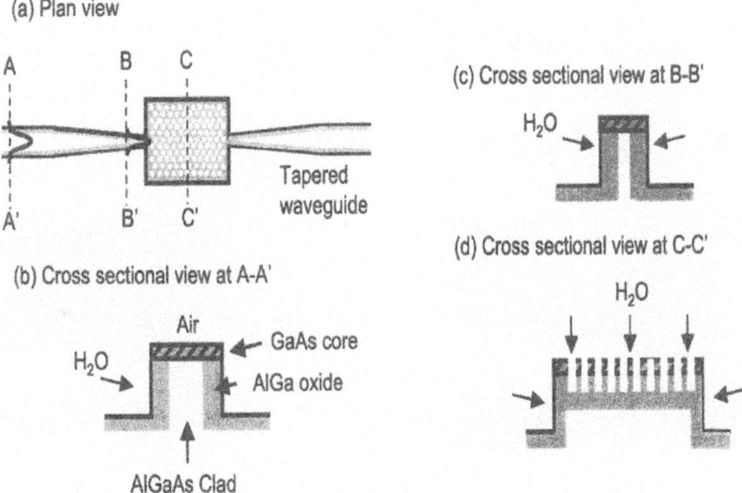

Fig. 6.8. Schematic illustration of fabrication process of an OC (oxidized cladding layer) PC sample sandwiched between a pair of tapered ridge waveguides; quoted from [10]

the air-bridge structure, the sample for steam oxidation was grown with a thin (250–300 nm thick) $Al_{0.1}Ga_{0.9}As$ core layer on top of a 2-μm-thick AlGaAs layer with a high Al content, i.e., $Al_{0.93}Ga_{0.07}As$, which was subsequently oxidized (see Fig. 6.8).

Following the RIBE dry-etching process for air-hole and waveguide mesa fabrication, the sample was introduced into a special apparatus for steam oxidation[10]. The oxidation process that was 15 min in duration was sufficient to completely oxidize the $Al_{0.93}Ga_{0.07}As$ remaining in the sample. This translates to a nominal oxidation rate of approximately 0.14 μm/min. The result of the sample fabrication is shown in Fig. 6.9. The cross-sectional view shown there was taken from a cleaved facet of one of the stripe waveguides. This image indicates that the samples could be of high quality, since no indication of delaminating or the like can be seen.

So far we have described the fabrication method for AlGaAs-based PC slabs. We would like to briefly describe PC slabs made of other materials. Currently, Si- and GaInAsP-based PC slabs have also been fabricated successfully and their optical properties have been intensively studied. The fabrication method for GaInAsP-based slabs is very similar to the case of AlGaAs except that a different dry-etching technique is employed. As for the undercut method to fabricate the air-bridge sample, it is much easier in the GaInAsP case than in the AlGaAs case, since a solution with better selectivity can be used [11]. In the case of Si-based slabs, fabrication of the samples is also easier than in the AlGaAs-based case, as already described; a SOI wafer is available as a starting material [12, 13]. Moreover, the SiO_2 layer can be easily removed by dissolution in hydrofluoric acid. In addition, the directionality of etching is almost vertical.

Here we briefly mention the reasons that the above three materials are currently used for PC slabs. One of the reasons is that semiconductor microfabrication technology can be applied to fabrication of these materials, and the refractive index is also large. In this sense the situation for Si- and GaInAsP-based PC slabs is almost the same as that for AlGaAs-based slabs

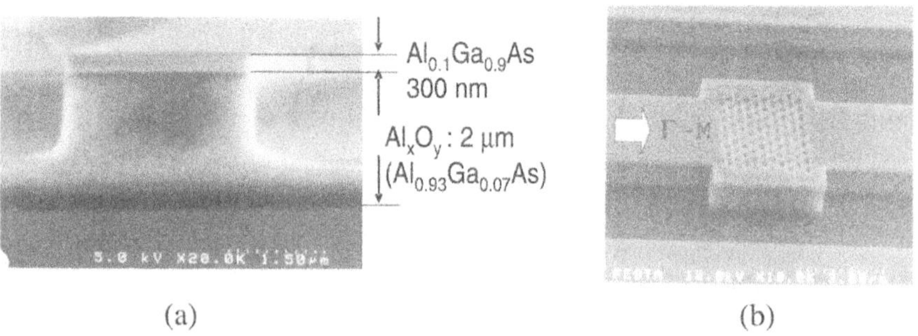

Fig. 6.9. SEM pictures of an OC sample with the oxidized cladding layer: cross-section (left) and top view (right)

described thus far. Another reason is that PC-waveguide devices based on these are considered to serve for future planar ultra-compact light circuits for telecommunication use. The optically transparent ranges for the above two materials are smaller than that of AlGaAs, but nicely cover the wavelength range from 1.3 to 1.55 μm used for present optical communication; for example, the range for Si is from 1.2 to 1.7 μm. However, it is remarked that use of AlGaAs as a slab material, more exactly, use of AlGaAs-based PC waveguides and related devices, may also be attractive in a few respects, because it may also be used for interconnection between the IC or LSI chips. Furthermore, GaInAsP-based quantum wells or quantum dots can be easily embedded in AlGaAs-based slabs, which may work as a nonlinear material to enhance the signal, making use of resonant effects, for example, in an all-optical switch. This subject will be discussed in a later chapter.

6.4 Optical Properties

We experimantally examined in detail the optical properties, in particular, the transmittance (T) spectra of three types of AlGaAs-based PC slabs. We present below a few examples of the T-spectra observed by using a cw wavelength-tunable Ti-sapphire laser [9], which will be described in Chap. 9. First, we show an example of the T-spectrum for an AB type of sample; actually, we observed many spectra for a variety of samples fabricated with different parameters. The spectra shown in Fig. 6.10(a) were observed in the $\Gamma - M$ direction for the electric field of the incident light perpendicular (for TM-like modes) and parallel (TE-like modes) to the slab plane [9]. The ab-

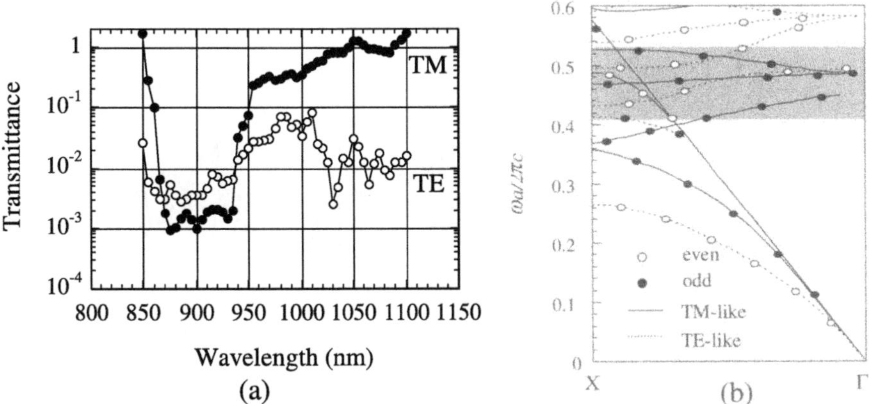

Fig. 6.10. An example of transmission spectra of an air-bridge PC slab sample (**a**), and the corresponding photonic band structure (**b**), where the hatched area corresponds to the observed wavelength region, and the straight line is the air light-line; quoted from [9]

solute T-value is determined by normalizing the transmission with use of a reference measured with a nominally identical waveguide without PC. The dimension of this sample is such that a, d (slab thickness), f (air-filling factor), and N are 450 nm, 270 nm, 0.52, and 5, respectively. For comparison, the corresponding band structures are also shown in (b), which were calculated by solving Maxwell's equations using an admixture of plane-wave expansion and real space decomposition [6].

First, we would like to state that the observed spectra for a variety of samples are essentially well explained in terms of the corresponding band structures. In the case of Fig. 6.10(a), for example, it is seen that for the TM-mode a deep drop in T manifests itself from 850 to 950 nm, and importantly, the absolute T-value reaches almost 100% above 1020 nm. This opaque region concerning the former is interpreted as arising from a stop band between the second- and third-lowest TM-like bands of odd symmetry in Fig. 6.10(b). In this connection, it is remarked that for a similar sample but with $f = 0.60$ and $a = 400$ nm, a narrow but definite T-dip due to the lowest stop band for TM-like modes is observed. As for the latter, it is quite reasonable, since the lower energy side of the second-lowest band is below the light line. For the TE-like modes, on the other hand, T is seen to be low over the observed energy region. This result is also consistent with the band structure. Namely, considering that the energy region of the second-lowest band below the light line is too narrow to cause a high T, and postulating that the band above the light line is leaky enough in this case, T should be low over the observed region, which corresponds basically to the stop band.

The above interpretation is nicely supported by the T-spectra calculated by using the finite-difference-time-domain (FDTD) method. In Fig. 6.11 is presented a comparison between the observed T-spectrum (solid circles) and the calculated one (solid line), where the observed spectrum is shifted in

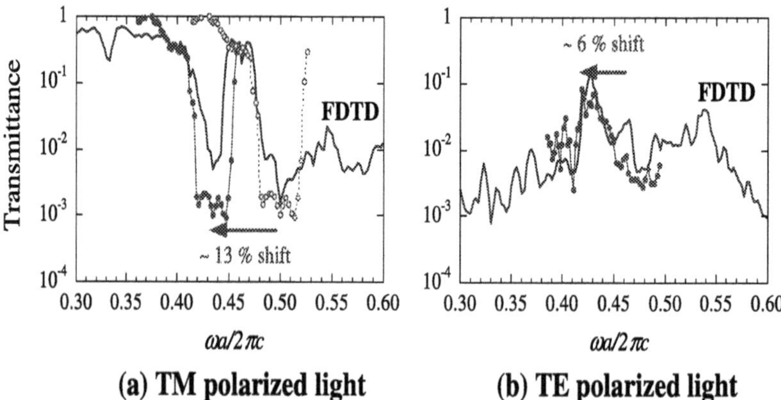

Fig. 6.11. Comparison of the observed tansmittance spectra with those calculated by using the 3D FDTD method; quoted from [9]

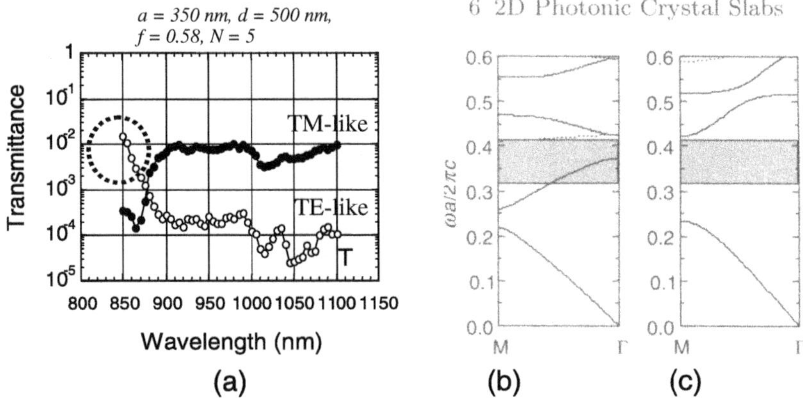

Fig. 6.12. (**a**) An example of the observed tansmittance spectrum in the $\Gamma - M$ direction for a SC-type sample and the corresponding calculated band structures. (**b**) TM-like mode and (**c**) TE-like mode. The solid and dotted lines represent the coupled and uncoupled bands, respectively

energy by 13% and 6% for (a) and (b), respectively. Obviously, the correspondence is very good.

Next, an example of the spectrum for an SC-type slab sample is shown in Fig. 6.12, where the calculated band structures are also presented for comparioson. The parameters for this sample are such that a = 350 nm, f = 0.58, and N = 10, and the thichness of the core layer is 500 nm while the total depth of air holes is 900 nm. In this case, however, the band calculation for quasi-guided modes that may be approximately treated as being similar to those in the case of a symmetric PC slab [14], was made in the following way: the PBS was calculated for an infinitely long air-hole 2D PC, embedded in a background (core) material, assuming an effective refractive index of 3.39, instead of 3.52 for the homogeneous case [2]. Overall features of the TM-like spectra observed for many samples with different parameters can be explained rather well in terms of the PBS calculated in this way. The absolute T is very small for both modes, which is reasonable. More exactly, for TM-like modes the observed energy region corresponds to the first stop band and the second-lowest band above the light line, while for the TE-like mode, it corresponds to the stop band, causing the very low T. For the latter T increases to a certain extent on the shorter wavelength side, because the wavelength approaches the second band. For the former, on the other hand, the drop in T is likely to be explaind by assuming that the PBS should actually shift to some extent to the higher energy side relative to the hatched area.

Finally, the issue as to what extent the SC or heterostrucural PC slab of air-hole type can be tolerable in practical use or in future devices arises. The answer is, of course, "it depends". In Fig. 6.13 is shown a plot of the

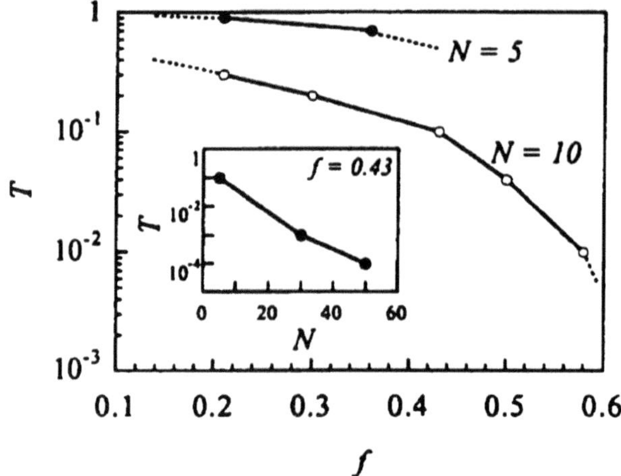

Fig. 6.13. Dependence of transmittance T (propagation loss) of the second band in a SC PC slab of air-hole type on f (air-filling factor). The closed and open circles indicate the data points for a series of PCs with $N = 5$ and 10, respectively. The inset shows the dependence on N for three samples with the same f-value of 0.43; quoted from [15]

transmittance (T) of the second band lying below the light line as a function of N and f [15]. As is clearly seen there, T decreases rapidly as either N or f increases, which is caused by the radiation loss, i.e., by light escaping from the slab. This fact suggests that SC type samples should not work well for large N, except for the lowest band.

6.5 Non-Bulk PC Slabs

In the preceeding sections we have described only bulk PC slabs. Other than the bulk cases, research on defect-involved PC slabs such as the so-called PC slab waveguides and small cavities (point-defect type) has been extensively performed both experimentally and theoretically. This is because these are expected to be used for compact, ultrafast planar optical integrated circuits in future. This work has been treated independently in Chaps. 11 and 12.

Part II. An Array of Spheres: Leaky Modes

In this part we present experimental results that show that the leaky PBs cause sharp resonance signals in the transmission and reflection experiment which agree quite well with the theoretical calculation. The PC used in the experiment is a monolayer array of dielectric spheres of millimeter range.

6.6 Q-values of Leaky Photonic Band Modes

Photonic band modes in a slab PC inevitably leak out of the PC to have a finite lifetime, when they are lying in the leaky region where the mode frequency ω and its wavevector $\boldsymbol{k}_{\parallel}$ parallel to the surface of the slab satisfies the relation $\omega > c|\boldsymbol{k}_{\parallel}|$ (see Sect. 2.3). Because of their finite lifetime, PBs in the leaky regions have so far attracted relatively little attention, as compared to PBs of the waveguide region of $\omega < c|\boldsymbol{k}_{\parallel}|$, where the escape is completely inhibited. However, light extraction in the direction perpendicular to the surface of a slab is anticipated in many ways and related more or less to the excitation of leaky PBs. Therefore, the Q-values involved in the excitation of the leaky modes play a crucial role in determining their technological usability. Here we are concerned with the data obtained by the transmission experiment from a monolayer PC of arrayed spheres and show that the leaky PBs are usable in enhancing optical signals by their resonant excitations.

In the experiment, which was reported in [16], a 2D array of Si_3N_4 spherical balls with radius r of 1/16 inch (=1.588 mm) was set up. They have a highly perfect spherical shape and quite uniform size with inaccuracy in diameter of less than 1/2000 inch (= 0.0127 mm). The Si_3N_4 spheres were placed manually on the xy plane in a monolayer triangle lattice of lattice constant $a = 2r$ in a 50 mm 42 mm-sized aluminum frame. The p-polarized terahertz (THz) wave was emitted from a bow-tie-shaped low-temperature grown GaAs photoconductive antenna (LT-GaAs antenna) [17]. It was incident on the sample at an incidence angle θ. The 2D wavevector $\boldsymbol{k}_{\parallel} = (k_x, k_y)$ of the incident THz wave was chosen to be along the symmetry axis Γ-M of the 2D Brillouin zone. The distance between the Γ and M points was $|\boldsymbol{k}_{\Gamma \to \mathrm{M}}| = 2\pi/a\sqrt{3}$. The transmitted p-polarized THz wave was detected by a bow-tie-shaped LT-GaAs antenna. The spot size of the incident THz beam on the sample was about 40 mm in diameter and the frequency resolution is estimated to be 0.75 GHz. The magnitude of the transmittance $T(\omega)$ and the phase shift were measured by sub terahertz (sub-THz) time domain spectroscopy (TDS).

Figure 6.14 (upper) shows the measured and calculated $T(\omega)$ at an incidence angle $\theta = 0°$. In the case of normal incidence, the first-order Bragg diffraction occurs at 109.1 GHz. The theoretical T was obtained by the vector KKR method [18, 19] (see Sect. 4.1.1). The refractive index of Si_3N_4 was taken as $n = 2.99$ and its imaginary part κ was neglected. The lattice constant a was the only other parameter used in the calculation.

The measured spectrum shows a number of distinct dips. This fine structure will be shown to be caused by the excitation of the optically active 2D PBs, that is, PBs that are coupled to the external light. The capability of responding to the external light implies that these modes are leaky. The overall pattern of the observations agrees well with the calculation. Agreement is particularly good in the range from 40 GHz to 75 GHz. However, the absolute value of the calculated T is larger than that found experimentally. Further-

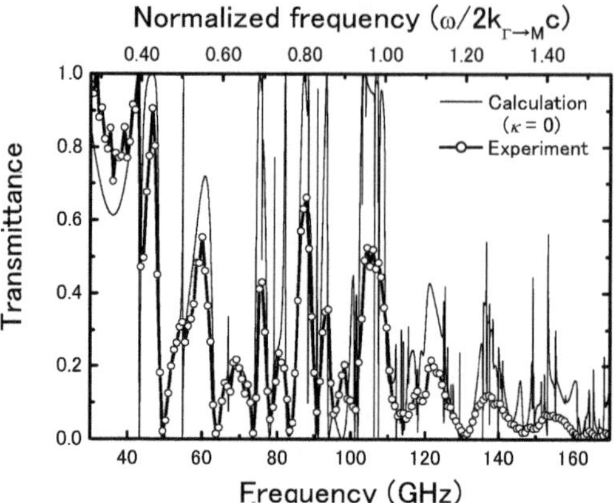

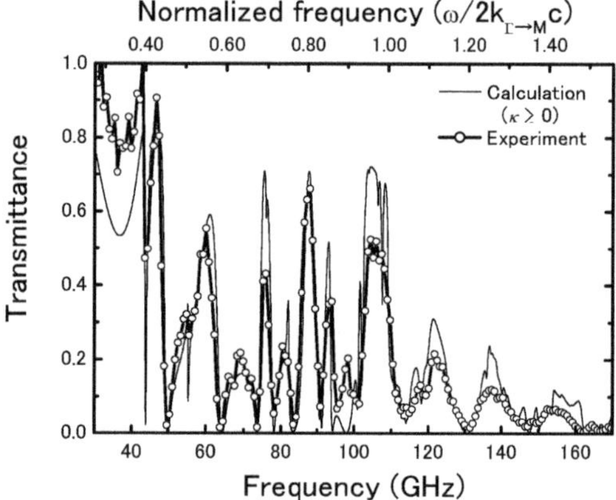

Fig. 6.14. Experimental transmission spectrum of Si_3N_4 spheres arrayed in a triangular lattice as compared with the calculated result (normal incidence). Circles with the solid line represent data obtained by TDS with a resolution of 0.7 5 GHz. The fine solid line represents the calculated result (upper) with $n = 2.99$, $\kappa = 0$ and (lower) with $\kappa > 0$

more, some of the asymmetric spiky features of the calculation, accompanied by rapid falls to zero in the calculation, which are characteristic of the Fano effect [20], are absent in the experimental results.

Figure 6.14 (lower) presents $T(\omega)$, which is calculated by allowing for the absorption of Si_3N_4 spheres and the light transmission through the thin vinyl films used to fix the system on both sides. The experimental data of the upper panel is also reproduced. From the transmission data of a plate sample of Si_3N_4, we found that the complex refractive index $(n(\omega), \kappa(\omega))$ of Si_3N_4 has a slight dispersion in the frequency region of this study, from (2.96, 0.0038) at 50 GHz to (2.99, 0.0054) at 100 GHz. To obtain the theoretical result of Fig. 6.14 (lower), we have fitted the measured ω dependence of the refractive index using a third-order polynomial. The T of the thin vinyl films was measured to be 93% in the sub-THz frequency range. We multiply the calculated T by $(0.93)^2$ to take account of the two films. The interference effect between the two films and the PC can be neglected because of the low reflectivity of the films.

Compared to the numerical curve of Fig. 6.14 (upper), T is much reduced and a lot of spiky structure disappeared. As a result, the agreement is greatly improved across a wide frequency region. Agreement is especially good in the low frequency region below 95 GHz. This implies that the PC used in this study has an ideal lattice and good uniformity in the size of spheres and that the absorption by Si_3N_4 is mainly responsible for the discrepancy between theory and experiment seen in Fig. 6.14 (upper). The importance of the absorption coefficient for interpreting the transmission spectra of PCs has been pointed out by Hase et al. for a 2D photonic crystal made of dielectric rods [21]. Close agreement over a wide frequency range involving as many PBs as presented here is very remarkable.

The phase of the complex transmission amplitude has information about the density of states (DOS) of PBs (see Sect. 4.3.3). We note that the reciprocal lifetime of leaky PBs is proportional to the full-width at half maximum (FWHM) of the DOS Lorentzian peak [19]. The phase shift is defined by

$$\Delta\phi(\omega) = \phi_{\text{sam}}(\omega) - \phi_{\text{ref}}(\omega) \tag{6.1}$$

for the transmitted electric field expressed by $E(\omega) = |E(\omega)| \exp[-i\phi(\omega)]$. The quantity $\phi_{\text{ref}}(\omega)$ is the phase of the waveform measured without the sample. The change of DOS, relative to that of free space without the sample is obtained by differentiating $\Delta\phi(\omega)$ with respect to ω:

$$\text{change of DOS} = \frac{1}{\pi}\frac{\partial}{\partial\omega}\Delta\phi(\omega). \tag{6.2}$$

We confirmed (not shown here) that the change of DOS agrees very well between theory and experiment.

6.7 Dispersion Relation and Lifetime

According to the analysis of [19], PBs show up mostly as dips in the transmission spectra. Correspondingly, we take the dips of T to indicate the existence of the 2D PBs, and plot these as θ varies to obtain the experimental dispersion relation of the 2D PBs. The angle of incidence θ and the 2D wavevector $\boldsymbol{k}_\parallel$ in the Γ-M direction of the incident light is related as $k_{\Gamma\to\text{M}} = (\omega/c)\sin\theta$. Because of the continuity of the tangential components of wave vectors, this value just gives the wave number of the excited PBs.

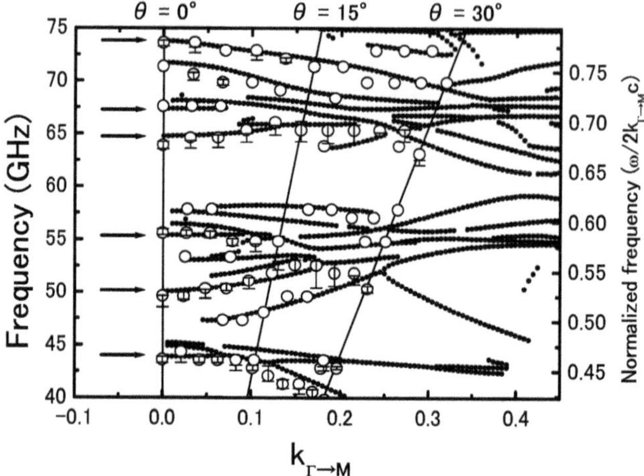

Fig. 6.15. Experimental and theoretical dispersion curves of 2D photonic bands. Only the dispersion curves of the PBs active to the p-polarized light are shown. The horizontal axis is normalized so that $k_y = 0.5$ at the M point of the first Brillouin zone. Solid circles are the theoretical dispersion curves, and open circles are the observed values. The bars on the data points show the measured FWHM. The three straight lines are the $k - \omega$ curves for three incident

Figure 6.15 shows the observed band dispersion of the p-active PBs. The open circles and the accompanying bars show, respectively, the dip positions in the transmission spectrum and the FWHM of the sharp peaks, obtained from the measured phase shift $\Delta\phi(\omega)$, respectively. For the dips with FWHM larger than 3 GHz, we did not include the bars. Superimposed on them are the solid circles which show the theoretical dispersion relations obtained from the theoretical dip positions, calculated for $\kappa > 0$.

Figure 6.15 demonstrates good agreement between theory and experiment including the PBs located in a higher-frequency region, showing that many photonic bands are indeed excited in the experiment. In particular, on the PBs marked by the solid arrows, which appear as distinct dips of the transmittance at the point $k = 0$, the theoretical dispersion curves are traced unambiguously by the transmission experiment throughout the region above the light line. Furthermore, Fig. 6.15 shows that the Q-values of the PBs are roughly 100.

Because of the scaling property (see Sect. 2.1.2) set up between the lattice parameters of PCs and the wavelength of light, the agreement obtained here in the millimeter size system guarantees automatically a promising availability of PBs in the leaky region of the visible range. In the regular array of spheres, it is well-established theoretically that resonant excitation of leaky PBs reduces the lasing threshold (see Sect. 10.5), enhances the near-field intensity (see Sect. 4.4.2), enhances the resonant Smith–Purcell radiation (see

Sect. 10.7) and so on. We have shown here that the fine structures in the transmittance $T(\omega)$, too, are caused by the same effect.

References

1. S. Johnson, S. Fan, P. R. Villeneuve, J. D. Joannopoulos, and L. A. Kolodziejski, Phys. Rev. B**60**, 5751 (1999)
2. T. F. Krauss, De LaRue, and S. Brand, Nature (London) **383**, 699 (1996)
3. D. Labilloy, H. Benisty, C. Weisbuch, T. F. Krauss, R. M. De La Rue, V. Bardinal, R. Houdre, U. Oesterle, D. Cassagne, and C. Jouanin, Phys. Rev. Lett. **79**, 4147 (1997)
4. T. Baba and T. Matsuzaki, Electron. Lett. **31**, 1776 (1999)
5. H. Benisty, D. Labilloy, C. Weisbuch, C. J. M. Smith, T. F. Krauss, A. Béraud, D. Cassagne, and C. Jouanin, Appl. Phys. Lett. **76**, 532 (2000)
6. N. Carlsson, T. Takemori, K. Asakawa, and Y. Katayama, J. Opt. Soc. Am. B **18**, 1260 (2001)
7. T. Ochiai and K. Sakoda, Phys. Rev. B**63**, 125107 (2001)
8. Y. Sugimoto, N. Ikeda, N. Carlsson, K. Asakawa, N. Kawai, and K. Inoue, J. Appl. Phys. **91**, 922 (2002)
9. N. Kawai, K. Inoue, N. Carlsson, N. Ikeda, Y. Sugimoto, K. Asakawa, and T. Takemori, Phys. Rev. Lett. **86**, 2289 (2001)
10. Y. Tanaka, Y. Sugimoto, N. Ikeda, T. Yang, K. Asakawa, Y. Watanabe, K. Inoue, T. Maruyama, K. Miyashita, and K. Ishida, Jpn. J. Appl. Phys. **42**, 7331 (2003)
11. T. Baba, K. Inoshita, H. Tanaka, J. Yonekura, M. Ariga, A. Matsutani, T. Miyamoto, F. Koyama, and K. Iga, J. Lightwave Technol. **17**, 2120 (1999)
12. T. Baba, A. Motegi, T. Iwai, N. Fukaya, Y. Watanabe, and A. Sakai, IEEE J. Quantum Electronics, **38**, 743 (2002)
13. M. Tokushima, H. Kosaka, A. Tomita, and H. Yamada, Appl. Phys. Lett. **76**, 952 (2000)
14. K. Inoue, N. Kawai, Y. Sugimoto, N. Carlsson, N. Ikeda, and K. Asakawa, in *MRS Symp. Proc.*, vol. 637, 2001, E3.2.1
15. N. Kawai, K. Inoue, N. Ikeda, N. carlsson, Y. Sugimoto, K. Asakawa, S. Yamada, and Y. Katayama, Phys. Rev. B**63**, 153313 (2001)
16. T. Kondo, M. Hangyo, S. Yamaguchi, S. Yano, Y. Segawa and K. Ohtaka, Phys. Rev. B**66**, 033111 (2002)
17. M. Tani, S. Matsuura, and K. Sakai, Appl. Opt. **36**, 7853 (1997)
18. K. Ohtaka, Phys. Rev. B **19**, 5057 (1979)
19. K. Ohtaka, Y. Suda, S. Nagano, T. Ueta, A. Imada, T. Koda, J. S. Bae, K. Mizuno, S. Yano and Y. Segawa, Phys. Rev. B **61**, 5267 (2000)
20. U. Fano, Phys. Rev. **124**, 1866 (1961)
21. M. Hase, et al., *Proceedings of Smart Structures and Materials 2000, Newport Beach, 2000*, ed by V. K. Varadan (SPIE, Washington, 2000) p. 314

7 Three-Dimensional Photonic Crystals

S. Noda, T. Kawashima, and S. Kawakami

Introduction

Much interest has been taken in photonic crystals (PCs) [1] in which the refractive index n changes periodically. A photonic band gap (PBG) is formed in the crystals, and the propagation of electromagnetic waves is prohibited for all wavevectors. Various important scientific and engineering applications such as control of spontaneous emission, a zero-threshold laser, very sharp bending of light, and so on, are expected by utilizing the PBG and the artificially introduced defect states and/or light emitters.

To develop these potential of PCs as much as possible and put them in the real world, the satisfaction of the following requirements is considered one of the most important issues: (i) a three-dimensional (3D) PC with a complete PBG should be constructed in the optical wavelength region; (ii) the introduction of an arbitrary defect state into the crystal should be possible at an arbitrary position; (iii) the introduction of an efficient light-emitting element should also be possible; and (iv) an electronically conductive crystal is preferable for the actual device application. Although various important approaches such as a self-assembled colloidal crystal [2], a GaAs-based three-axis dry-etching crystal [3], and a silicon-based layer-by-layer crystal [4, 5] have been proposed and developed to construct 3D PCs, it is considered difficult for these methods to satisfy the above requirements simultaneously. For example, in the case of a PC based on silicon with indirect band gap, it is difficult to apply it to an active photonic device.

Noda et al. proposed and demonstrated a new method for fabricating a complete 3D PC by stacking III–V semiconductors with a wafer-fusion and a very precise alignment technique [6–8]. In this method, the crystal is constructed with III–V semiconductors which are widely utilized for optoelectronic devices, and thus the above requirement (iii) is satisfied. In addition, since the wafer-fusion technique enables us to construct an arbitrary structure and to form an electronically active interface, the above requirements (i)–(iv) will be satisfied. Thus, once the 3D PC is implemented according to the proposed method, it will open the door for various applications including an active quantum device such as a zero-threshold laser. Recently, another fabrication technology which potentially satisfies the above requirements was reported by Aoki et al. [9], which is the stacking of very thin plates with

2D periodical structure by using micromanipulation under SEM observation. The important issue in this method is the limited size in area of PC.

In addition, 3D PCs have other attractive properties, such as strong isotropy or relatively high dispersion in the pass band frequency. By utilizing these properties, many kinds of applications, such as light collimators, dispersion compensators and so on, can be achieved. In this case, not only 3D PCs with a complete PBG described above but also 3D PCs without a complete PBG may be useful. Therefore, 3D PCs constructed by a simple fabrication method are attractive although the structure does not support a complete PBG. Kawakami et al. proposed and demonstrated an available and practical fabrication method for fabricating 3D structures, which is named autocloning [10].

In this chapter, two important fabrication technologies for 3D PCs are described. One is the technology using the wafer fusion technique, which is proposed by Noda et al. The fabrication method and optical properties of 3D PCs with a complete PBG developed at infrared to near-infrared wavelengths are described comprehensively and in detail. The other is the autocloning technology proposed by Kawakami et al. The mechanism and features of autocloning and its application to optical devices are reported in detail.

Part I. Fabrication and Optical Properties of 3D Photonic Crystals at Infrared to Near-Infrared Wavelengths

In this part, the fabrication method of 3D PCs with a complete PBG using the wafer fusion and alignment technique and their optical properties are described.

7.1 3D Photonic Crystal by Wafer Fusion and Alignment and its Band Structure

Figure 7.1 shows a schematic drawing to show the proposed fabrication procedure of a 3D PC by utilizing a wafer-fusion and alignment technique. The material system employed in this work is a III–V semiconductor system such as GaAs and InP. First of all, an AlGaAs (or InGaAsP) etching stop and a GaAs (or InP) PC layer are grown on a GaAs (or InP) substrate as shown in Fig. 7.1(a). Then, a stripe pattern is formed on the PC layer as shown in Fig. 7.1(b), where the period, the width, and the thickness of the stripe pattern are determined by the PBG wavelength. A pair of striped wafers is stacked with a crossed configuration and wafer-fused in H_2 atmosphere as shown in Fig. 7.1(c). One of the substrates and the etching stop layers are selectively and sequentially etched off by wet-chemical etching as shown

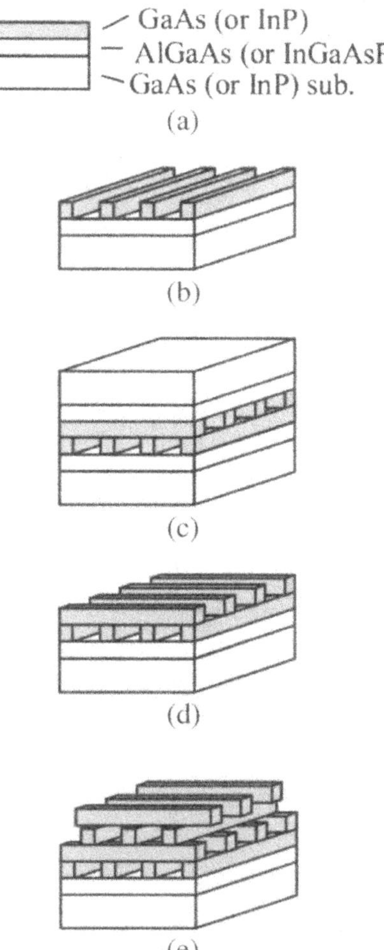

Fig. 7.1. Schematic drawing to show the proposed fabrication procedure of a 3D photonic crystal by utilizing a wafer-fusion and alignment technique

in Fig. 7.1(d). Then the wafer is cleaved into two pieces, and processes (c) and (d) are repeated. Figure 7.1(e) illustrates a crystal structure with four stacked layers constructed like this. When the processes (c) and (d) are repeated again, a crystal structure with eight stacked layers can be formed.

The constructed crystal structure is an asymmetric face-centered cubic (a-fcc) structure. Figure 7.2 shows the first Brillouin zone (BZ) and the calculated band diagram of the structure [11]. The calculation was made with a plane-wave expansion method, where the following parameters are assumed: the refractive index (n) of the semiconductor is 3.2, the filling fraction of the semiconductor to the air is 0.25, and the ratio of thickness to period of each

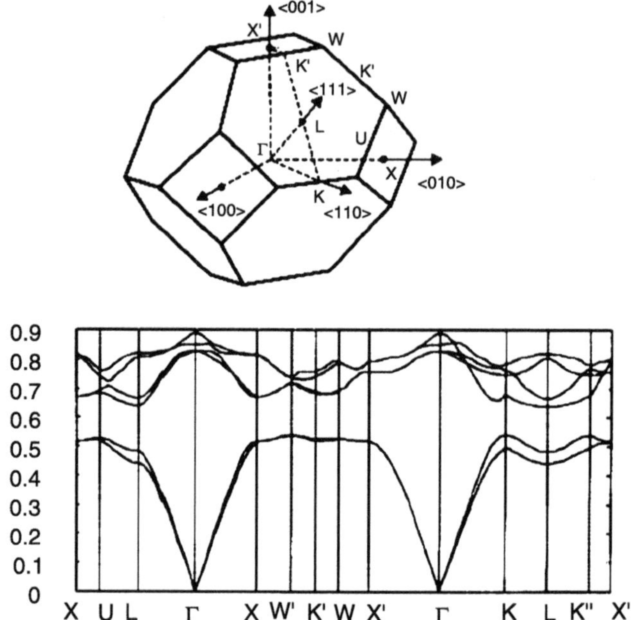

Fig. 7.2. First Brillouin zone and the calculated band diagram of the photonic crystal developed in this work

stripe is 0.3. As can be seen from the figure, the structure has a complete PBG for all wavevectors.

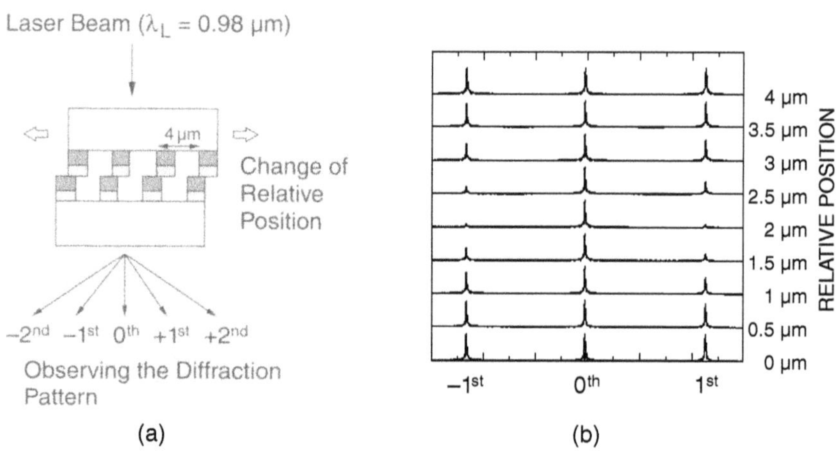

Fig. 7.3. (a) Schematic drawing to show the principle of alignment by utilizing the laser beam diffraction pattern observation technique, where the stripe period of each stripe is assumed to be 4 μm. **(b)** Calculated intensities of the ±first-order diffraction spots as a function of the relative position between two parallel stripes

The most important point in the construction of the PC according to Fig. 7.1 is the precise alignment between the first and third stripes (and between the second and fourth stripes) at step (d), where individual stripes should be shifted by a half period to construct the a-fcc structure. For this purpose, we have developed an alignment method [6, 12] by utilizing a laser beam diffraction pattern observation technique as follows. The individual stripe layers work as diffraction gratings, and a laser beam incident on the diffraction gratings is diffracted as shown in Fig. 7.3(a). When the relative position between two parallel stripes changes, the intensities of the higher-order diffraction spots also change. The result of the theoretical calculation is shown in Fig. 7.3(b), where the ±first-order diffraction spots become minimal when the relative position between the parallel stripes is shifted by just half a period. By utilizing this phenomenon, we can achieve a very precise alignment [12].

7.2 3D Photonic Crystals at Infrared Wavelengths

The PCs at infrared (5–10 μm) wavelengths were first created by using GaAs according to Fig. 7.1, where the structure with four stacked layers (one-unit structure) was constructed. Figure 7.4 shows the top view of the crystal with four stacked layers, where the stripe period, width, and depth are 4 μm, 1 μm, and 1.2 μm, respectively. It is seen that the individual stripe layers are excellently well-positioned. The transmission spectrum of the crystal was measured with a system composed of a Globar lamp, a monochromator, and an HgCdTe detector. The measurement result is shown in Fig. 7.5 (solid line),

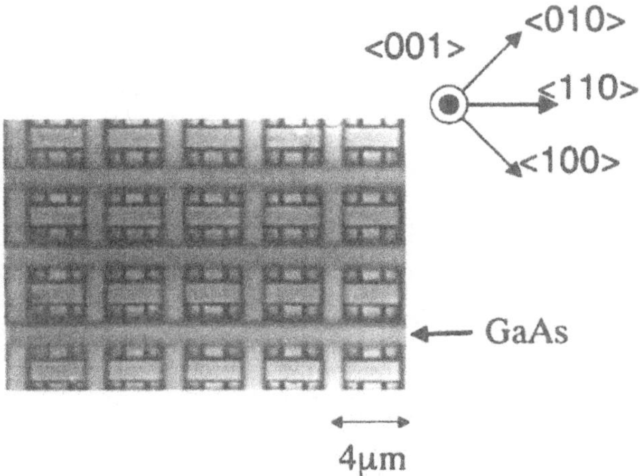

Fig. 7.4. Top view of the crystal with four stacked layers, where the stripe period, width, and depth are 4 μm, 1 μm, and 1.2 μm, respectively

where the light is incident normally on the sample surface (⟨001⟩ direction). A clear transmittance (T) dip is seen in the 5.5–9 μm wavelength region, which indicates the formation of a band gap (BG) in the 3D PC. The magnitude of the maximum attenuation is as large as ~15 dB for a one-unit structure. The broken line in Fig. 7.5 shows the theoretical result calculated

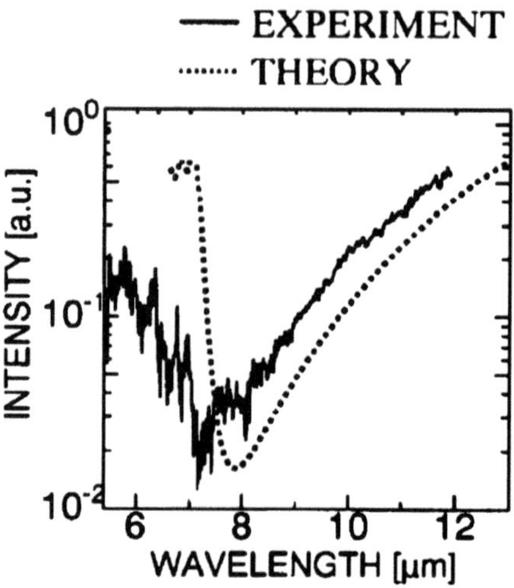

Fig. 7.5. Transmission spectrum of the crystal with four stacked layers (one-unit), where the light is incident normally on the sample surface (⟨001⟩ direction). The theoretical result by transfer matrix theory is also plotted with broken line in the figure. Relatively good agreement is seen between the experimental and theoretical results

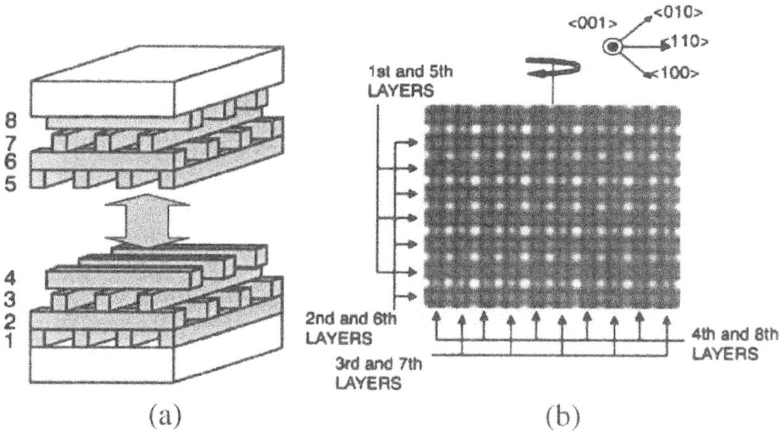

Fig. 7.6. **(a)** Schematic drawing to show the construction method of a photonic crystal with eight stacked layers (two-unit) by stacking a pair of one-unit crystals. **(b)** Infrared transmission image of the crystal thus constructed

by transfer matrix theory [13, 14]. A relatively good agreement is seen between the experimental and theoretical results, and it indicates that the fabricated structure works well as the PC.

Then, a PC with a two-unit (eight stacked layers) structure was constructed by stacking a pair of one-unit PCs as shown in Fig. 7.6(a). The infrared transmission image of the crystal is shown in Fig. 7.6(b), where it is seen that the individual layers are well positioned. The transmission spectrum of the crystal is shown in Fig. 7.7, where the result of the one-unit (four stacked layers) crystal is also shown for reference. As can be seen in the figure, an attenuation as large as 30 dB, which is twice that of one-unit crystal, has been successfully achieved. The 30 dB attenuation corresponds to 99.9% reflection and is considered to be enough for strong photon localization with the BG. It is also noteworthy that the band edges become very clear and sharp in the case of an eight stacked structure.

The angular dependence of the transmission spectrum was also investigated. The sample was rotated as shown in Fig. 7.6(b) to let the incident direction change from ⟨001⟩ toward ⟨110⟩, which corresponds to the change in the BZ from Γ-X' toward Γ-K (see Fig. 2). The experimental results are shown in Fig. 7.8. The attenuation of more than 20 dB (99%) is clearly seen at around 6.5–9 μm for various incident angles. The result is consistent with the theoretical band diagram (Fig. 7.2), which indicates that the crystal has a complete PBG. It is also seen that a dip appears at the band edge of the longer wavelength side (~9.2 μm) when the incident angle increases. This shows the degenerate first and second bands at the Γ-X' point split when the incident angle increases and the incident light direction changes from Γ-X' to Γ-K. These results indicate that the two-unit crystal offers us good performance as the photonic band gap material.

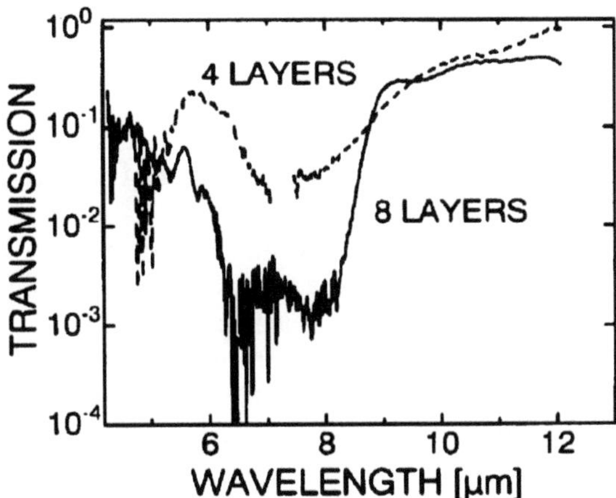

Fig. 7.7. Transmission spectrum of the photonic crystal with eight stacked layers. For comparison, the result for the crystal with four stacked layers is also shown

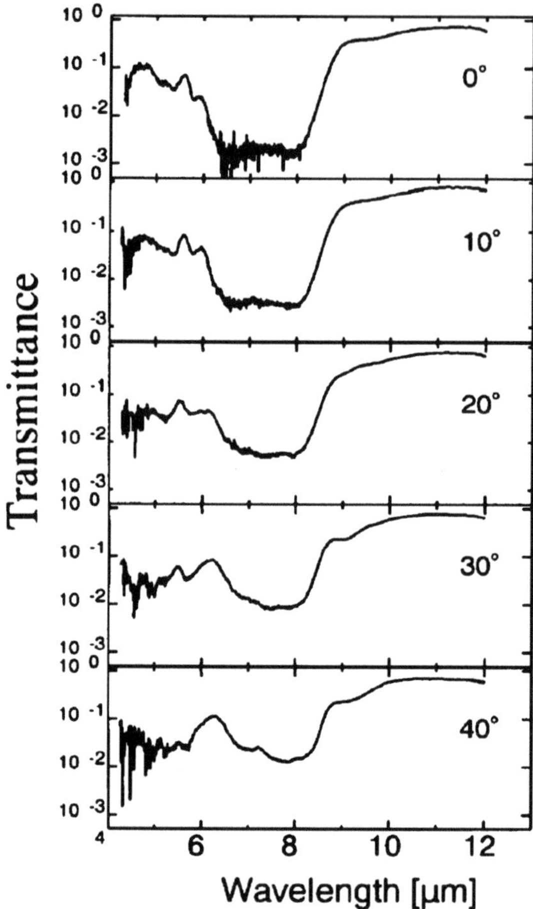

Fig. 7.8. Angular dependence of the transmission spectrum of the photonic crystal with eight stacked layers. The sample was rotated as shown in Fig. 7.6(b)

7.3 3D Photonic Crystals at Near-Infrared Wavelengths

Encouraged by the above excellent results at infrared wavelengths, we next constructed PCs at near-infrared wavelengths (1–2 μm). The material system was changed from GaAs to InP [8] since InP is widely utilized for optoelectronic devices in the optical communications field.

The important issue in this case is that the lattice constant (a) of the crystal should be reduced to as small as 1/5–1/6 of that of the PC in the infrared wavelength region. Figure 7.9 shows the top view of the two stacked layers. In the figure, the center square part corresponds to the PC area, and as seen in the expanded view, submicron-order stripes are well stacked. The area with a relatively large stripe period, which surrounds the PC area, is used as

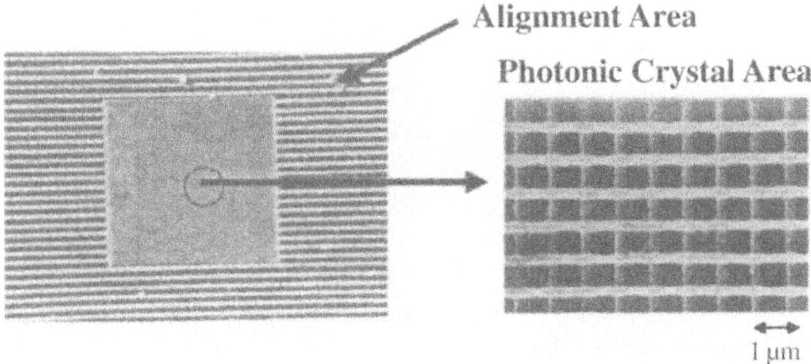

Fig. 7.9. Top view of the two stacked layers with submicron stripe period. The center square part corresponds to the photonic crystal area, and as seen in the expanded view, submicron-order stripes are well stacked. The area with relatively large stripe period, which surrounds the photonic crystal area, is utilized as the alignment area

an alignment area for the following reason. When we use the PC area itself for the alignment by the laser beam diffraction pattern observation technique, the diffracted laser beam cannot pass through it since the diffraction angle becomes too large. To avoid this situation, alignment stripes with an odd-multiple period of that of the stripes of the area were prepared. Figure 7.10(a) shows the diffraction spots after the alignment is accomplished, where the intensities of ±first-order spots are found to become minimal, and Fig. 7.10(b) shows the resultant cross-sectional view of the aligned parallel stripes.

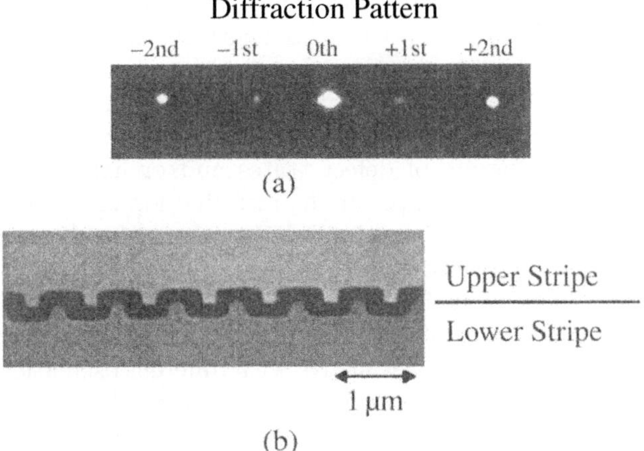

Fig. 7.10. (a) Diffraction spots after the alignment is accomplished, where the intensities of ±first-order spots are found to become minimal. **(b)** Resultant cross-sectional view of the aligned parallel stripes

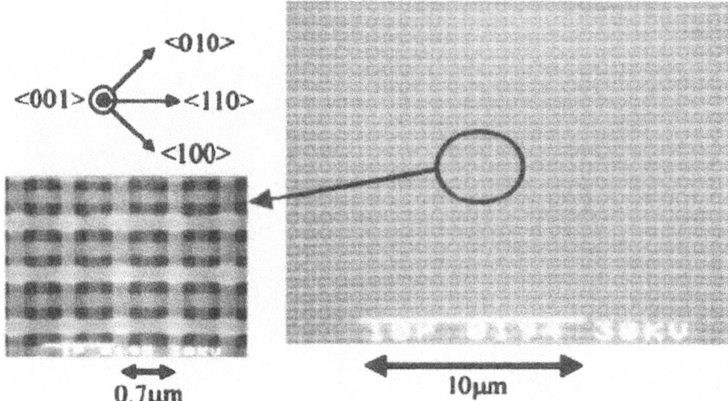

Fig. 7.11. SEM image of 3D photonic crystal fabricated with advanced processing techniques

It is seen that excellent alignment has been achieved between two parallel stripes. The accuracy was estimated to be within 50 nm, and in this case, the band gap width can be theoretically estimated to be as large as 100 meV [11, 14], which is large enough for the control of spontaneous emission from a light emitter whose spectral width is almost determined by the room temperature energy. Based on the above results, the PC with four stacked layers was constructed as shown in Fig. 7.11. For the sample, clear attenuation has been observed in the 1–1.5 μm wavelength region, and the maximum attenuation is ~23 dB. The result is consistent with the theoretical calculation by using the transfer matrix method where the alignment accuracy is taken into account. These facts clearly show that III–V semiconductor-based PCs become realistic in the optical wavelength region.

In summary, semiconductor 3D PCs have been successfully developed by wafer fusion and alignment in the infrared and near-infrared wavelength regions. It has been shown that by this method the crystals have a complete PBG and the introduction of defect states and/or light-emitting elements may be possible. For an example of the introduction of a light-emitting element, a surface-emitting laser with 2D PC structure has been demonstrated and unique lasing characteristics have been obtained [15–17]. Our success should open the door for various applications to quantum optics and devices. Figure 7.12 indicates an example of a future very compact quantum optical circuit, where nano-ampere laser arrays with different oscillation frequencies, optical modulators, wavelength selectors, and very sharp-bend waveguides are integrated in a very small area only by introducing appropriate artificial defects inside the crystal.

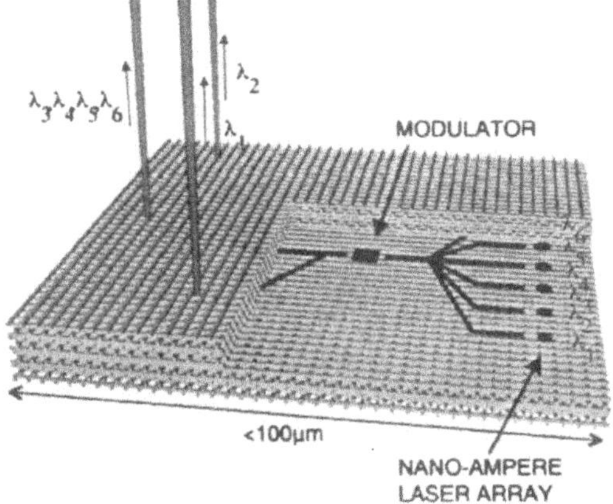

Fig. 7.12. Example of future very compact quantum optical circuits, where nano-ampere laser arrays with different oscillation frequencies, optical modulators, wavelength selectors, and very sharp-bend waveguides are integrated in a very small area only by introducing appropriate artificial defects inside the crystal. Note that parts of the figure (for example, a bent waveguide part) are not so accurate since the figure is drawn mainly to show the concept

Part II. Autocloning Technology and its Applications

In this part, a unique method for fabricating 3D PCs, named autocloning, is presented. It is based on the deposition of multilayers by a sputtering process. By using autocloning, 3D PCs are formed automatically without a precise alignment process. Also several novel applications of autoloning are described.

7.4 Autocloning Technology

Autocloning was developed in Tohoku University for the purpose of industrialization of PCs. It is based on stacking mulitilayers by a sputtering process, and is similar to conventional optical multicoating processes. The fabrication process is as follows. After the formation of the periodic corrugation pattern on a substrate by lithography, such as electron beam lithography, layers are stacked on the substrate by a combined sputtering deposition and sputter etching process. Under appropriate conditions, the deposition of the multilayer progresses with the preservation of the surface corrugation pattern which is reflected in the substrate pattern. In this way, a multidimensional periodic structure (a 3D periodic structure or a 2D structure with a periodicity along the stacking direction) can be formed sequentially (Fig. 7.13).

Figure 7.14 shows a schematic diagram of the apparatus for the autocloning process. It is a conventional bias sputtering apparatus. The bias sputtering process is often used for deposition of a passivation layer for burying

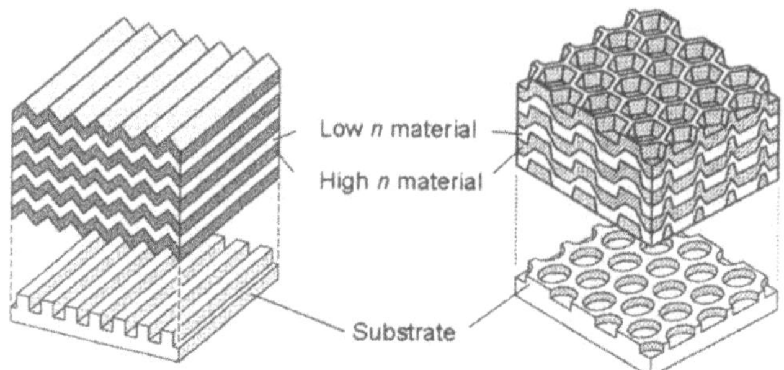

Fig. 7.13. Schematic illustrations of autocloned structures

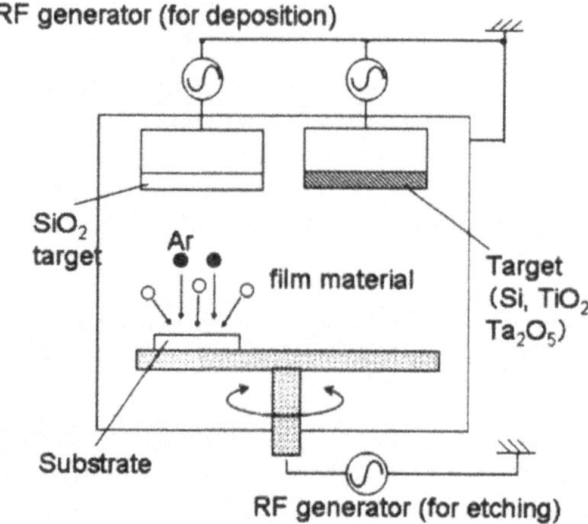

Fig. 7.14. Schematic diagram of the apparatus for autocloning

patterned metal wiring in LSI (large-scale integration). In the process, rf (radio frequency) power is applied to both the target and the substrate electrode. As a result, ions are accelerated not only to the target but also to the substrate; sputter deposition and sputter etching take place on the surface of the film simultaneously.

7.5 Mechanisms of Autocloning

We discuss the mechanism of the preservation of the shape in the autocloning process. First of all, we discuss the mechanisms of the flattening process by bias sputtering.

As mentioned above, in the bias sputtering process, sputter deposition and sputter etching take place simultaneously. The etching rate of the sputter etching depends on the angle between the direction of incidence of the ions and the normal direction to the etched surface, and the rate has a maximum value at a specific angle [18]. In general, the ions are launched to the substrate electrode normally. Namely, we should consider only the angle θ of the surface (see Fig. 7.15(b)). The net deposition rate combined with the deposition rate and the etching rate also depend on the surface angle. By continuing the bias sputtering process, the surface at the angle of which the etching rate is maximum is formed dominantly. If the net rate at the angle is lower than $b \cos\theta$, the convex part is eroded away and a flat surface is obtained [19]. (b is the net deposition rate at the horizontal surface.)

Let us consider autocloning. Roughly speaking, when we adjust the balance between the deposition and the etching appropriately, the growth rate at the slope becomes close to $b \cos\theta$, and it looks like the slope does not grow in lateral directions. Namely, we can suppose the autocloning process is based on stacking multilayers by sputter deposition and shaping by sputter etching. To verify this, we assume that the following three phenomena mainly influence the film shaping process, and we introduce a computer simulation for calculating the influence. In the simulation, a film is considered as a group of tiny square pixels. The reaction quantity for each phenomenon can be calculated for each pixel, and the surface shape evolution can be obtained by superposing these results [20].

A. Deposition of Particles Coming from a Target with a Distribution of Incident Angles

The particles coming from a target to a substrate have a distribution of incident angles caused by the mutual collisions of the particles. Because the dimensions of the surface shape are much smaller than the mean free paths of the particles, a shadowing effect is caused, and the deposition rate at a point depends on the surrounding shape. Owing to this effect, the surface evolution starting from ridges is calculated as shown in Fig. 7.15(a). The rate

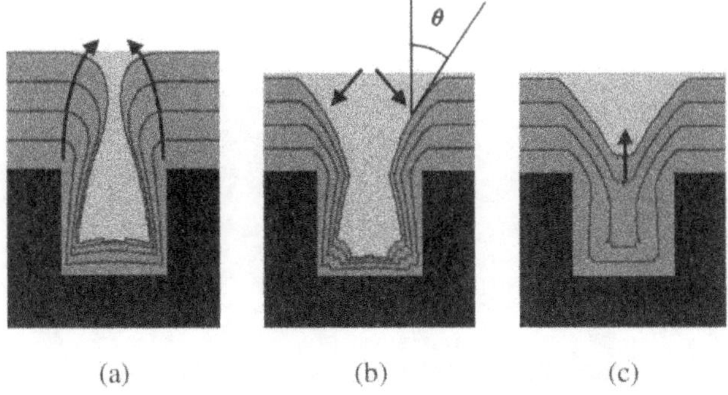

Fig. 7.15. (a) A simulation result of surface evolution starting from a rectangular groove caused by deposition of diffuse incident particles. **(b)** The result of superposing the effect of sputter etching on (a). **(c)** The result of superposing the effect of redeposition on (b). The gray lines are markers to make the evolution clear

at a convex part is higher than that at a concave part. As a result, the valleys become deeper and narrower.

B. Sputter Etching with an Angular Dependent Etching Rate

When rf power is fed to the substrate electrode, electrical potential to plasma becomes highly negative. As a consequence, Ar ions are accelerated towards the substrate, where they cause sputter etching. By continuing sputter etching, a constant slope at the angle which has the maximum sputtering yield is formed. When this effect is superposed on that of A, the calculated surface evolution is as shown in Fig. 7.15(b). The valleys tend to become wider and straight slopes are formed.

C. Redeposition of Film Materials

When the film surface is sputter etched, or particles from the target rebound on the film surface, redeposition by secondary generated particles is observed. The rate of redeposition also depends on the surrounding shape. Contrary to the effect of A, the rate at a concave part is higher than that at a convex part. The simulation result of superposing the effects of A, B and C is shown in Fig. 7.15(c). The valleys become filled up.

To verify whether the autocloning phenomenon is interpreted as a superposition of the above three effects, we simulate the autocloning process, which involves repetition of deposition and etching. The simulation result and the corresponding experimental result are shown in Fig. 7.16. The results agree well, suggesting that autocloning is a superposition of the three elementary phenomena. When deposition and etching are carried out in an appropriate ratio for a certain time, the changes of shape caused by each phenomenon cancel each other out and the stationary shape is preserved. Moreover, the

Fig. 7.16. On the right is a cross-sectional SEM image of a 2D a-Si/SiO_2 structure. On the left is the calculated result of the 2D structure. The parameters used in the calculation correspond to the conditions of the experiment

agreement in Fig. 7.16 also means that we can search the conditions and predict the shape utilizing only the simulator [21].

Consequently, the autocloning process is based on a balance between sputter deposition and sputter etching. We can understand the autocloning phenomenon as a result of the combination of the three primitive processes. When appropriately balanced, the constant slope is preserved and it looks like the wavy shape grows only in the perpendicular direction.

7.6 Features of Autocloning

In this section, we show some features of autocloning, and summarize the feasibility for industrialization.

7.6.1 Self-healing Effect

Autocloning possesses a self-healing effect for the preserved shape. Figure 7.17 shows a cross-sectional SEM photograph of a 2D structure stacked on a substrate whose pattern is partly disordered. After a few cycles of stacking, the defect is repaired and a uniform structure is formed. This phenomenon supports the stability of the preserved shape. This stability against perturbation is important to achieve a high tolerance for fabrication error of the substrate. Consequently, we can conclude that this feature is suitable for mass production [22].

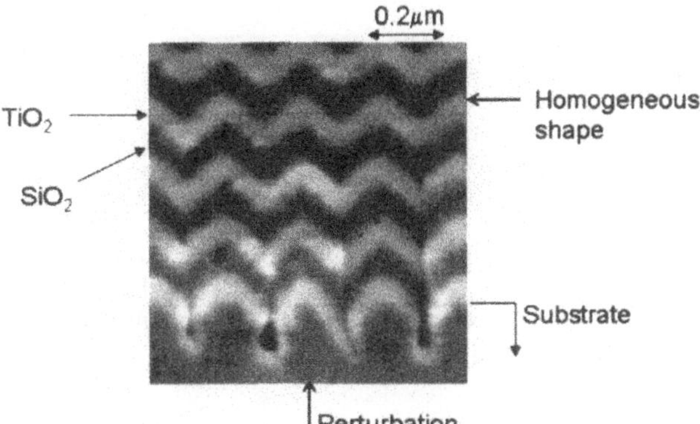

Fig. 7.17. A cross-sectional SEM image of the 2D TiO_2/SiO_2 structure. The influence of the perturbation on the substrate is canceled immediately after a few layers are stacked, and the corrugated shape becomes homogeneous. This effect does not depend on the material

7.6.2 Productivity

The phenomena observed in the autocloning process are basically the same as those in the bias sputtering process. The bias sputtering process is well defined, and various techniques for raising productivity have already been developed. We can utilize such established techniques. Now we can produce large-uniform 2D PCs (20 mm square). A 2D PC sample thus fabricated serves as a polarization beam splitter [23].

7.6.3 Flexibility of Materials

In the autocloning process, in principle, we can use any materials that can be sputtered. Consequently we can use a pair of materials whose refractive indices are very different from one another. We have verified that we can fabricate PCs using a-Si, SiO_2, TiO_2, or Ta_2O_5. TiO_2 and Ta_2O_5 are transparent materials in the visible wavelength region [24].

7.6.4 Flexibility of Lattice Type

In the case of autocloned structures, the lattice type and the lattice constant depend on the substrate patterns. Accordingly we can select the symmetry of the structure as we desire. In Fig. 7.18, surface AFM (atomic force microscope) images of autocloned 3D structures are shown. The left image is the case of a triangular arrangement of substrate patterns, and the center is that of a square arrangement. Moreover the right is the case that the pattern has an aperiodic part.

Fig. 7.18. Surface AFM images of 3D autocloned structures stacked 20 periods (8 µm thickness). The left is the case of a triangular arrangement of substrate pattern, the center is that of a square arrangement, and the right is the case that the pattern has an aperiodic part

7.6.5 Scaling Law

Except for the limitation that the mean free path of sputtered particles is much longer than the period of the structure, the above primitive phenomena do not depend on the dimension of the structure. Namely, the law of scaling can hold in the autocloning phenomenon as shown in Fig. 7.19.

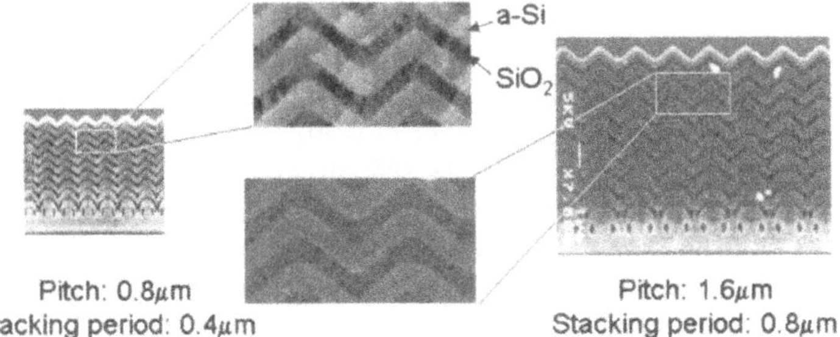

Fig. 7.19. On the left is a 2D structure whose pattern pitch is 0.8 μm and stacking period is 0.4 μm. On the right is half of that on the left

7.6.6 Lattice Modulation

We have succeeded in stacking PCs on a substrate having several patterns whose periods are different from each other. Figure 7.20 shows a cross-sectional SEM photograph of the interface between the structures having different lattice constants. By comparison with the two areas, we can see that the levels of the ridges in each structure are the same, and the depths of the valleys are different. It is very important for "heterostructured PCs" that we can fabricate structures having several kinds of lattice constants simultaneously.

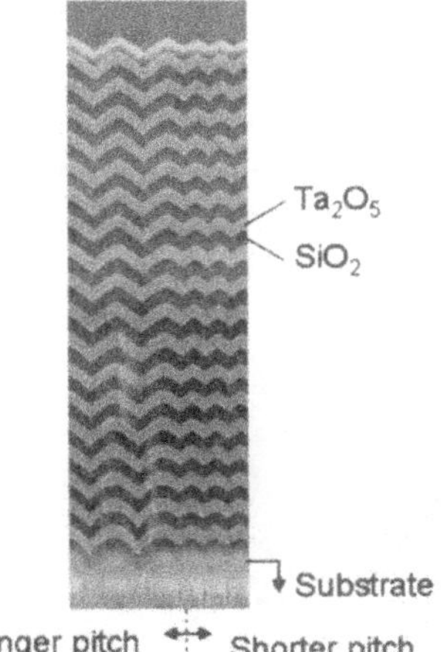

Fig. 7.20. Cross-sectional SEM image of the interface between areas having different horizontal periods

7.7 Concept of Lattice Modulation of Photonic Crystals and its Application

By using autocloning technology, several applications have been proposed, such as polarization beam splitters [23], waveplates [24], superprisms [25], lattice-modulated waveguides [26], and resonators [27]. Since autumn 2002, samples of polarization beam splitters have been shipped by Photonic Lattice Inc. [28]. Among these, the concept of lattice modulation of photonic crystals and its applications is described in the following.

As mentioned above, we can modify the lattice constant a easily. If lattice modulation can be introduced into PCs, the degree of freedom in designing application devices increases dramatically. For example, we will be able to use it for mode matching between waveguides. Thus, the concept of lattice modulation is one of the keys to making practical PC devices. We have proposed a new type of device utilizing lattice modulation.

7.7.1 Channel Waveguide by Using Lattice Modulation

Figure 7.21(a) shows schematic diagrams of the channel waveguides constructed by the lattice modulation of autocloned PCs. When light propagates in the z direction, the effective n of each region depends on the lattice constant period or the lattice orientation as shown in Fig. 7.21(b), so we can control n by adjusting the pitch or the axis, and the stacking period. The index of the center region A is higher than the neighboring regions, and consequently the center region A acts as the core of the waveguide. The cross-section and the near-field pattern of the a-modulated waveguide fabricated from a Ta_2O_5/SiO_2 system are shown in Fig. 7.22 [26]. The boundary between the areas with a different in-plane pitch is found to grow in the direction perpendicular to the substrate. The spot diameter is about 4 μm.

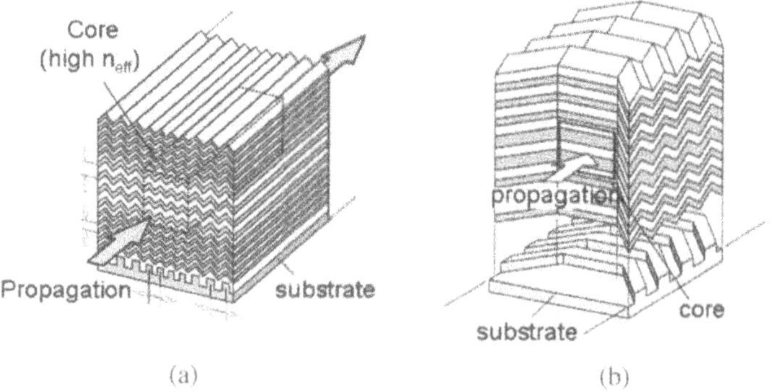

Fig. 7.21. Schematic illustration of PC waveguide. **(a)** Lattice-constant modulation and **(b)** lattice-orientation modulation

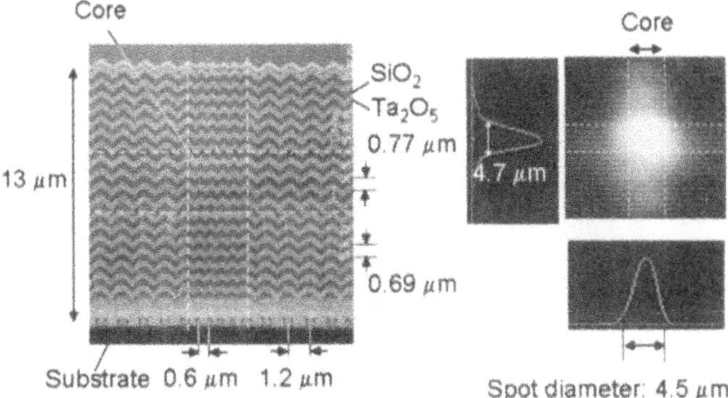

Fig. 7.22. Cross-sectional SEM image of a lattice-modulated PC waveguide and near-field pattern of a lattice-modulated waveguide ($\lambda = 1.55$ μm)

This is the first time that fiber-to-fiber transmission measurement of a PC waveguide by butt-jointing has been accomplished, involving the direct connection of the waveguide to a high Δ single-mode fiber which is spliced by the TEC (thermal extended core) technique to a standard SMF (single-mode fiber). A propagation loss of 0.1 dB/mm is obtained. The coupling losses can be estimated to be about 0.43 dB in total, which includes the coupling loss due to the Fresnel reflection and the mode mismatch. The good connectability with fiber and the low propagation loss are thus experimentally verified, which cannot be attained in the line-defect type PC waveguide.

7.7.2 Optical Resonator

Various functions can be realized in an optical circuit based on a lattice modulation waveguide. As an example, the optical resonator shown in Fig. 7.23 was fabricated and evaluated [27]. The resonator is composed of mirrors and a cavity, which operates as a BG for the incident wavelength and a pass band, respectively. In the resonator, the light is confined within the core area with a high effective refractive index which is formed by the large period of stacking layers, similar to the lattice modulation waveguide. The fabrication is simple since the 2D or 3D periodic structure with various periods and directions, which is a heterostructured PC, is formed by stacking multilayers once on the patterned substrate. There is no need for alignment at all. The transmission spectrum of an optical resonator with a waveguide of 1.5 mm length was measured by the fiber-to-fiber system. As shown in Fig. 7.24, a resonating peak value of $Q = 270$ was obtained at a wavelength of 1498 nm. The center frequency and the Q-value are adjustable in the initial patterning process. As conventional PLCs (planar lightwave circuit) cannot implement such an in-line type resonator structure, this example shows the rich functionality of autocloned PC circuits.

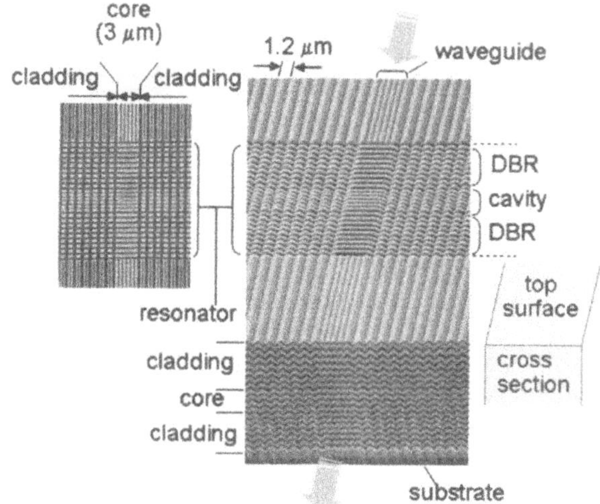

Fig. 7.23. SEM image of an in-line resonator (perspective)

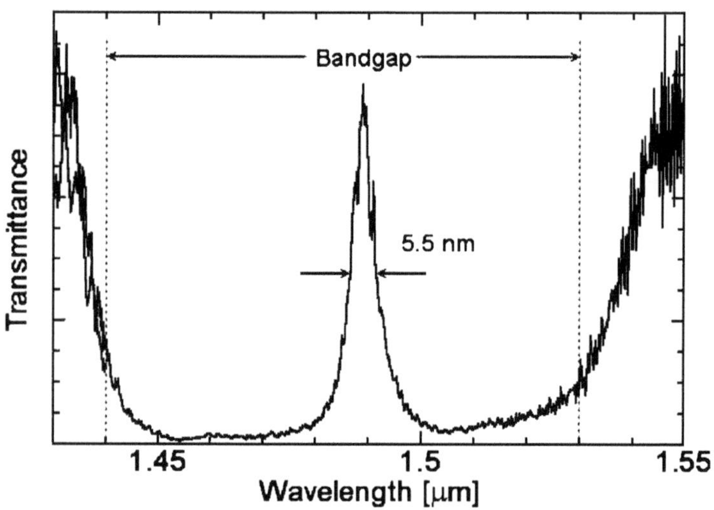

Fig. 7.24. Transmission spectrum of an in-line resonator ($Q = 270$)

The functional circuit which utilizes propagation along the in-plane direction is also promising as a next-generation PLC because it is easy to connect to a fiber and offers low loss and rich functionality. We will continue to improve the fabrication technique and to propose and develop new functions, such as a directional coupler or dispersion compensator, to expand the range of application of autocloning type PCs.

7.8 Expansion of Band Gaps

Autocloned structures are basically wavy structures whose thicknesses are constant. Such structures support large BGs in the normal direction to the film. On the other hand, in the lateral direction, the average of n is almost constant. This means that the periodicity in the lateral directions gives rise to only small BGs. In the case of applications as planar devices, the larger BGs in the lateral directions lead to a greater variety of applications, such as reflectors, resonators, or dispersion control. Thus we are investigating new processes for increasing BGs in the lateral directions.

7.8.1 Introducing Reactive Ion Etching

The autocloned shape is formed from sputter etching, and does not much depend on the material. As a result, the shape settles into a similar one in many cases. So we introduce RIE (reactive ion etching) instead of sputter etching into the autocloning. In the case of RIE, the etching rate depends very much on the etched materials. In addition to that, the rate also depends on the sparseness of the etched film. Figure 7.25 shows a cross-sectional SEM image of a 2D structure which was fabricated by using RIE. In this case, the etching rate at concave parts is different from that at convex parts, which is caused by the difference of sparseness of the etched parts. The thickness of each film is highly modulated, and it causes expansion of BGs in the lateral directions [29].

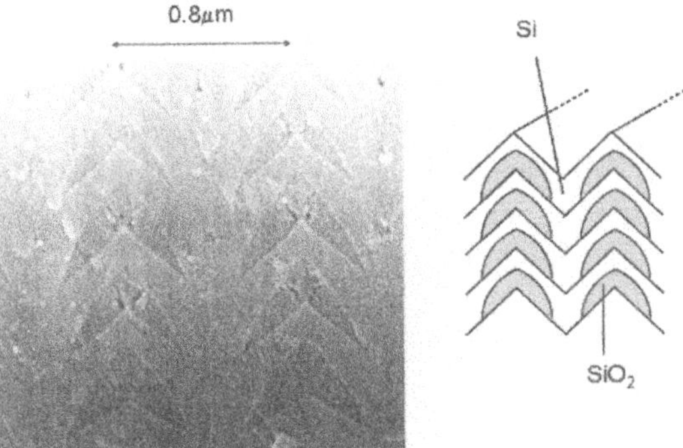

Fig. 7.25. Cross-sectional SEM image of a 2D structure fabricated by the combination of autocloning and RIE. The horizontal SiO_2 beams with arrowhead like cross-section are fabricated in Si. The thickness of each layer is highly modulated

7.8.2 Autocloned Structures Supporting FBG

The realization of 3D full BGs is one of several objectives of our research on PCs. We have proposed a new method of fabricating PCs having 3D full BGs

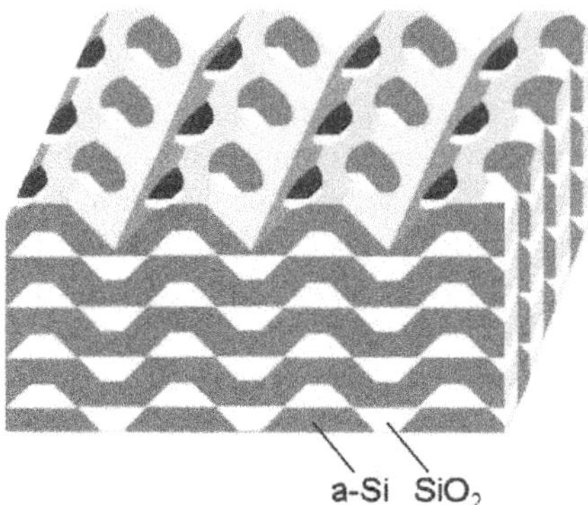

Fig. 7.26. Schematic illustration of a structure having 3D full band gaps. After a 2D structure is fabricated using autocloning, periodic holes are drilled normal to the substrate with reactive ion etching

utilizing autocloning technology [30]. A schematic figure of the structure is shown in Fig. 7.26. First we fabricate a 2D periodic structure by autocloning, and then drill periodic holes with lithography and dry etching. This method is based on a technique proposed by an MIT group [31]. In the case of our method, the 2D structure in the early stage can be fabricated much more easily. Our collaborator, NTT research group, succeeded in observing the BGs in the some representative directions [32].

In summary, autocloning is a fabrication method for multidimensional structures by stacking the corrugated films keeping the periodicity. It has robustness against perturbation, and is tolerant of materials and lattice types. Accordingly, we can fabricate various PCs with high reliability by using autocloning. Since it is based on a sputtering process which is an industrialized technology, there is little obstacle to industrializing the autocloning process. So it is obvious that autocloning is a promising candidate for mass-production of PC devices.

References

1. E. Yablonovitch, J. Opt. Soc. Am. B **10**, 283 (1993)
2. For example, V. N. Astratov, V. N. Bogomolov, A. A. Kaplyanskii, A. V. Prokoev, L. A. Samoilovich, S. M. Samoilovich, Yu. A. Vlasov, Nuovo Cimento D (Italy), **17D**, ser.1, no.11, 1349 (1995)
3. V. Arbet-Engels, E. Yablonovitch, C. C. Cheng, and A. Scherer, *Micro-cavities and photonic bandgaps*, (Kluwer Academic, Dordrecht, 1996), pp.125-131
4. E. Ozbay, A. Abeyta, G. Tuttle, M. Tringides, R. Biswas, C. T. Chan, C. M. Soukoulis, and K. M. Ho, Phys. Rev. B **50**, 1945, (1994)
5. S. Y. Lin, J. G. Fleming, D. L. Hetherington, B. K. Smith, R. Biswas, K. M. Ho, M. M. Sigalas, W. Zubrzycki, S. R. Kurtz, and J. Bur, Nature **394**, 251 (1998)

6. S. Noda, N. Yamamoto, and A. Sasaki, Jpn. J. Appl. Phys. **35**, L909 (1996)
7. N. Yamamoto, S. Noda, and A. Chutinan, Jpn. J. Appl. Phys. **37**, L1052 (1998)
8. S. Noda, N. Yamamoto, H. Kobayashi, M. Okano, and K. Tomoda, Appl. Phys. Lett. **75**, 905 (1999)
9. K. Aoki, T. Miyazaki, H. Hirayama, K. Inoshita, T. Baba, N. Shinya, and Y. Aoyagi, Appl. Phys. Lett. **81**, 3122 (2002)
10. S. Kawakami, Electron. Lett. **33**, 1260 (1997)
11. A. Chutinan, and S. Noda, J. Opt. Soc. Am. B **16**, 240 (1999)
12. N. Yamamoto and S. Noda, Jpn. J. Appl. Phys. **37**, 3334 (1998)
13. J. B. Pendry and A. MacKinnon, Phys. Rev. Lett. **69**, 2772 (1992)
14. A. Chutinan and S. Noda, J. Opt. Soc. Am. B **16**, 1398 (1999)
15. M. Imada, S. Noda, A. Chutinan, T. Tokuda, M. Murata, and G. Sasaki, Appl. Phys. Lett. **75**, 316 (1999)
16. S. Noda, M. Yokoyama, M. Imada, A. Chutinan, and M. Mochizuki, Science **293**, 1123 (2001)
17. D. Ohnishi, K. Sakai, M. Imada, and S. Noda, Electron. Lett. **39**, 612, (2003)
18. H. Bach, J. Non-Cryst. Solids, **3**, 1 (1970)
19. C. Y. Ting, V. J. Vivalda, and H. G. Schaefer, J. Vac. Sci. Technol. **15**, 1105 (1978)
20. S. Tazawa, S. Matsuo, and K. Saito, IEEE Trans. Semiconductor Manufacturing, **5**, 27 (1992)
21. S. Kawakami, T. Kawashima, and T. Sato, Appl. Phys. Lett. **74**, 463 (1999)
22. T. Kawashima, K. Miura, T. Sato, and S. Kawakami, Appl. Phys. Lett. **77**, 2613 (2000)
23. Y. Ohtera, T. Sato, T. Kawashima, T. Tamamura, and S. Kawakami, Electron. Lett. **35**, 1271 (1999)
24. T. Sato, K. Miura, Y. Ohtera, S. Kawakami, and T. Tamamura, QELS 2000, QTuC5, San Francisco, May 2000
25. H. Kosaka, T. Kawashima, A. Tomita, M. Notomi, T. Tamamura, T. Sato, and S. Kawakami, Phys. Rev. B, **58**, 16, R10096 (1998)
26. Y. Ohtera, T. Kawashima, Y. Sakai, T. Sato, I. Yokohama, A. Ozawa, and S. Kawakami, Opt. Lett. **27**, 2158 (2002)
27. T. Sato, Y. Ohtera, T. Kawashima, H. Ohkubo, K. Miura, N. Ishino, and S. Kawakami, The 28th European Conference on Optical Communication (ECOC02), 04.4.2, Sept. 2002
28. http:///www.photonic-lattice.com/.
29. T. Kawashima, T. Sato, Y. Ohtera, and S. Kawakami, IEEE J. Quantum Electron. **38**, 899 (2002)
30. M. Notomi, T. Tamamura, T. Kawashima, and S. Kawakami, Appl. Phys. Lett. **77**, 4256 (2000)
31. S. Fan, P. R. Villeneuve, R. D. Meade, and J. D. Joannopoulos, Appl. Phys. Lett. **65**, 1466 (1994)
32. E. Kuramochi, M. Notomi, I. Yokohama, J. Takahashi, C. Takahashi, T. Kawashima, and S. Kawakami, Optical and Quantum Electronics, **34**, 53 (2002)

8 Other Types of Photonic Crystals

Y. Segawa and K. Ohtaka

Other than the PCs discussed so far, a number of interesting systems have been expoited and studied. Here the descriptions are given for double-periodic PCs, quantum-well PCs, isotropic-band PCs and metallic PCs.

8.1 Double-Periodic Photonic Crystals

What kind of feature will be expected if the 1D optical lattice has a dual periodicity, in other words, if the dielectric function of a dielectric multilayer structure is modulated with another periodicity different from that of the basic unit cell? Such an optical superlattice structure may be regarded as being analogous to the electronic system in semiconductors, either having a superlattice structure or being subject to a strong static magnetic field. In the former case, the electrons are under the influence of 1D dual-periodic crystal potentials, while in the latter, the electrons are under the influence of the Coulomb potential and of a periodic phase factor defined as $a^2H/2\pi(\hbar c/e)$ in the 2D plane perpendicular to the magnetic field H [1]. As is well known, the resulting electronic energy spectra exhibit a fantastic "butterfly diagram", as demonstrated by Hofstadter [1]. A fantastic butterfly-like diagram indeed appear even more clearly in the 1D dual-periodic PC [2]. Also a strong localization of Bloch modes is predicted. The electromagnetic field of these modes is shown to be strongly localized around the troughs of undulated periodic dielectric structures, which leads to an extraordinary high-Q-value effect in dual-periodic 1D PCs [3]. See an experiment described in Sect. 13.3. For the model of dual-periodic dielectric multilayers, we have assumed a sequence of square-well optical "potential" barriers composed of the layer "A" (the thickness being a) with the dielectric constant ε_a and the layer "B" (thickness b) with dielectric constant ε_b ($< \varepsilon_a$). For simplicity, we assume that $\varepsilon_b = 1$ and we take the unit of length as $a = 1$. The multilayer is composed of a sequential stack of the unit cell made of the A and B layers with the basic lattice constant of $a + b$. Let us define the position of each unit cell by the site index i. To introduce dual-periodicity, we assume that the dielectric constant of the A layer at the i-th site is given by the following site-dependent mode function:

$$\varepsilon_a(i) = \frac{\varepsilon_0 + 1}{2} + \frac{\varepsilon_0 - 1}{2} \cos(2\pi p i). \tag{8.1}$$

Here, ε_0 and p are the constants which characterize the energy spectra for photonic states in the model dual-periodic multilayers.

When $p = m/n$ where m and n are integer numbers, the multilayer becomes dual-periodic. In this case, the multilayer is analogous to a superlattice semiconductor with period $n(a + b)$. Examples of such dual-periodic multilayer structures are illustrated in Fig. 8.1 for $p = 1/10$ and $p = 3/7$, both assuming $\varepsilon_0 = 8$, $a = 1$ and $b = 3$. The dashed curves indicate the sinusoidal modulations given by (8.1) for the respective cases. Note that the superposition of two periodicities, with the spacings of $a+b$ and $(a+b)m/n$ respectively, leads to a supermultilayer structure composed of a larger unit cell with the period of $(a + b)n$. This supercell period is m times longer than the period of modulation, as seen in the case of $p = 3/7$ shown in Fig. 8.1(b).

In order to calculate the optical transmittance (T) or reflectance (R) for a dual-periodic multilayer, we have to deal with a superlattice structure having a finite length L. In the present calculation, we assumed that $L = 4n$, that is, a length which is four times larger than the superlattice periodicity. Then, the numerical calculation of T is straightforward, if we prepare a number of transfer matrices for these finite dual-periodic multilayers.

Numerical calculations have been performed for the PBS, the transmittance T, and the energy spectra (the density of states spectra) upon assuming specific values for a, b, ε_0 and p.

In Fig. 8.2, we show the typical results for $b/a = 1, 3$ and 5. As seen, each diagram in the three cases consists of close vertical black lines which are separated by the white regions in between. The black lines correspond

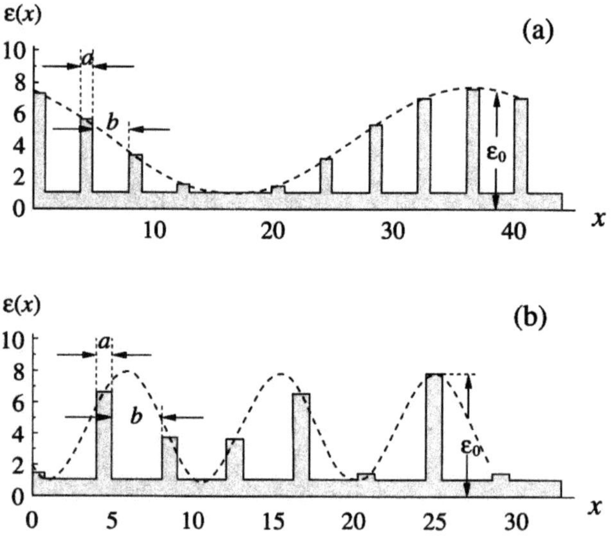

Fig. 8.1. Model of dual-periodic multilayers: **(a)** $\varepsilon_0 = 8$, $b/a = 3$ and $p = 1/10$; **(b)** $\varepsilon_0 = 8$, $b/a = 3$ and $p = 3/7$

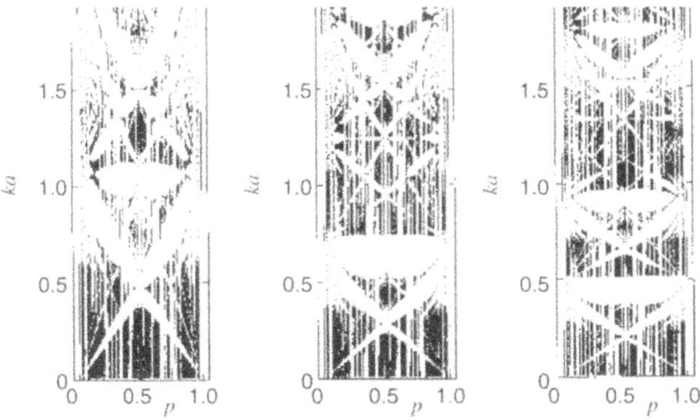

Fig. 8.2. Typical energy spectra of dual-periodic multilayers: **(a)** $\varepsilon_0 = 8$, $b/a = 1$, **(b)** $\varepsilon_0 = 8$, $b/a = 3$ and (c) $\varepsilon_0 = 8$, $b/a = 5$

to the transmission bands, which are the energy regions where more or less k-dependent PBs are distributed, whereas the white regions between the respective black line regions at a fixed value of p are the energy regions of $T = 0$, that is the regions corresponding to the PBGs.

The most remarkable feature in Fig. 8.2 is no doubt the fantastic butterfly shapes which are developed in the black-and-white patterns in the respective diagrams demonstrated in Fig. 8.2(a), (b) and (c). Although the band distributions have been calculated only for rational numbers for p, the change of patterns looks like smooth functions of p, so that one can trace how the respective bands or BGs are continuously changing as p is varied, that is, when the dual-periodicity is introduced into the monoperiodic multilayer structures corresponding to the cases for $p = 0$. The patterns as a whole are dependent on the value of p, as seen from comparison of Fig. 8.2(a), (b) and (c), yet the butterfly-like features are seen in common, indicating that these features are not specific characters of the PB diagram for particular p-values, but are evidently inherent to the band structures of the dual-periodic multilayer structures.

It should also be remarked that the characteristic patterns shown in Fig. 8.2 for dual-periodic multilayers are reminiscent of the well-known clustering patterns for the energy spectra of Bloch electron bands in a crystal under uniform magnetic fields (a typical example is given by Hofstadter [1]). The physical interpretation of this remarkable feature has been given in terms of the rational and irrational magnetic field parameter with respect to the periodicity of the crystal lattice. We note that if there is any relation in common between the diagrams of optical band structures plotted in Fig. 8.2 and those for the Bloch electron energy spectra in magnetic fields, the only reasonable and realistic explanation for that would be the effect of the dual-periodicity on the wave propagation modes in the 1D multilayer PCs and the 2D electronic systems of solids in uniform magnetic fields, respectively. According to this interpretation, the present system of dual-periodic dielectric multilayers is considered to be unique as an optical analogue of the electronic system in crystals under a magnetic field. It is possible to reveal some interesting features by introducing the concept of a dual-periodicity in dielectric multilayer structures: (1) the perturbation of the original system in the form of the slowly varying modulation leads to the appearance of several very narrow photonic Bloch modes, and (2) an extraordinarily large enhancement of the EM field intensity is predicted to occur due to the effective localization of Bloch photons in the heavy-photon-like bands in agreement with the discussion by Ohtaka [4–6]. In contrast to the effect of photon localization on the defective layer, the present effect is essentially attributed to a bulk property, characteristic of the dual-periodic system. The photon localization and EM field enhancement takes place not in the particular spatial region, but in the extended regions over the whole multilayer structure. Thus, a dual-periodic multilayer structure may act as an anomalously high-Q optical cavity for the specific photonic Bloch modes. It is also notable that these features can be controlled by means of fine tuning of the dual-periodic modulation parameter p and the modulation function shape. There is a large degree of freedom for improving the characteristics according to the requirements of specific optical applications and sample preparation technology.

8.2 Quantum-Well Photonic Crystals

The quantum confinement effects of electrons in semiconductors have been studied over the past few decades and the basic features are well understood. For example, in the case of a quantum-well structure, the wavefunctions of the electrons and holes in the quantum well are changed from plane waves to standing waves. The confined electrons and holes have their own quantum numbers and form sub-band structures. In this section we treat the quantization of photonic states in the quantum-well PCs and discuss how the quantization manifest itself in the light transmission property. We deal with quantum-well PCs of arrayed dielectric spheres both experimentally and theoretically.

Periodic arrays of dielectric spheres provide us with prototypical PCs. The PB effect in such systems has been studied both theoretically [4, 7–11] and experimentally [12–16]. In a spherical system, accurate theoretical analysis is possible using the group theory and the spherical vector expansion method for a PC system consisting of a dielectric sphere. Experiments have been performed in the millimeter-wave region, and the results show good agreement with the results of theoretical analysis [16].

Recently, the concept of a "photonic quantum well", which consists of a quantum-well structure similar to the structure of a semiconductor, has been proposed [17, 18]. In these studies, the sharp peaks in the stop band (SB) were calculated and considered as the effect of the bound state or resonant tunneling. The calculated peaks were not explained as the quantized state of PBs but as impurity states of the PC. But, in this photonic quantum well, the quantization of the PB will occur in a similar way to the quantization of the electron band in semiconductors. However, no experiments on the quantum-well structure using a PC have been performed.

We measured the transmittance spectra and the intensity of electrical fields in the well crystal of a quantum-well structure fabricated by a PC arranged in millimeter-sized Si_3N_4 spheres. The quantization of PBs is discussed in comparison with a semiconductor quantum well [19].

To fabricate the quantum-well structure, two kinds of 3D PCs were prepared. They were made by a layered 2D periodic array of Si_3N_4 beads of mm size as constituent spherical dielectric particles. The 2D periodic array is a close-packed hexagonal lattice, stacked to make a hexagonal crystal. The building blocks for the 3D PC we used were spherical balls of Si_3N_4 with a diameter $d = 1/8$ inch. Si_3N_4 has a fairly high ε-value of 8.67 in the millimeter wavelength region investigated. Figure 8.3 shows illustrations of the quantum-well structure made with PCs. PBG can be controlled by changing the air gap of layered 3D PCs. The frequency region of the BG of the barrier is different from that of the well due to the different air gaps. The PB of the well crystal overlaps the BG of the barrier crystal, and the PB of the well crystal is confined and quantized. The reason why we considered that this configuration corresponds to four-layered quantum-well crystals will be discussed later. It is expected that sharp transmitted peaks corresponding to the quantized PB in the BG of the barrier crystal would be observed.

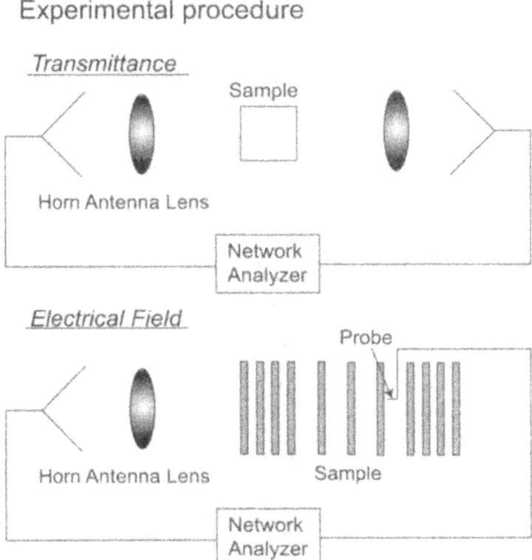

Fig. 8.3. Illustrations of a quantum well structure and measurement systems. Here, we have changed the distances of the air gap

The transmittance (T) spectra and the intensity of the electric field in the air gap of the PCs were measured by using a network analyzer (WILTRON 360B) as a function of millimeter wavelength for normal incidence. For the measurement of T, two horn antennas were used to produce a probe EM wave and to detect T. We used two lenses ($f = 300$ mm) to make a plane-wave incident EM wave and to focus its transmitted component. A semi-rigid cable 0.8 mm in diameter was used to probe the electric field in the air gaps of the well crystal.

Figure 8.4 shows the transmission spectra when the number of layers in the quantum well crystal is 3–9, D_w (air gap width of the well region) is 3.2 mm and D_b (air gap width of the barrier region) is 1.1 mm. Figure 8.4(a) and (b) respectively show the experimental T-spectra and the results of theoretical calculations. The lowest trace of the figure is drawn for a well width of three layers, while the topmost one is for nine layers. The number of well lay-

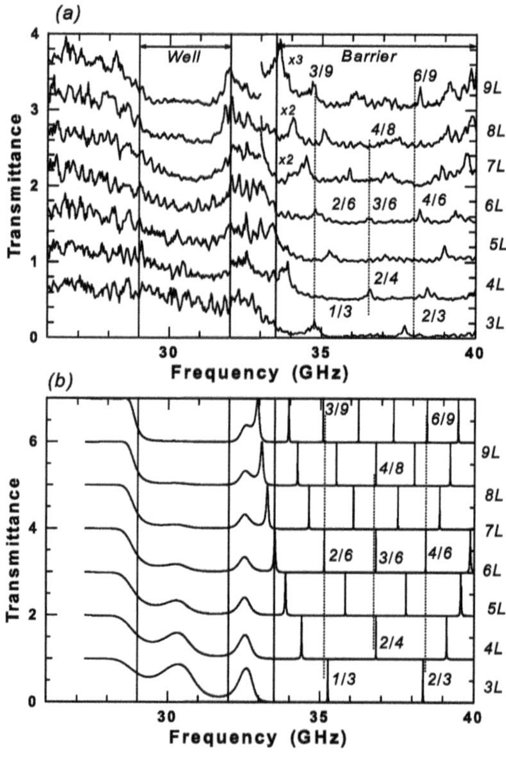

Fig. 8.4. Transmittance spectra when the air gap of the well crystal (B) was 3.2 mm and that of the barrier was 1.1 mm. The well is 3–9 layers. **(a)** and **(b)** show the experimental and theoretical transmittance results, respectively. The two regions shown by the horizontal arrows are the PBG regions of the well and barrier crystals, to be found when the layers are stacked infinitely

ers is represented on the right-hand side of the figure as $3L$. If both the PCs of the barrier and the well are perfect crystals, they will have BGs in different frequency regions. The range of the BG of the well crystal is 29–32 GHz, and that of the barrier crystal is 33.5–40 GHz. Horizontal arrows show the range of each band gap. Therefore, in our experiment, the PB of the well crystal in the range of 33.5 to 40 GHz is confined by the band gap of the barrier crystal. Based on theoretical calculations, the confined PB corresponds to the third PB. As shown in Fig. 8.4(a), some sharp peaks appeared in the BG of the barrier crystal. In the case of a four-layered crystal, whose shape corresponds to that shown in Fig. 8.3, three peaks are observed at 33.9 GHz, 36.5 GHz and 38.4 GHz. We think that these structures correspond to the quantized states of photonic bands in a well crystal. The quantized wavenumber of the states is determined similarly to the quantized states of electrons in a semiconductor quantum well. Let the lattice constant of the unit cell be a and the width of the quantum well be $L = na$. The wavenumber of the BZ boundary is given by $\pi/a(\equiv k_B)$, and the quantized wavenumber of the first quantized state is $\pi/L = \pi/(na) = k_B/n$. That is, if the well width is four layers, the wavenumbers of all quantized states are $\frac{1}{4}k_B$, $\frac{2}{4}k_B$ and $\frac{3}{4}k_B$. Therefore, the second quantized level of four layers, the third of six layers and the fourth of eight layers should have the same quantized wavenumber and energy. The first (second) level of three layers, the second (fourth) of six layers and the third (sixth) of nine layers will appear at the same positions. Thus, the wavenumbers of the observed quantized levels are shown in the figure by dotted lines and fractions such as 2/4. The number of the layers was determined to be four for the spectra of $L = 4$ and so on. Therefore, the boundary layers between the regions of the air gap width D_w and D_b (see the lower panel of Fig. 8.3) are considered to belong to the well region.

The experimental T was compared with the theoretical calculations shown in Fig. 8.4(b). The results of the calculation agreed well with the experimental results in terms of the frequency of sharp peaks, although the transmitted intensity of sharp peaks is not unity in the experiment. This discrepancy is due to the loss of the electromagnetic wave in the case of a thick sample. Because our sample was slightly distorted, it is thought that the loss occurred by the scattering or diffusion of the EM wave on the 2D lattice plane. Therefore, it is difficult to observe sharp peaks such as those seen in the spectra of a large well number, for example, 7–9 layers.

The intensity of the electric field $|E|^2$ was measured in the air gaps of the well region of the system in order to study the symmetry of each quantized structure. The results are shown in Fig. 8.5 for the system with $D_w = 3.2$ mm, $D_b = 1.1$ mm, and the number of the wells equal to six. The bottom of the figure depicts the structure of the measured quantum well with the gray boxes showing the plane of the 2D lattice of beads. The measured intensity $|E|^2$ was shown by the squares. The theoretical results are given by solid curves both for Re E and $|E|^2$ as functions of the distance from the first layer. Im E is

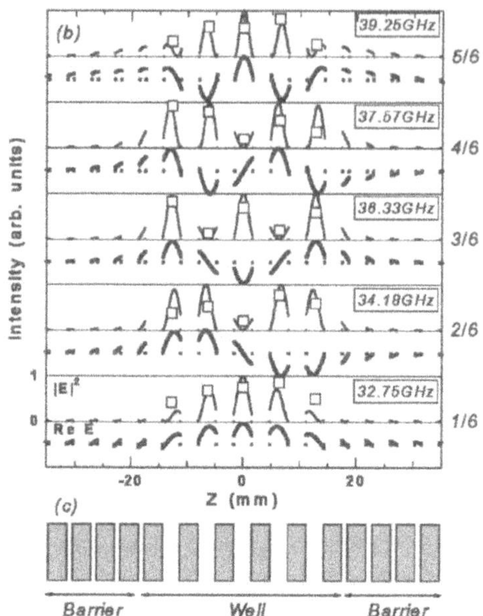

Fig. 8.5. (b) Intensity of the electric field in the air gaps as a function of the distance from the surface. The large squares show the experimental results of $|E|^2$ and the solid curves show the of theoretical calculations, for $|E|^2$ in the upper half and ReE in the lower half. **(c)** Diagram of the measured quantum well structure. The gray boxes show the plane of the 2D lattice

not given as it has the same symmetry as Re E. Obviously, the experimental results are in good agreement with the theoretical calculations. Moreover, the symmetry of fine peaks will be discussed. From the calculation of Re E, one can see the symmetry, that is, the first, third and fifth quantized states are even modes, and the second and fourth are odd modes. The intensity at the third air gap, which is at the center of the well crystal, is almost zero for the second (34.18 GHz) and fourth (37.57 GHz) peaks, but it is nonzero for the first, third and fifth states. Considering that the center is the loop for even parity and the node for odd parity, the experimental results clearly show that the mode of the fine peaks changes alternately between even and odd parity. This indicates that the envelope function of the quantized electromagnetic wave is determined as $\sin(n\pi/L)$. Thus, the difference of the electrical field pattern between the fine peaks is well explained by the various even and odd quantized states in a quantum well.

8.3 Isotropic Band Photonic Crystals

It is now well known that the origin of photonic gaps is not limited to the periodicity of systems [20]. This is best clarified from the observation of isotropic photonic gaps in 2D quasi-periodic PCs composed of identical dielectric cylinders [21–27]. All of the quasi-periodic PCs proposed and examined so far have the common feature of rotational symmetry around certain rods. By extending this feature, we study in this work the optical properties of rotationally symmetric 2D photonic system. This would not only be interesting from the viewpoint of the understanding of the physical origin of photonic gaps but also open up the fascinating possibility for optical technology applications such as 2D optical waveguides.

We measured the transmission of a circular PC (CPC), the non-periodic but systematic arrangement of dielectric cylinders. The CPC is shown schematically in Fig 8.6. The principle of this system is a dense arrangement of cylinders, which we distribute so that the distance between the neighboring pairs is nearly the same. Then, the distribution of the cylinders is designed by using concentric circles. The positions of the cylinders are taken to be

$$x = dN \cos\left(\frac{360m}{6N}\right) \tag{8.2}$$

$$y = dN \sin\left(\frac{360m}{6N}\right) \tag{8.3}$$

where N, d and m denote the number of concentric circles, the difference of radii of neighboring concentric circles (called the radial spacing) and the number of cylinders ($0 \leq m < 6N$), respectively.

By changing the radial spacing d, the electric field power was calculated at the center of the CPC of $N = 5$. One remarkable drop was found to appear around a frequency of 10 GHz in the transmission spectrum. With decreasing d, we found that the frequency of the transmission drop shifted upwards and the depth of the drop became smaller. Based on the numerical data, the sample size for the experiments was decided to be $r = 2$ mm and $d = 8$ mm. We fabricated our CPC using alumina cylinders whose refractive index is 3.1. The transmission spectrum was measured by a probe antenna, because the CPC had no translational symmetry. The polarization of the millimeter wave was set parallel to the cylinders (TM mode).

The result is shown in Fig. 8.7. When the millimeter wave propagates along the y-axis defined in Fig. 8.6, there is a large drop and some fine drops in the experimental spectrum. The large drop was always found around 10 GHz irrespecctive of the position of the probe antenna in the CPC. The depth of the drop in the transmission of the CPC shown in Fig. 8.7 was as small as −20 dB, comparable to the minimum detectable limit of our probe antenna. On the other hand, the appearance of the fine drops was different at the same measurement points when the direction of the probe antenna insertion was

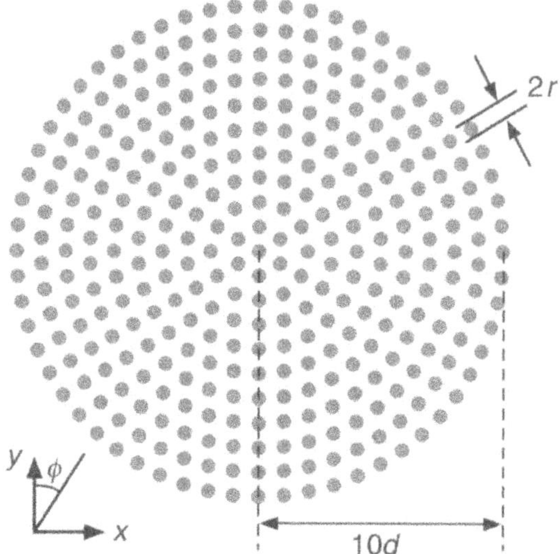

Fig. 8.6. Scheme of CPC with six-fold symmetry. A millimeter-wave propagates along the y-axis. The width of the wave was as large as the sample size

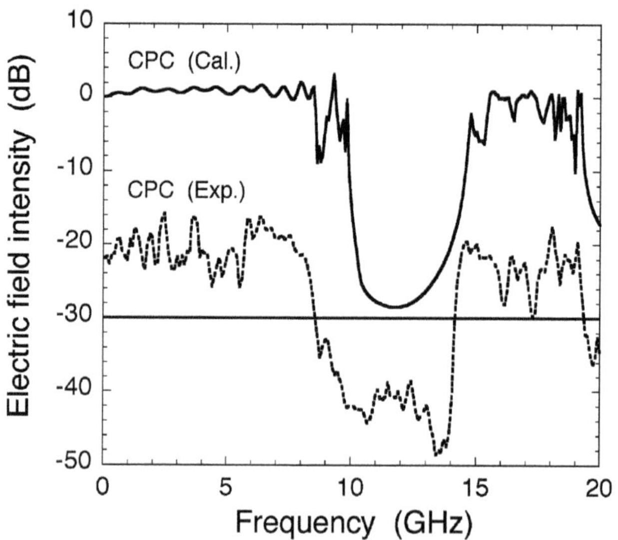

Fig. 8.7. Spectra of the electric field power calculated and measured at the center of the CPC. The spectra are offset at invervals of 10^2

changed. From these results, the drop around 10 GHz was confirmed to show the optical character of the CPC, that is the photonic gap.

The relation between the PG and the direction of the incident wave was examined. Here, the upper and lower band edges were defined at the frequency where the drop of transmission crossed the level of −10 dB. The incident angle ϕ from the y axis was varied from $-90°$ to $+90°$. The measurement points were fixed at the center of the CPC. The upper and lower band edges were found around 9 GHz and 14 GHz, respectively. Though the arrangement of dielectric cylinders had six-fold symmetry, there seemed to be no correlation between the band edges and the incident angle ϕ. From this result, it was concluded that the CPC had an isotropic PG. We have proposed the CPC as a novel dielectric structure and have shown the isotropic PG. In general, two possible origins for the PG are suggested; the coherent interference of scattering waves from the periodic cylinders (like Bragg diffraction) and the interaction of modes between the bonding and anti-bonding states within each cylinder (like Mie resonance) [28–30]. By taking the nonperiodic structure of dielectrics and nearly the same distance between the cylinders into account, the origins of the PG for a CPC may be due to Mie resonance. Further study is in progress on the optical features of CPCs.

From the viewpoint of the distance between cylinders, there are two other types of dielectric structures that show an isotropic PG. One is the structure of quasicrystals which possess the same distance between the cylinders and two kinds of unit cells [21–27]. Though there is no translational symmetry, the resonance of diffracted light is observed in photonic quasicrystals (PQCs). This phenomenon indicates that a PQC exhibits not only short-range, but also long-range order. Hase et al. showed that the isotropic PG could be obtained by five-fold and eight-fold symmetric PQCs [31]. These PQCs were fabricated by epoxy resins ($\varepsilon = 2.4$). The radius r of the dielectric cylinders and the tile side length a of the five-fold PQC (eight-fold PQC) were $r = 22\,\mu\text{m}$ and $a = 85\,\mu\text{m}$ ($69\,\mu\text{m}$), respectively. The transmission spectra were obtained in the far-infrared region. Even when the propagation direction of light was changed, the drop of the transmission was always found at the same normalized wavenumber $ka/2\pi \sim 0.54$ (0.50).

The other type of dielectric structure is uniformly distributed photonic scatterers (UDPSs), proposed by Miyazaki et al [32]. In this system the cylinders are distributed freely except that the distance between cylinders is larger than a certain value $D_{\min}$, i.e., $|\boldsymbol{R}_i - \boldsymbol{R}_j| \geq D_{\min}$ for any pair of cylinder sites $\boldsymbol{R}_i$ and $\boldsymbol{R}_j$. It should be noted that the UDPSs have neither long-range nor short-range order. It has been numerically and experimentally confirmed that the UDPS can have a clear cut PG, depending on the distance between the cylinders, not on how they are distributed topologically. Also, the PG is found to be relatively robust against the variations of various parameters of the UDPS. The frequency region of the PG is shifted insignificantly with the change of the radial distribution of cylinders, but significantly with the change of area fraction of the cylinders. For the design of a PG positioned at $\lambda = 1.55\,\mu\text{m}$, they fabricated the UDPS with $r = 0.11\,\mu\text{m}$. The drastic change of the optical features of UDPS was demonstrated by controlling the randomness of the distance. The UDPS is also a useful dielectric structure for obtaining an isotropic PG.

8.4 Metallic Photonic Crystals

Light of frequency ω cannot penetrate into a metal matrix when ω is less than ω_{p}, the plasma frequency of the metal. Since the plasma frequencies of ordiary metals lie in the frequency region of visible or shorter wavelengths of light, metallic PCs must have lattice constants of that order in order for their PBs to have peculiar effects originating from plasma oscillations. Also, in order to lower the resonant frequencies, we should preferably use PCs of regular arrays of *isolated* metal units, for the plasma oscillations thereof are usually found below ω_{p} as seen below.

There seems to be two motives in studies of metallic PCs. The first is to make use of the large frequency separations between two adjacent plasmon resonances to introduce a wide PBG, hopefully an omni-directional BG. It is to be noted that the band gap in question has nothing to do with Bragg reflection of the PC [33–35]. The second motive is to take advantage of the plasmon resonances to enhance optical signals. Since the plasmon polariton modes in a metallic PC are folded back at the BZ edge in ω-k space, the portion pulled into the leaky region (see Sects. 2.3 and 3.4) works as an origin of sharp resonant signals having a high quality factor Q. One typical example is Raman scattering from a molecule adsorbed on the surface of a metal PC (see Sect. 4.6 for the enhanced Raman signal involving PCs).

In the calculation of the PBs related to plasmon oscillations, the straightforward use of the plane-wave expansion method suffers the difficulty of slow convergence [36]. This is caused by the fact that surface plasmon oscillations involved in metal units are those of the surface charge and hence the treatment to express the localized events in terms of plane waves is more delicate than the dielectric cases.

Since the properties of metallic PCs of arrayed cylinders are discussed by Sakoda [37], we describe in this article a number of characteristic features of PBS of plasmon polaritons, obtained by the vector KKR method (see Sect. 4.1.1) for an array of metallic spheres. This method was extensively applied in [33, 34] to metallic PCs. All the results presented here are new and based on unpublished works of Ohtaka and his collaborators.

8.4.1 Band Structure of a Metallic Photonic Crystal

Sphere Plasmon Modes. To understand the PBS of metallic PCs, it is useful to start from the plasmon oscillations of an isolated metal sphere. Neglecting the retardation effect, its plasma oscillation is covered completely by the Laplace equation, $\Delta V(\boldsymbol{r}) = 0$ everywhere, for the scalar potential $V(\boldsymbol{r})$. To avoid the confusion with the radial variable r, we use the symbol r_0 for the radius of spheres. We then find that $V(\boldsymbol{r})$ of symmetry l, m has the form

$$V(\boldsymbol{r}) = \begin{cases} \sum_l A_l r^{-l-1} Y_{lm}(\theta,\phi) & r > r_0 \\ \sum_l B_l r^l Y_{lm}(\theta,\phi) & r < r_0 \end{cases} \tag{8.4}$$

where $Y_{lm}(\theta, \phi)$ is the spherical harmonic of the angular momentum index $(l\, m)$ for the polar angles θ and ϕ of the vector $\boldsymbol{r}$. The boundary condition at $r = r_0$ is the continuity of $V(\boldsymbol{r})$ and the continuity of the radial component of the displacement field defined by $\boldsymbol{D}(\boldsymbol{r}) = -\varepsilon_0\varepsilon(\boldsymbol{r})\boldsymbol{\nabla}V(\boldsymbol{r})$. For $\boldsymbol{r}$ inside the sphere, we use the metallic dielectric constant $\varepsilon(\omega)$ for $\varepsilon(\boldsymbol{r})$. For a metal sphere in free space, these boundary conditions using (8.4) lead to

$$l\varepsilon(\omega) + (l+1) = 0. \tag{8.5}$$

This equation gives the frequencies of the (non-retarded) plasma oscillations of a metal sphere. There is a $(2l+1)$-fold degeneracy with respect to m.

When the Drude form

$$\varepsilon(\omega) = 1 - \frac{\omega_{\mathrm{p}}^2}{\omega^2} \tag{8.6}$$

is used for $\varepsilon(\omega)$, (8.5) leads to

$$\omega_l = \omega_{\mathrm{p}}\sqrt{\frac{l}{2l+1}}. \tag{8.7}$$

These plasmon modes, called hereafter sphere plasmons, therefore have eigenfrequencies of $\sqrt{1/3}\omega_{\mathrm{p}}$, $\sqrt{2/5}\omega_{\mathrm{p}}, \cdots$ for $l = 1, 2, \cdots$, respectively, and exist in the frequency range $0 < \omega < \omega_{\mathrm{s}}$, where $\omega_{\mathrm{s}} = \omega_{\mathrm{p}}/\sqrt{2}$ is the surface plasmon frequency. Consequently the resonant frequencies exist in the frequency region lower than ω_{s} and they concentrate indefinitely when the resonant frequency approaches ω_{s} in the limit $l \to \infty$. With these dipole or multipole moments induced in metal spheres, the electromagnetic energies are transported through a metallic PC, in spite of the fact that the frequency of light is well below the plasmon frequency.

As a becomes larger, the retardation effect enters, because it takes a longer time for an electromagnetic wave to traverse the sphere. Due to the retardation effect the sphere plasmons no longer have real eigenvalues but have finite lifetimes. The lifetime is induced by the energy dissipation due to irradiation from the excited multipole moments specified by (l, m). The outflow to infinity of the radiation is described by the spherical Hankel function of the solutions of the full Maxwell equations. Inside the sphere, the field is described by the spherical Bessel function j_l of pure imaginary argument, because $\varepsilon(\omega_l)$ is negative in the argument $(\omega/c)\sqrt{\varepsilon(\omega)}r$ of j_l. Therefore the field decays monotonically towards the sphere center with the maximum amplitude realized at the wall of the sphere. In short, sphere plasmons are surface modes localized to the surface region of a sphere with a finite lifetime. We note that the radiative origin of the lifetime is exactly the same in character as that of the whispering gallery modes (WGM) of the Mie resonance discussed in Sect. 3.5.

The frequency and lifetime of a sphere plasmon are well covered by the increment of the density of states (DOS) of electromagnetic modes, which are obtained by the frequency derivative of the calculated scattering phase shift $\delta_l^{\mathrm{N}}(\omega)$ of the N type partial vector-wave (TM modes) (see Sect. 3.5 for the mode classification, and see Sect. 4.3.2 for the DOS calculation). The reason why only N type modes have plasma oscillations is that they alone have nonzero radial components of the associated electric field and can induce surface charge oscillation due to the strong depolarization effect. In contrast to the Mie resonance of a dielectric sphere, the modes with the M symmetry (TE mode) do not have any resonance. A larger confinement effect is expected for a higher l, which leads to a longer lifetime, as in a dielectric WGM.

Array of Metallic Spheres. In the periodic array of metallic spheres, the PBs are specified by the wavevector $\boldsymbol{k}$. Neglecting the interaction between spheres, the $(2l+1)$-fold degenerate sphere plasmons of multipole l are degenerate for all spheres, i.e., they have a degeneracy of $N_{\mathrm{s}}(2l+1)$, N_{s} being the number of spheres. A rough sketch of the PBs in the arrayed metallic spheres is obtained by drawing the flat dispersion curves at $\omega = \omega_l$ for each l, crossing them with the light line $\omega = c|\boldsymbol{k}|$ and then taking into account the polariton splittings.

Honest numerical calculation is, however, necessary to take the retardation effect into account and to cover the interaction between spheres which causes the modes to hop between the neighboring spheres (see Sect. 3.6 for the formation of the tight-binding PB). The vector KKR equation is very powerful to incorporate all these effects.

With $l \leq 6$ taken into account, the example of the PB structure is given in Fig. 8.8 for the wavevector $\boldsymbol{k}$ in the Γ-X direction. The spheres are arrayed in the fcc lattice with the ratio of the radius of the spheres to the fcc cube size being 0.18 (volume fraction of spheres being 0.1). We take $\omega_{\mathrm{p}} = 0.5$ in normalized units. The $(2l+1)$-fold degeneracy of sphere plasmons is split into several optically active bands of E irreducible representation (see Sect. 4.2.3) and other optically inactive bands. Figure 8.8 displays only the band structure of optically active bands. How the degeneracy is dissolved for each l is analyzed in [5].

Figure 8.9 shows the PBS of the frequency near the sphere plasmon frequency. The bands shown by solid circles are optically active bands of E irreducible representation, reproduced from Fig. 8.8. Open circles show the optically inactive bands, which were removed in Fig. 8.8. The left figure shows the band structure of $l = 2$ sphere plasmons (mainly $l = 2$, strictly, because there is always a mixing with the other l in the lattice) and the right figure shows that from $l = 3$. In the case of $l = 2$, the five-fold degeneracy of sphere plasmons is split into $2+1+1+1$, i.e., one E band and two nondegenerate bands, while in $l = 3$ we see the splitting $7 = 2 \times 2 + 1 + 1 + 1$. It is to be noted that the dispersion curve of the light-plasmon coupled bands of $l = 3$ is not monotonic. This is an interesting example of a nonmonotonic dispersion relation of a surface-plasmon polariton.

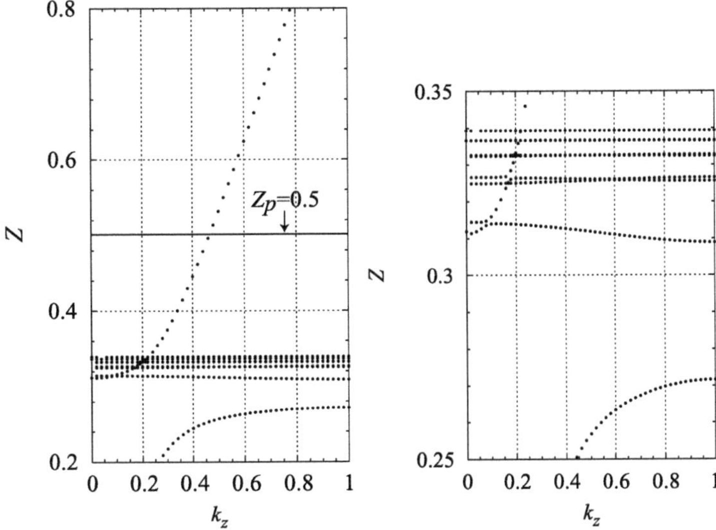

Fig. 8.8. Band structure of fcc array of metallic spheres for the wavevector $\boldsymbol{k}$ in the (001) direction. Frequency Z and wavenumber k_z are normalized by $2\pi c/a$ and $2\pi/a$, respectively, a being the size of the fcc cube. The line of $Z_{\mathrm{p}} = 0.5$ shows the plasmon frequency ω_{p}. Sphere plasmons of $\ell \leq 6$ are taken into account in the calculation. The right figure is an enlargement of the left

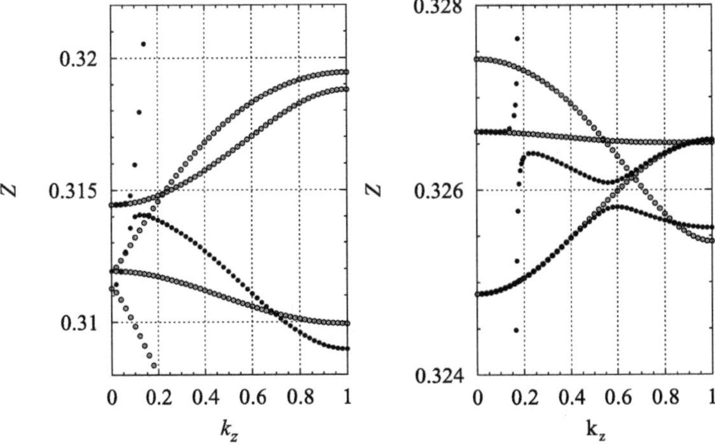

Fig. 8.9. Band structure of fcc array of metallic spheres for the wavevector $\boldsymbol{k}$ in the (001) direction. All the bands, optically active or inactive, are plotted. The bands originating mainly from $l = 2$ are shown in the left panel and those of $l = 3$ are shown in the right panel. In the right panel nonmonotonic parts of the dispersion relations are to be noted

8.4.2 Transmittance and Local Field Enhancement

The layer-doubling method described in Sect. 4.4 is a very powerful method of calculation for the transmittance and near-field intensity of the incident light. All we need in this method is the monolayer scattering data obtained by the vector KKR method. There is nothing special in treating a metallic PC other than that we employ the Drude form (8.6) for $\varepsilon(\omega)$.

The transmittance in the frequency region of the band $\ell = 3$ is given in Fig. 8.10 for a slab of fcc PC with N, the number of stacked layers, equal to 8. Because of the nonmonotonic behavior shown in Fig. 8.9 (right), two modes (maybe, more than two) can be excited at a time when a horizontal line at one frequency intersects twice or more the dispersion curves. We find there are very sharp resonant dips the right panel. Judging from their marked asymmetries, we may presume that the effect is caused by the excitation of a series of quantized levels in the sense of Sect. 3.4, which are embedded in the continuum of the band states of smaller confinement, i.e., a kind of Fano resonance is the origin of the asymmetric and sharp resonant features.

Figure 8.11 shows the frequency dependence of the local field excited at the top of the spheres at the first layer. We find very good correlation between the frequency of the remarkable enhancement with the BS of optically active PBs. It is notable that the field intensity just at the resonance is as large as 10^4, relative to the amplitude of the incident light. This is about two factors larger than the enhancement expected in the Mie resonance of the dielectric spheres (see, e.g., Fig. 4.16).

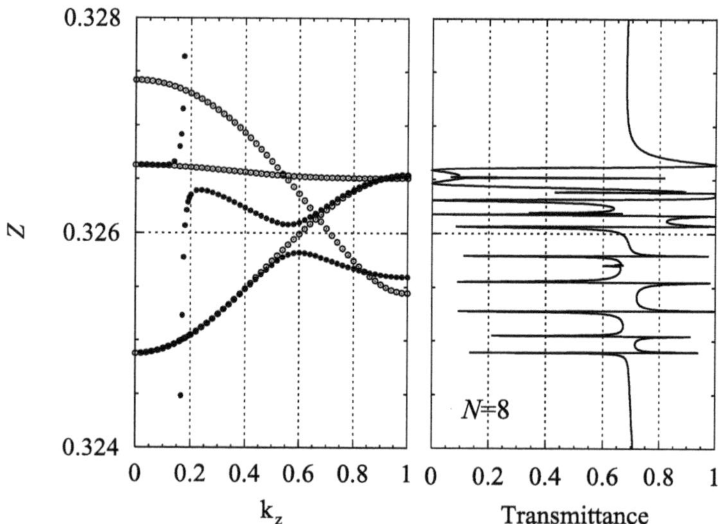

Fig. 8.10. Transmittance involving the nonmonotonic band of $\ell = 3$. The BS is a reproduction of Fig. 8.9, and the transmittance is calculated by the layer-doubling method for the $N = 8$ stacked layer

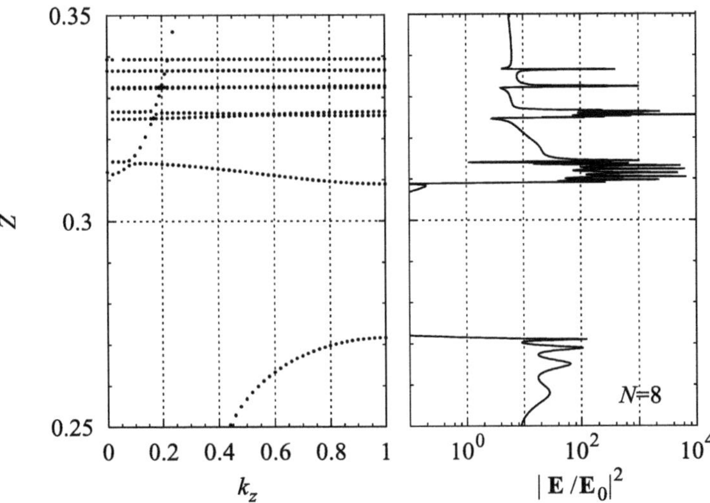

Fig. 8.11. Near-field intensity versus band structure. The intensity of the near field at the top position of a sphere in the first layer is plotted. The result is for a slab with $N = 8$

Since the Raman signal of the adsorbed molecule is related to the enhancement of the probe light at the molecular position, the huge enhancement of the local field shows that the surface-enhanced Raman scattering is very promising in a PC of metal spheres.

8.4.3 Effect of Absorption on the Local Field Enhancement

So far we found that the plasmon resonance gives a large band gap effect in the optical response. Since the dielectric function of free electrons must involve an imaginary part, it is important to see how the absorption effect modifies the results of the real dielectric function. We examine this effect on the enhancement of the local field in this section. The effect of the absorption on T was investigated in detail by Yannopapas et al. [34] and El-Kady et al. [38].

Figure 8.12 shows the enhancement of the local field at the top of the spheres of the surface layer. The parameters in the Drude form of the dielectric function is the same as used to obtain Fig. 8.11 except that the form

$$\varepsilon(\omega) = 1 - \frac{\omega_\mathrm{p}^2}{\omega(\omega + \mathrm{i}\gamma)} \tag{8.8}$$

is used for the metallic ε. The result of $\gamma = \omega_\mathrm{p}/50$ shows that the absorption effect influences the magnitude of the near-field very profoundly. It is therefore crucial to use a practical dielectric constant to obtain a quantitatively reliable results.

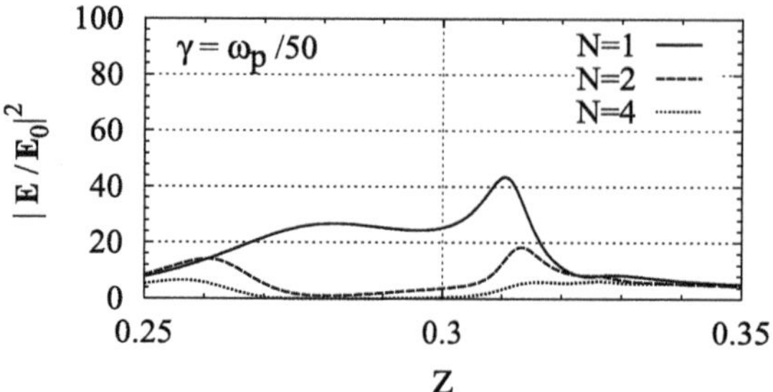

Fig. 8.12. Near-field intensity when absorption is taken into account. The intensity is calculated for three values of N at the top position of a sphere of a fcc PC. The effect of the imaginary part γ in the Drude form is quite remarkable. The case of $\gamma = 0.02\omega_{\mathrm{p}}$ is plotted

In summary, we have seen the characteristic features of PBs of metallic PCs. We have analyzed a fcc array of metallic spheres using the Drude form of the metallic dielectric constant. We find that sphere plasmons give rise to a band structure of nontrivial dispersion curves and T and the intensity of the induced near-fields can be analyzed in terms of the calculated BS. It is also shown that the absorption effect remarkably affects the quality of the resonant enhancement of the near-field intensities.

References

1. D. R. Hofstadter: Phys. Rev. B **14**, 2239 (1976)
2. R. Shimada, T. Koda, T. Ueta and K. Ohtaka: J. Phys. Soc. Jpn. **67**, 3414 (1998)
3. R. Shimada, T. Koda, T. Ueta and K. Ohtaka: J. Appl. Phys. **90**, 3905 (2001)
4. K. Ohtaka and Y. Tanabe: J. Phys. Soc. Jpn. **65**, 2265 (1996)
5. K. Ohtaka and Y. Tanabe: J. Phys. Soc. Jpn. **65**, 2276 (1996)
6. K. Ohtaka and Y. Tanabe: J. Phys. Soc. Jpn. **65**, 2670 (1996)
7. K. Ohtaka: Phys. Rev. B **19**, 5057 (1979)
8. N. Stefanou, V. Karathanos and A. Modinos: J. Phys. Condens. Matter **4**, 7389 (1992)
9. Xindong Wang, X.-G. Zhang, Qingliang Yu and B. N. Harmon: Phys. Rev. B **47**, 4161 (1993)
10. T. Suzuki and P. K. L. Yu: J. Opt. Soc. Am. B **12**, 570 (1995)
11. K. Ohtaka, H. Miyazaki and T. Ueta: Mater. Sci. Eng. **B48**, 153 (1997)
12. E. R. Brown and O. B. McMahon: Appl. Phys. Lett. **67**, 2138 (1995)
13. I. I. Tarhan, M. T. Zinkin and J. H. Watson: Opt. Lett. **20** 1571 (1995); Phys. Lett. **68**, 3506 (1996)
14. I. I. Tarhan and G. H. Watson: Phys. Rev. Lett. **76**, 315 (1996)
15. T. Fujimura, K. Edamatsu, T. Itoh, R.Shimada, A. Imada, T. Koda, N. Chiba, H. Muramatsu, and T. Ataka: Opt. Lett. **22**, 489 (1997)
16. K. Ohtaka, Y. Suda, S. Nagano, T. Ueta, A. Imada, T. Koda, J. S. Bae, K. Mizuno, S. Yano, and Y. Segawa: Phys. Rev. B **61**, 5267 (2000)

17. H. Miyazaki, Y. Jimba, C. Y. Kim, and T. Watanabe: J. Phys. Soc. Jpn. **65**, 3842 (1996)
18. Y. Jiang, C. Niu, and D. L. Lin: Phys. Rev. B **59**, 9981 (1998)
19. S. Yano, Y. Segawa, J. S. Bae, K. Mizuno, H. Miyazaki, and K. Ohtaka: Phys. Rev. B **63**, 153316 (2001)
20. E. Lidorikis, M. M. Sigalas, E. N. Economou and C. M. Soukoulis: Phys. Rev. B **61**, 13458 (2000)
21. Y. S. Chan, C. T. Chan, and Z. Y.Liu: Phys. Rev. Lett. **80**, 956 (1998)
22. S. S. M. Cheng, L. -M. Li, C. T. Chan and Z. Q. Zhang: Phys. Rev. B **59**, 4091 (1999)
23. C. Jin, B. Cheng, B. Man, Z. Li, D. Zhang, S. Ban and B. Sun: Appl. Phys. Lett. **75**, 1848 (1999)
24. M. E. Zoorob, M. D. B. Charlton, G. J. Parker, J. J. Baumberg and M. C. Netti: Nature **404**, 740 (2000)
25. C. Jin, B. Cheng, B. Man, Z. Li and D. Zhang: Phys. Rev. B **61**, 10762 (2000)
26. X. Zhang, Z. -Q. Zhang and C. T. Chan: Phys. Rev. B. **63**, 081105(R) (2001)
27. M. Bayindir, E. Cubukcu, I. Bulu and E. Ozbay: Phys. Rev. B **63**, 161104(R) (2001)
28. C. F. Bohren and D. R. Huffman: *Absorption and Scattering of Light by Small Particles.* (John Wiley, New York, 1983), Chap. 4.
29. M. M. Sigalas, C. M. Soukoulis, C. -T. Chan and D. Turner: Phys. Rev. B **53**, 8340 (1996)
30. E. Lidorikis, M. M. Sigalas, E. N. Economou, C. M. Soukoulis; Phys. Rev. Lett. **81**, 1405 (1998)
31. M. Hase, H. Miyazaki, M. Egashira, N. Shinya, K. M. Kojima and S. Uchida: Phys. Rev. B **66**, 214205 (2002)
32. H. Miyazaki, M. Hase, H. T. Miyazaki, Y. Kurokawa and S. Shinya: Phys. Rev. B (in print).
33. A. Moroz: Phys. Rev. Lett. **83**, 5274 (1999); Phys. Rev. B**66**, 115109 (2000)
34. V. Yannopapas, A. Modinos and N. Stefanou: Phys. Rev. B**60**, 5359 (1999)
35. Zhenlin Wang, C. T. Chan, Weiyi Zhang, Naiben Ming and Ping Sheng: Phys. Rev. B**64**, 113108 (2001); Phys. Rev. B**66**, 115109 (2000)
36. V. Kuzmiak, A. A. Maradudin and F. Pincemin: Phys. Rev. B**50**, 16835 (1994)
37. K. Sakoda: *Optical Properties of Photonic Crystals*, (Springer, Berlin Heidelberg New York 2001) pp 151–176
38. I. El-Kady, M. M. Sigalas, R. Biswas, K. M. Ho and C. M. Soukoulis: Phys. Rev. B**62**, 15299 (2000)

9 Spectroscopic Methods for Characterization

K. Inoue

The unique band structure of a PC can, in principle, be obtained by calculation, as has been described so far. Any kind of optical property is governed through the DOS by PBS. In this chapter, we explain how to experimentally examine the optical properties. First of all, how to experimentally confirm the calculated PBS is a problem. We first describe this, which is followed by a discussion. Any PC sample is artificially fabricated, so it is subject to fluctuations of the parameters such as the lattice constant. So, it is remarked that this kind of inhomogeneity causes uncertainty in the energy position of the band or BG to be determined, while the calculated band or band gap is free from this kind of uncertainty, but still involves a different kind of uncertainty. Hence the importance of the measurement.

9.1 How to Characterize a Sample

The most typical method for characterzing a PC sample or almost equivalently, for knowing experimentally the PBS, is to measure either the transmission or reflectance spectrum, or hopefully, both. The transmittance (T) spectrum is also important in another sense. Namely, it also provides us with information about the coupling strength or coefficient between an individual band, more exactly, a particular energy position of a band and an external light wave; the coupling strength depends not only on the symmetry of the band, but also on the group velocity (v_g). So long as the PC sample is not extremely small, there is no essential problem with the spectroscopic methods used for this purpose. However, the situation in a thin PC slab differs completely from those normal samples; until very recently, it has been almost impossible for one to observe such a spectrum over a broad near-infrared spectral range, which is badly needed not only for characterizing the optical properties, but also for developing devices in relation to optical telecommunications. In view of the importance of PC slabs, as will be described later, we describe in detail how to unambiguously observe such a transmission spectrum; it is remarked that roughly speaking, more than a third of all papers published up until now are related to 2D PC slabs.

Next, the propagation characteristics of an ultrashort light pulse are also important for two major reasons. One is that we need to directly estimate the

v_g-value as well as its dispersion (GVD), and the other, to get information about the spectral shape of the transmitted pulse, which is crucially important in developing an ultrafast and compact optical integrated circuit in the future. As for the latter, only a few experiments have been done up until now, though.

9.2 Spectroscopy in the General Case

As described above, if a PC sample is not small, it is not difficult to observe both transmission and reflectance (R) spectra. Namely, in the case of a sample with flat incident and exit surfaces much larger than the relevant wavelength in air, a commercially available spectrometer can be applied. Generally speaking, the R-measurement is easier and better to get information about the energy position of a stop band or a coupled band in that direction. Unlike the R-spectrum, it may happen that it is impossible to observe the transmission spectrum in some cases. It is well known that fluctuations involved in a PC sample cause serious scattering loss, giving rise to a large attenuation of the transmitted light, and therefore, one can observe the spectrum only for a sample of good quality. Of course, this is not the case with the R-spectrum. In the latter case, R is unity in a range corresponding to the stop band in the propagation direction, and it becomes small whenever external light can go inside a sample, i.e., at a coupled band.

Transmission- and R-spectrum thus obtained by varying the incident angle should provide, in principle, information about the PBS under study. For example, the energy range corresponding to a PBG, i.e., the energy width as well as the central energy position, shift as a function of incident angle. However, strictly speaking, in the case of non-normal incidence a problem arises in correctly determining the energy position of a particular band. That is, since Snell's law does not hold for a PC in the normal sense, we cannot identify the propagation direction inside the PC without information about the PBS. Furthermore, the situation becomes more complicated whenever more than two modes coexist with the same energy.

Although the R-spectrum is very useful in determining the PBS, it does not necessarily serve for quantitatively estimating either T or the coupling efficiency between an external light wave and the PB. This is because some factors involved for this estimation are difficult to extract; for example, it is impossible to estimate the scattering loss from the R-spectrum.

Here we would like to point out that care must be taken of the correspondence between the observed R or T, and a BG or a band. Namely, high R or low T does not necessarily correspond to a stop band or a band gap [1], because it is also possible to correspond to an uncoupled band of odd symmetry. In 2D and 3D PCs of high symmetry there are many such bands that cannot couple, by symmetry, to external light [2, 3].

In some cases, emission spectroscopy may also be useful. It is not difficult to observe an emission spectrum from a light source embedded in a PC sample. From a comparison of emission efficiency as a function of wavelength between the cases with and without a PC, information about the PBS can be extracted.

9.3 Spectroscopy of a PC Slab

9.3.1 Transmission and Reflection Spectroscopy

There are two major reasons why measurement on a 1D or 2D PC slab sample is so difficult. First, the sample is so thin that the incident surface for external light becomes extremely small in size, typically $0.250 \times 1.0\,\mu m^2$, as is the case where a PC part is sandwiched between a pair of ridge type waveguides for the input and output signal or the case with single-line-defect PC-slab waveguides which will be described later. Notice that the thickness of the slab is much smaller than the wavelength of light in air. Consequently, we need to employ a powerful and wavelength-tunable light source suited for this purpose. Second, currently, scientists are mainly interested in a PC sample showing a marked feature in the near-infrared region; we have already explained the reason thus far. Unfortunately, however, there are no such laser sources available in this region. In short, until recently it had been thought that it was next to impossible to observe a transmission spectrum over the broad near-infrared wavelength region, as far as a PC slab was concerned. Nowadays, it turns out that this is not necessarily the case. We explain two methods separately below.

Laser-Based Method. In the near-infrared region, two kinds of wavelength-tunable cw lasers are available at present. One is the Ti-sapphire laser that covers the range from 800 nm to 1070 nm in wavelength. The other is laser diodes (LDs). As for the latter, combining three or four such LDs with different central wavelengths, one can cover a range from 1280 nm to 1650 nm. Figure 9.1 shows a typical experimental setup for observing the transmission spectrum in a slab sample, by using the former laser (Ti-sapphire). There are two ways to couple an external light wave from such a laser to a thin sample, i.e., one is to use an optical fiber, and the other, a lens with a large numerical aperture (NA), e.g., of 0.5 in the near-infrared region. Here, we explain how to make optical alignment in the former case, which is rather easier than the latter [4, 5].

First, we need to prepare a pair of optical fibers such that single-mode use can be guaranteed in the concerned wavelength region with polarization maintained (the so-called polarization-preserving fiber). Furthermore, the top of one side is shaped to be a lens by polishing [5], called a lensed fiber. First, by using a fiber coupler we introduce laser light into the other side of such

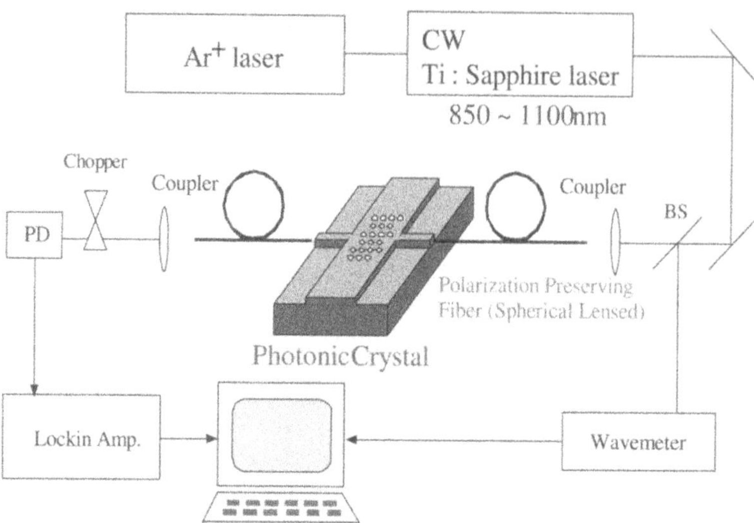

Fig. 9.1. A schematic drawing of the experimental setup for observing the transmission spectrum in a PC slab

an optical fiber. Then, light appearing on the other side is carefully focused onto the cleaved edge of the sample, i.e., either the edge of the stripe (ridge) waveguide connected to a PC sample, or the edge of a PC otherwise. The signal light appearing at the output edge after being transmitted through the sample is picked up by using another identical fiber, and detected by a detector, i.e., either an infrared detector such as a Ge detector, or an optical analyzer. By scanning the wavelength of the laser light, a transmission spectrum can be observed. The T-spectrum can be obtained in such a way that the transmission spectrum is normalized by the T-spectrum observed for a reference of an otherwise nominally identical structure without a PC. In the LD case, the method is essentially the same as the above.

Next, we describe how to get information about reflectance from the interface between a PC and the external material, including the case of air. As already described, it is, generally speaking, difficult to estimate correctly the R-value in the slab case. However, this information becomes very important in some cases. This is because information about T alone is not sufficient to distinguish a stop band or an uncoupled band from a coupled band.

Suppose, for example, very low T is observed over a wavelength range. The observed range does not necessarily correspond to a stop band, because T may also be very low due to radiation loss in the case of a leaky band. In this case, if R is high, then the low T range arises from either a stop band or an odd (uncoupled) band, while if R is not high, the range corresponds to a leaky band.

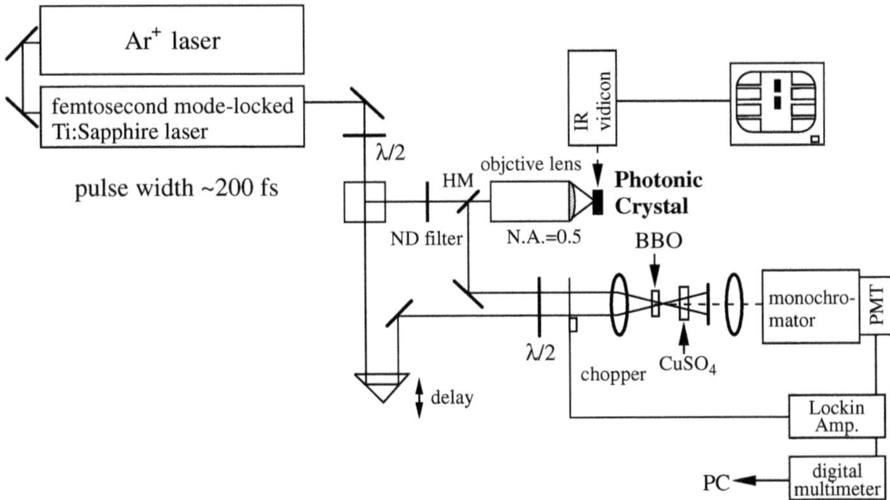

Fig. 9.2. Experimental setup for observing reflectance or the reflectance spectrum in a PC slab

Let us adopt as an example the case of a PC sample equipped with ridge waveguides. The reflectance from the interface between a PC and the outside material or the ridge waveguide can be estimated unambiguously with use of the time-of-flight method using an ultrashort light pulse [6]. The experimental setup is presented in Fig. 9.2. The principle of this method relies on the fact that the signal pulses reflected from the respective interfaces can be identified, since these appear at different times. Figure 9.3 shows how muptiple delayed pulses appear together with their origins. Here, in order to obtain the

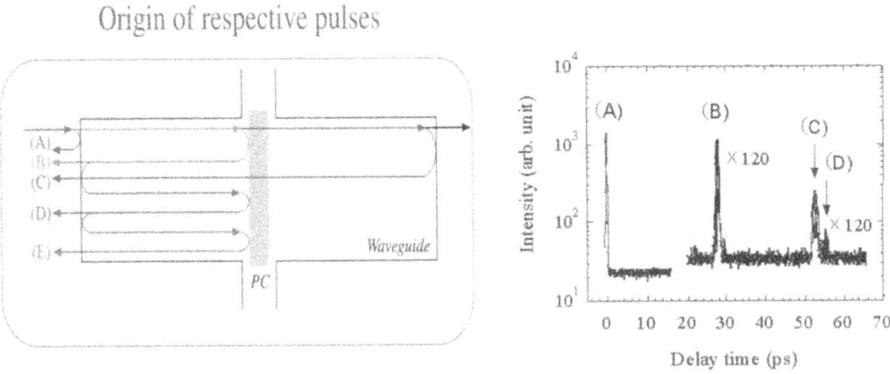

Fig. 9.3. *Left:* schematic explaining multiple pulses appearing with the respective delay times. *Right:* an example of such a spectrum of multiple pulses in the time domain, quoted from [6]

delayed signal pulses, we employ a cross-correlation method between these and the original one, utilizing the second-harmonic generation (SHG) signal that emerges only when these two pulses overlap in the time domain. One of the nice things about this method is that not only the reflectance in question, but also T can also be estimated, in principle, by comparing the respective intensities of the multiple-reflected signal pulses [7]; consequently, propagation loss such as radiative or out-of-plane loss may also be estimated in priciple.

Examples of such a reflected spectrum in the time domain are presented in Fig. 9.4. The sample used is the heterostructure (SC) type of PC slab as explained in Chap. 6; the sample parameters of a, f and N are described in the figure. The spectra (a) and (b) were obtained at the respective wavelengths marked by an arrow in the observed T-spectrum, which correspond to the fifth band and the second-lowest stopband in the model band structure (not shown), respectively. From comparison of the relative intensities of the pulses, R and T at the respective wavelengths are estimated as around 0.15 and 0.01 for the former, whereas more than 0.5 and less than 0.01 for the latter, respectively. These values indicate that loss of the present PC sample

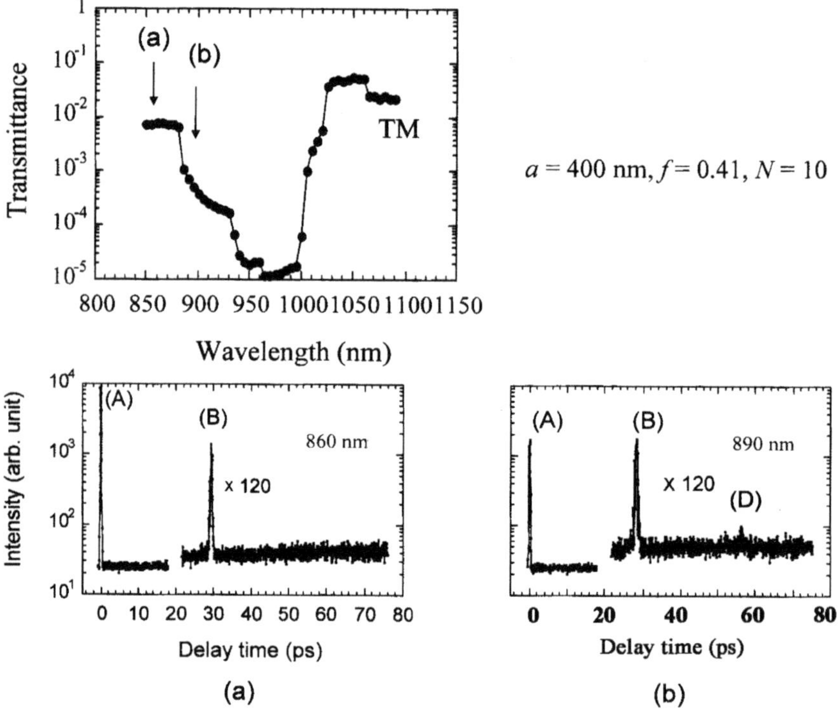

Fig. 9.4. An example of the reflection pulse in a SC sample at a wavelength corresponding to a coupled band (**a**) and the band gap (**b**) in the TM-polarized transmittance spectrum shown above

amounts to 0.8 for the former, and is smaller than 0.5 for the latter, which are both reasonable. In this connection, it is noted that for the former the wavelength corresponds to the energy position inside the light cone, and for the latter, the R-value is shown by calculation not to be high enough even in the stop band in the case of a SC type of PC slab [8].

Halogen-Lamp-Based Method. Recently, we have developed a novel spectroscopic method using a white light as a source, and a monochromator as well, which turns out to work very well for characterizing a PC slab sample [9–11]. We explain this method below in a bit more detail. In Fig. 9.5 is shown the experimental setup for observing the transmission spectrum for both a bulk PC slab and PC waveguides (PCWs) in the range from 850 to 1600 nm. A commercially available white light source (AQ-4303B, Ando Electric Comp. Ltd.), which comprises a halogen lamp (50 W), a condenser lens, and optical filters, is employed as the present source. Output light ranging from 400 to 1800 nm can be extracted through a standard optical connecter (FC) of ferrule type for an optical fiber.

In the present experiment, two types of polarization-preserving optical fibers were used, i.e., both for single-mode use, one above 800 nm and the other above 1300 nm, respectively. One end of the fiber was equipped with a ferrule to couple to the source, and the other end was formed into a spherical lens by polishing. To fix the polarization, a special polarizer made from metal particles (Melles-Greo Ltd. Co.,) is inserted into the light source. Thus, light from a halogen lamp via the fiber was coupled to a PC sample through the input ridge waveguide, the core of which is typically 250 nm in thickness and 0.6–2.4 μm in width. In the case of a sample without ridge waveguides, light from the fiber is directly focused onto the cleaved edge of the sample.

The light signal from the output waveguide or directly from a PC sample was collected by using another identical fiber, and the light coming out

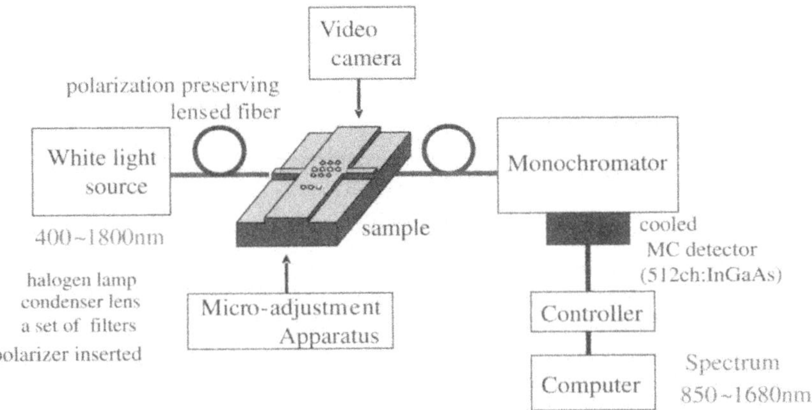

Fig. 9.5. Experimental setup for observing the transmission spectrum in a PC slab

from a ferrule on the other side was introduced into a monochromator. The monochromator employed is specially designed with an F-number of 4.2, so that one can easily select one of three gratings mounted at the same time by electrically rotating the mount-holder. As a photo-detector we employed a 1D multichannel type of InGaAs infrared detector consisting of 512 pixels (OMA-V; InGaAs, Princeton Instruments, Inc.); the size of each pixel is 50 μm in width and 500 μm in length, so that the total dimension is 25.6×0.5 mm^2. The detector, mounted in a metal dewar to cool down to $-100°$C, was set just behind the position where the exit slit of the monochromator was usually located, after the slit was removed.

The InGaAs chip has sensitivity from 850 to 1680 nm, and the quantum efficiency is almost flat in the wavelength range between 1000 and 1600 nm, being 90%. As a grating we employed a custom-ordered grating of 56×56 mm^2 in size, characteristic of 150 grooves/mm and having a braze-wavelength at 1250 nm. By using this grating, we can observe at one time a spectrum over the broad range of 600 nm. Depending on the case, we have also used another grating with 300 grooves/mm, which enables us to observe a spectrum with twice better resolution.

It is crucially important to precisely set the fiber in front of the sample for input, or behind it for the output. A special apparatus (Advantest Corp. Ltd.) was employed to attain this purpose. The apparatus is also equipped with two video cameras to monitor images of the sample and fibers, one from

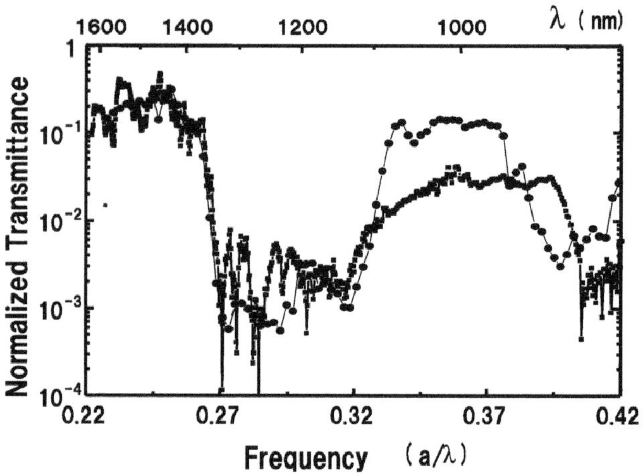

Fig. 9.6. An example of the TE-polarized T-spectrum in the $\Gamma - K$ direction for an air-bridge PC slab sample observed using the spectrometer shown in Fig. 9.5, revealing that a broad non-transmission region from 1070 to 1370 nm is observed due to the lowest-bandgap; notice that a comparison of the observed spectrum (*thick line*) with the theoretical one (*thin line*) calculated by using the 3D FDTD method reveals good correspondence. A sample of $d = 250$ nm, $a = 360$ nm, $R = 230$ nm, and $N = 10$ with a pair of ridge waveguides was used

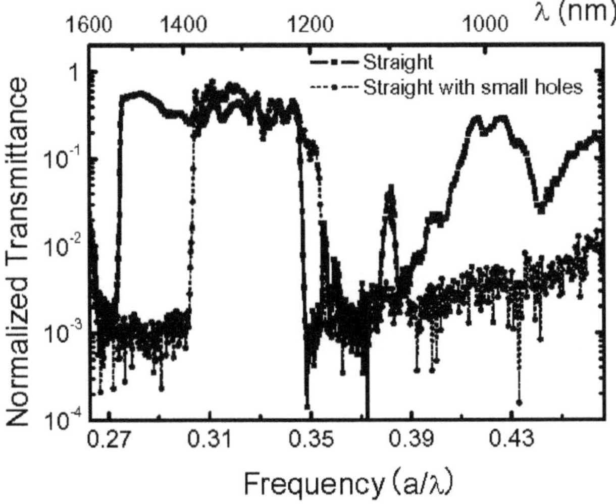

Fig. 9.7. An example of the transmission spectrum observed using the spectrometer shown in Fig. 9.5. Two kinds of single-line-defect (called W1) PC-slab (air-hole type) straight waveguides sandwiched with a pair of ridge waveguides are used for comparison; one is the normal type (*thick line*) with single-line of missing air-holes, and the other the type (*thin line*) with single line of air-holes replaced by smaller air-holes with the diameter $R' = 154\,\text{nm}$. The same parameters of $d = 250\,\text{nm}$, $a = 420\,\text{nm}$, $R = 230\,\text{nm}$, and $N = 50$ are used between the two types as the bulk PC slab. It is noted that the line-defect guided modes (below the air light-line) and quasi-guided modes (above the light-line) are responsible for a high transmission region observed from 1210 to 1530 nm for the normal type. The coresponding region for the small-hole type is found to shift to the shorter-wavelength side with the reduced spectral-width. All these features are consistent with the calculated result

the top and the other from the back side. An infrared video camera attached to a video-microscope unit has enabled us to see the shape of the sample from the top or to monitor the relative position between the tip of one fiber and the input edge. At the same time we can also monitor the scattered light at the input and output edge, at the interfaces between the waveguides, and from the PC itself.

Examples of the transmission spectra observed with use of this spectrometer are shown in Figs. 9.6 and 9.7. These spectra were taken with use of a grating of 150 grooves/mm, which covers a spectrum range of 600 nm at once, with the spectral resolution power intentionally low, such as 1.5 nm. It is, however, straightforward to improve, if necessary, the resolution, since all we have to do is to change the grating, i.e., to use a grating with better resolution, with the sacrifice of the wavelength range covered at one time. The accumulation time needed to observe these spectra is typically 1 min. The way to normalize the transmission spectrum is the same as that for the laser-based method. Finally, we make the comment that the reproducibility of the spectrum thus observed is very good, as it should be, because we employ a multichannnel detector.

9.3.2 Near-Field Spectroscopy

One of the nice things with a PC slab is that the near-field spectroscopy can apply also to this sample, unlike other samples. Pioneering work has already been successfully done in the literature [12, 13]. There is no doubt that in view of the importance of PC slabs, this spectroscopy will be more frequently used in future. The principle is rather simple. When a guided mode is excited by light source, one can get information about the distribution of the electric field strength along the surface of the slab, by shifting the tip, of extremely small diameter, of the special optical fiber, which is placed as close as possible to the surface. For example, we can get direct information on the pattern of a standing wave inside a sample for a band with a finite length, which should vary with changing the wavelength. Another example is to know how light propagates through, or is reflected from, the corner in the case of the PCW.

This spectroscopy can also be used to illuminate light from an optical fiber perpendicularly on the surface of a slab, and to detect light emerging from it by using the same fiber. White light is used for obtaining the reflection spectroscopy, while a laser with an appropriate wavelength is employed for observing the emission spectrum. For example, one can get information about the resonant frequency as well as the Q-value of a microcavity embedded in a PC waveguide. Therefore, this method is very powerful for designing such a cavity.

9.3.3 Resonant-Mode Spectroscopy

In some cases we also need to get information about the leaky (resonant) bands. It is not necessarily difficult to do so, since the leaky mode can couple to an extended mode in air. Suppose that white light with a broad wavelength range is incident on the upper surface of a slab with the incident angle fixed. At a particular wavelength for which the boundary condition is satisfied, the reflection spectrum shows a dip, or an anomaly, since light can couple resonantly to the leaky mode. The spectral width of the dip provides loss of the mode; many such anomalies appear within a broad wavelength range. By taking such a spectrum by changing the incident angle, we can depict the respective dispersions. Note that information about to what extent the Q-value depends on an individual band, or on the BZ position of the band, is very useful. An example has already been presented in detail in Part II of Chap. 6, where the sample is a monolayer of a 2D array of spheres [14]. This method, however, cannot apply to the guided mode.

9.4 Time-of-Flight Spectroscopy Using an Ultrafast Pulse

There are a few reasons why we need to perform propagation experiments with ultrafast light pulses. First, experiments can give information about to what extent the shape of such a pulse changes when passing through a PC sample, since the GVD is generally very large in a PC; this information is important when one develops an ultrafast and miniaturized optical circuit based on PCs, which rely on ultrafast pulse propagation. Second, it is necessary to get direct information about v_g from a time-of-flight (TOF) measurement of the pulse. In this connection, it is pointed out that importantly, v_g for a finite sample is not necessarily given by the slope of the band calculated usually for an infinitely long sample, so, direct measurement of v_g is inevitable. Furthermore, as already noted in the preceding chapters such as Chap. 2, that v_g is small is very important, since one can utilize the frequency region of low v_g to greatly enhance the signal intensity in a PC, by using external light.

One of the most typical TOF methods is shown in Fig. 9.8, which is essentially the same as Fig. 9.2. Here again, we employ a cross-correlation method (frequency up-conversion) for observing the pulse shape in the time-domain [15]. Namely, the intensity of the second-harmonic generation (SHG) signal arising as a result of overlapping in time of two incoming pulses, the signal and reference pulses, in the phase-matching direction is detected as a function of the time delay of the latter pulse. At the same time the spec-

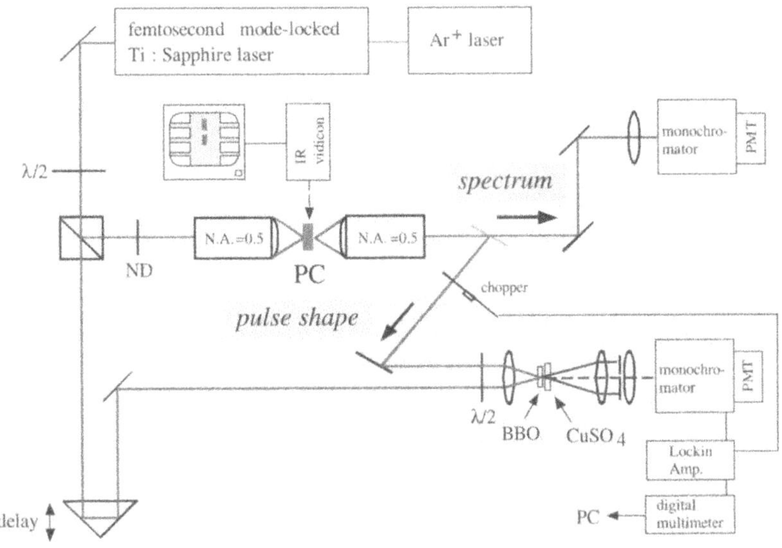

Fig. 9.8. Experimental setup for measureing the group velocity through the time-of-flight method

tral shape of the transmitted pulse is also observed in the path described as the *spectrum*. This is because the shape contains information about the transmission spectrum, which should also help to deduce the GVD.

Generally speaking, it is rather easy to perform such an experiment in the optical range, if the PC sample has a large input surface. However, it is not necessarily easy to do this kind of experiment in the near-infrared range, where currently most PC samples are fabricated in relation to future telecommunications. This is particularly true with a PC slab sample. The experimental setup shown in Fig. 9.8 is intended for such a sample in the near-infrared range. For this purpose, 200-fs pulses from a mode-locked Ti-sapphire laser operated at a rate of 76 MHz were employed as a light source, for which the wavelength can be varied from 680 to 1000 nm; it is also possible to utilize, in the region around 1550 nm, a wavelength-tunable LD with an ultrashort output pulse, e.g., of 2-20 ps, but the tunable range is not wide enough, typically only 60 nm. The spectral width of the pulse is typically 15 nm around 900 nm where the measurement was made.

The pulsed beam was focused on the cleaved edge-plane of the input waveguide by using a lens with numerical aperture (NA) of 0.5, and the output pulse, being collected with another identical lens, was measured in the way described above. The time delay is observed relative to a reference sample that is nominally identical but without a PC [6]. It is noted that in the case of a subpicosecond pulse, the spectral shape of the output pulse in the frequency domain changes from that of the incident pulse, as shown in Fig. 9.9, which indicates that the shape in the time domain is also more or less distorted. This is because T usually changes rapidly with wavelength,

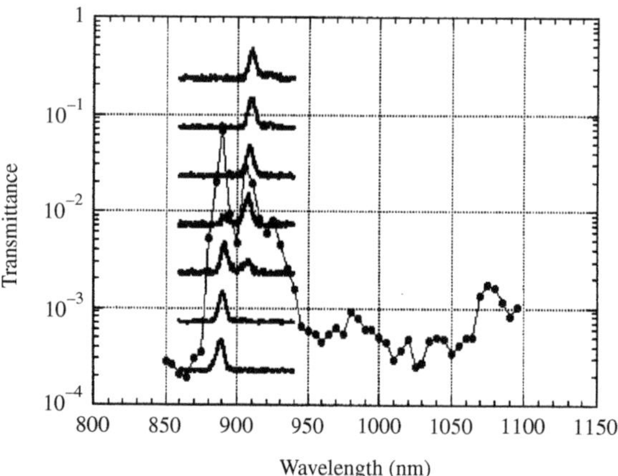

Fig. 9.9. Variation of the spectral shape of pulse with the central wavelength, which is transmitted through a PC slab with a pair of long stripe waveguides; the closed circles refer to the observed T-spectrum, quoted from[15]

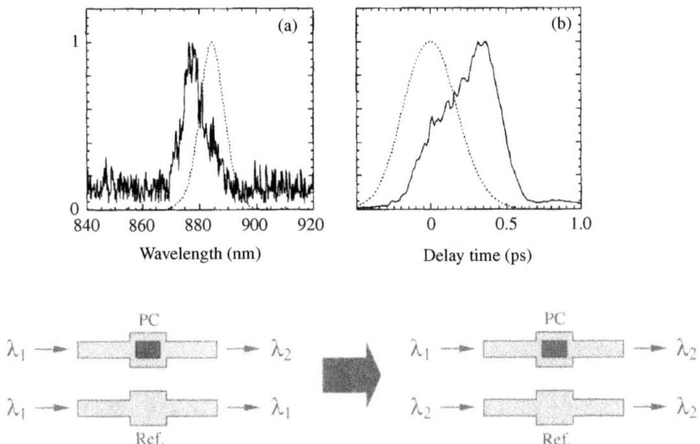

Fig. 9.10. *Lower:* a schematic explaining how to estimate the delay time of the ultra-short pulse passing through a PC slab sample in comparison to that through a reference one without a PC; for more details including (**a**), see the text. (**b**) An example of these pulses is shown in the case where the center wavelengths of these output pulses are adjusted to be the same; the signal and reference pulses are shown by solid and dotted lines, respectively

including the band edge region; a large GVD also causes distortion of the pulse passing through the PC sample. So, it is also possible that the spectral shape in the frequency domain differs between the pulses passing through the samples with and without a PC, as shown in the upper (a) of Fig. 9.10. In this case, as schematically shown in the lower part of the figure, the relative delay time has been estimated in such a way that the central wavelength of

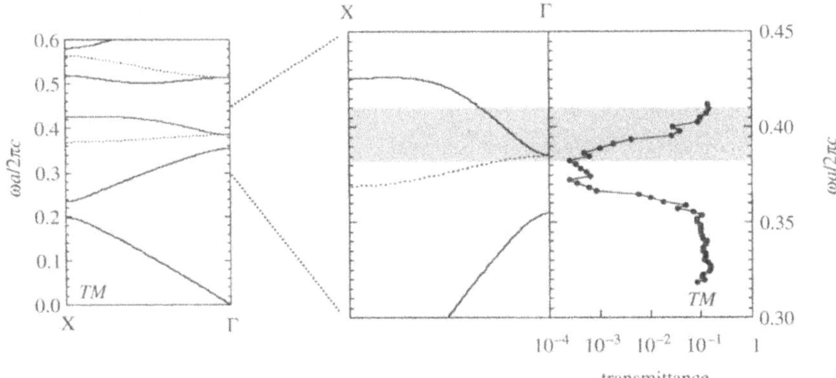

Fig. 9.11. The band (hatched area) used for subpicosecond pulse propagation measurement. The sample is a SC-type PC slab with $N = 10$ with a pair of long stripe waveguides

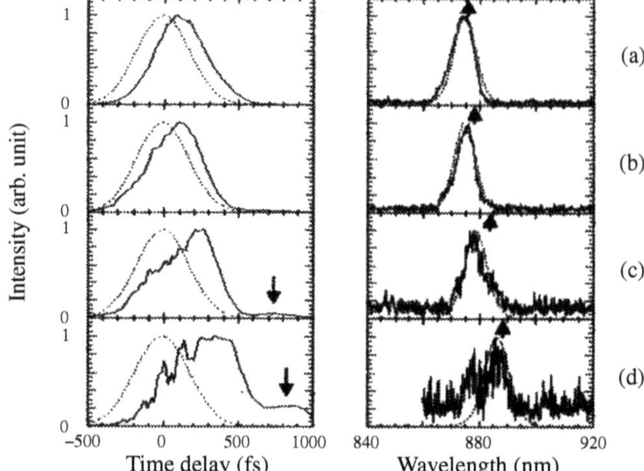

Fig. 9.12. The pulses (left) transmitted through the SC-type samples shown in Fig. 9.11 with (solid line) and without (dotted line) a PC as a function of delay time, where the arrow in (c) and (d) indicates the leak pulse through the substrate below 850 nm. Their spectra (right) are also shown, where the arrow indicates the central wavelength of the incident pulse for a sample with a PC: (a) 877 nm, (b) 879 nm, (c) 884 nm, and (d) 888 nm; quoted from[15]

the reference output pulse is adjusted to be the same as that of the signal one. An example obtained in this way is shown in Fig. 9.10(b).

An example of the v_g-measurement performed for a SC-type of PC slab sample with $N = 10$ is presented below. The third-lowest coupled band in the Γ-$X(M)$ direction for the TM mode shown in Fig. 9.11 was used; this band is relatively flat, so we are very interested in this band. In Fig. 9.12 are presented examples of the spectra in both domains, i.e., the frequency and time domains, where the signal pulse (solid line) passing through the sample with the PC is seen to be delayed as compared to the reference pulse (dotted line), indicating that v_g is smaller in the PC than in the reference.

References

1. See, for example, M. Wada, Y. Doi, K. Inoue, and J. W. Haus, Phys. Rev. B**55**, 10443 (1997)
2. K. Sakoda, Phys. Rev. B**51**, 4672 (1995)
3. M. Wada, K. Sakoda, and K. Inoue, Phys. Rev. B**52**, 16297 (1995)
4. K. Inoue, N. Kawai, M. Sasada, N. Ikeda, N. Carlsson, K. Asakawa, S. Abe, S. Yamada, and Y. Katayama, Pacific Rim Conference on Lasers and Electro-Optics'99 (Seoul, 1999-8); *Summary of Papers* (IEEE, New Jersey, 1999) pp.563-564.
5. N. Kawai, K. Inoue, N Carlsson, N. Ikeda, Y. Sugimoto, K. Asakawa, and T. Takemori, Phys. Rev. Lett. **86**, 2289 (2001)

6. N. Kawai, K. Inoue, N. Ikeda, N. Carlsson, Y. Sugimoto, K. Asakawa, S. Yamada, and Y. Katayama, Phys. Rev. B**63**, 153313 (2001)
7. K. Inoue, N. Kawai, Y. Sugimoto, N. Carlsson, N. Ikeda, and K. Asakawa, *MRS Symp. Proc.* vol. 637, 2001, E.3.2.1
8. B. D'Urso, O. Painter, J. O'Brien, T. Tombrello, A. Yariv, and A. Scherer, J. Opt. Soc. Am. B **15**, 1155 (1998)
9. K. Inoue, *Internat'l Workshop on PECS IV*, *Abstracts* ; pp. 18 (2002)
10. Y. Sugimoto, Y. Tanaka, N. Ikeda, K. Kanamoto, Y. Nakamura, S. Ohkouchi, H. Nakamura, K. Inoue, H. Sasaki, Y. Watanabe, K. Ishida, H. Ishikawa, and K. Asakawa, IEICE TRANS. ELECTRON. **E87-C**, 316 (2004)
11. K. Inoue, Y. Sugimoto, N. Ikeda, Y. Tanaka, K. Asakawa, H. Sasaki, and K. Ishida, Jpn. J. Appl. Phys. **43**, L446 (2004)
12. S. I. Bozhevolnyi, V. S. Volkov, T. Sondergaard, A. Boltasseva, P. I. Borel, and M. Kristensen, Phys. Rev. B **66**, 235204 (2002)
13. K. Okamoto, M. Loncar, T. Yoshie, A. Scherer, Y. Qui, and P. Gogna, App. Phys. Lett. **82**, 1676 82003)
14. T. Kondo, M. Hangyo, S. Yamaguchi, S. Yano, Y. Segawa and K. Ohtaka, Phys. Rev. B**66**, 033111 (2002)
15. K. Inoue, N. Kawai, Y. Sugimoto, N. Carlsson, N. Ikeda, and K. Asakawa, Phys. Rev. B**65**, 121308 (2002)

10 Interaction Between Light and Matter in Photonic Crystals

K. Inoue, K. Ohtaka, and S. Noda

Needless to say, all phenomena related to light are based on interaction between light and matter. In this chapter we examine how the interaction will change in a PC and how the optical signal intensity is drastically enhanced in several physical phenomena. In Part I we describe and discuss mainly direct interaction between light and matter, with emphasis placed on experimental research related to the control of the radiation field in a PC. The light propagation in a PC may also be included in this chapter, but we have restricted ourselves here to a subject related closely to the radiation and postpone a variety of new or unique phenomena of light propagation until Chap. 11. In Part II, we treat the indirect interaction or an interaction via high-speed running charged particles, i.e., a PC version of Smith-Purcell radiation effect. Since the topics of this chapter are closely related to the PC devices described in Chap. 11, some of them will appear there once again.

Part I. Direct Interaction

10.1 Suppression and Enhancement of Emission

Spontaneous emission from an atom or molecule placed in a 3D PC is inhibited, if the photon energy of emission $\hbar\omega_0$ lies within an energy range of PBG; this fact motivated Yablonovitch to propose a PC in 1987 as such a peculiar system. This behavior of quantum electrodynamics (QED) is physically interpreted as arising from the fact that the probability $W(\omega_0)$ for the emission to occur following the atomic transition vanishes, if the density of states (DOS) of photons $\rho(\omega)$ is zero identically. Namely, according to Fermi's golden rule in quantum mechanics [1], $W(\omega_0)$ is expressed as

$$W(\omega_0)\mathrm{d}\omega = \frac{2\pi}{\hbar}|\langle f|H_{\mathrm{int}}|e\rangle|^2\rho(\omega_0)\mathrm{d}\omega, \qquad (10.1)$$

where $H_{\mathrm{int}} = \boldsymbol{p}\cdot\boldsymbol{E}$ refers to the interaction Hamiltonian that couples the atomic dipole moment $\boldsymbol{p}\,(\boldsymbol{p} = e\boldsymbol{r})$ to the electric field (electric dipole interaction), and $|\mathrm{e}\rangle$ and $|\mathrm{f}\rangle$ are the excited (initial) and ground (final) electronic states, respectively. Consequently, no matter how large the magnitude of $|\langle\mathrm{f}|H_{\mathrm{int}}|\mathrm{e}\rangle|$ may be, $W(\omega_0)$ vanishes whenever $\rho(\omega) = 0$ in a 3D PBG region, indicating that the spontaneous emission is inhibited there. Conversely, can the probability of spontaneous emission be enhanced at the band-edge frequency of a 3D PBG ? Unfortunately, there is not a straightforward answer to this basic question of QED, because the DOS and spatial dependence of the electric field are a bit complicated in a PC. In the case of a dipole located in a homegeneous system, for example, in vacuum, the squared matrix element of (1.1) and hence the probability $W(\omega)$ depend neither on the position $\boldsymbol{r}$ nor the orientation of the dipole (except for a familiar orientational dependence expressed by cosine function). In this simple case, therefore, the spontaneous emission probability A (Einstein's A constant) is proportional to a constant $\omega_0^3|\boldsymbol{p}_{\mathrm{ef}}|^2$ [1], where the quantity $|\boldsymbol{p}_{\mathrm{ef}}|$ is the magnitude of the dipole moment associated with the transition from $|\mathrm{e}\rangle$ to $|\mathrm{f}\rangle$ states. When the field is inhomogeneous like that in a PC, however, the magnitude of A, or the radiative lifetime τ ($\tau = A^{-1}$) is not simply an intrinsic constant inherent to the eigenstates of the atom, but can be altered, depending upon where and how the atom is positioned. This fact was first pointed out by Purcell in 1946 [2]. In a PC, $\rho(\omega)$ depends critically on ω, and H_{int} is expressed as

$$H_{\mathrm{int}} = \sum_{n,\boldsymbol{k}} \boldsymbol{p} \cdot \boldsymbol{E}_{n,\boldsymbol{k}}(\boldsymbol{r}) \tag{10.2}$$

in terms of the normal modes $\boldsymbol{E}_{n,\boldsymbol{k}}(\boldsymbol{r})$ for the PB state $(n, \boldsymbol{k})$ at the atom position $\boldsymbol{r}$. Consequently, $W(\omega_0)$, A and τ all depend on not only $\rho(\omega)$ but also $\boldsymbol{E}_{n,\boldsymbol{k}}(\boldsymbol{r})$. As a result, the intensity of spontaneous emission observed with a finite solid angle in a particular direction, which is usually the case experimentally, depends on the position and orientation of the dipole. So, instead of

$$\rho(\omega) = \sum_{n,\boldsymbol{k}} \delta(\omega - \omega_{n,\boldsymbol{k}}), \tag{10.3}$$

it is more convenient to introduce the local density of states $\rho(\omega, \boldsymbol{r})$ defined as [3],

$$\rho(\omega, \boldsymbol{r}) = \sum_{n,\boldsymbol{k}} |\boldsymbol{p} \cdot \boldsymbol{E}_{n,\boldsymbol{k}}(\boldsymbol{r})|^2 \delta(\omega - \omega_{n,\boldsymbol{k}}). \tag{10.4}$$

Using $\rho(\omega, \boldsymbol{r})$, A is given by $(2\pi/\hbar)\rho(\omega, \boldsymbol{r})$.

Now, we are in a position to discuss the QED problem of whether the probability A can be enhanced at the band edge of a 3D PC, as compared to that in a uniform system. Near the band edge of a PB, the dispersion relation of photon may be expressed generally as

$$\omega = \omega_c + a(\boldsymbol{k} - \boldsymbol{k}_c)^2,$$

where ω_c and $\boldsymbol{k}_c$ refer to the frequency and wavevector of the band edge, respectively. Therefore, $\rho(\omega)$ increases in proportion to $(\omega - \omega_c)^{1/2}$ or $(\omega_c - \omega)^{1/2}$ depending on the sign of the prefactor a and the divergence of $\rho(\omega)$ never takes place at a band edge. The same is true also with $\rho(\omega, \boldsymbol{r})$, no matter how and where the dipole may be placed in a PC. However, $\rho(\omega, \boldsymbol{r})$ becomes relatively large in a particular frequency region near the band edge, if an atom is positioned appropriately [4]. Concomitantly, an enhancement of the spontaneous emission intensity should result in a particular direction, not just at the band edge but somewhere near the band edge. It is interesting to consider here the case of 1D PCs. We expect the situation to be quite d ifferent from the 3D one. First, $\rho(\omega)$ itself exhibits a singular behavior expressed by $|\omega - \omega_c|^{-1/2}$. This behavior at the band edge, however, does not imply the divergence of the emission probability. Besides, considering $\rho(\omega, \boldsymbol{r})$ in place of $\rho(\omega)$, we must take into account the 3D nature involved in using a finite solid angle to collect photons for observation. Namely, the off-axis photon modes must be taken into account. A simple $\rho(\omega)$ or $\rho(\omega, \boldsymbol{r})$ of 1D PC is not sufficient to explain the observation. As the result, the divergent behavior at the band edge should never happen. The enhancement, therefore, should occur somewhere near the band edge, not just at the band edge, similarly to the 3D case discussed above. The same is true also with a 2D PBG, where $\rho(\omega)$ become typically constant when $\omega \sim \omega_0$. However, there may be an exceptional case where this typical 2D behaviour cannot be expected. For example, when the band dispersion is almost flat in a particuar direction of the BZ, the mixture between 1D and 2D behavior shows up near the band edge.

Now, we turn to the experimental aspects. Actually, not many works have been reported on this QED problem. We examined in 1996 the case for a 2D PC, adopting a sample of the capillary plate described in Sect. 5.1 [5]. We utilized the spontaneous emission from optically-pumped dye molecules, which were injected into the air holes. We controlled the amount of dye molecules injected carefully so that they might not affect the PBS. The emission intensity was observed in the Γ-M direction and the intensity variation with wavelength was measured by collecting the radiation within a solid angle as small as 6×10^{-6} rad. The experimental setup for this measurement is shown in Fig. 10.1. The dyes (styryl-13) were injected only to the air holes at the sample center, and pumped by an argon laser along the direction normal to the 2D plane.

In the E-polarized spectrum observed in the Γ-M direction as shown in Fig. 10.1, the emission intensity is clearly seen to be suppressed in an energy range around 930 nm. This wavelength region agrees well with that of the second-lowest stop band (SB) of this sample (not shown), indicating that the suppression is caused by this SB; for the H-polarized spectrum, on the other hand, any clear attenuation is not observed in Fig. 10.1. A dip around 890 nm is likely to be caused by a stop band.

Next, we focus our attention on whether or not the normalized intensity of emission is enhanced at the band edge; notice that the spectrum is normalized by the emission intensity observed in a similar way from the PbO glass (the same PC sample without dyes). However, no matter how small a solid angle

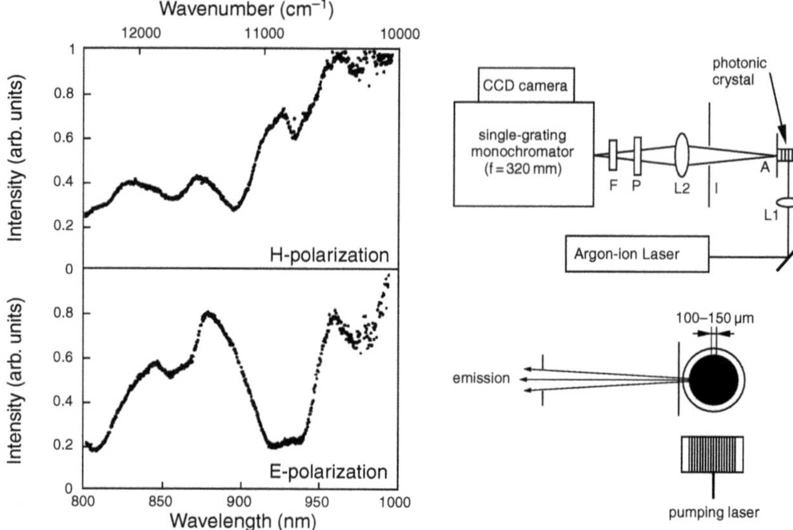

Fig. 10.1. The relative intensity (*left*) of spontaneous emission observed in the $\Gamma - M(X)$ direction within the 2D PC plane of a 2D capillary PC made of PbO glass, indicating the suppression due to the second stop gap. The emission arises from a small number of dye molecules embedded in the air holes that has been excited by an argon ion laser, as shown on the right. The solid angle for observation is set to be as small as 6×10^{-5} rad with use of an iris (I): L: lens, P: polarizer, F: filter, A: aperture

was adopted to observe the emission in the 2D plane, no enhancement was observed at the band edges. This result may be reasonable, because the out-of-plane (off-axis) modes join in the local DOS, as already explained.

Observation of enhancement at the band edge as well as suppression due to a band gap, of spontaneous emission is reported in a 1D PC in the literature [6]. Such suppression has also been observed in a particular direction in a 3D PC in [7]. Importantly, very recently Noda and his collaborators have observed the suppression effect of the emission in a few particular directions from an emitter placed in a 3D PC with a complete PBG, indicating a full suppression effect [8].

10.2 Extraction of Light from a Slab with High Efficiency

It has long been a problem as to how efficiently one can extract light from a small light source embedded in a thin slab. Theoretically, it has been proposed by Fan *etal.* that the extraction efficiency should be greatly improved by using a PC slab [9]. Suppose the case where the emission energy region is located above the air light-line. In this case, radiation emitted by such a source couples out-of-plane modes that possess a large component in free space, and therefore, goes into free space after propagating to some extent along the slab. They have demonstrated by simulation that this is indeed the case.

Experimentally, Baba *etal.* were the first to confirm this proposal in the optical region [10]. They measured the emission intensity from a light source embedded in an air-bridge type of PC slab made of InAs, and estimated experimentally the extraction efficiency by comparing the intensity with that in the homogeneous slab. As a result, they found that the efficiency was improved by a factor of more than 20. The result is shown in more detail in the next chapter.

10.3 Direct Observation of Small Group Velocity

In brief, control of the radiation field is based on a distinct feature of the DOS in a PC or, equivalently, the anomalous dispersion relation of light. Making use of the small group velocity v_g is very attractive as a tool for controlling the radiation field. For example, it causes a drastic effect in many physical phenomena described in the subsequent sections. In other words, that v_g is small corresponds generally to the large density of photon states, which indicates that the electric field strength is quite large. So, the effective interaction per unit length is regarded as being large there.

In this section, we describe a few measurements where v_g is directly observed in the time domain, i.e., by using the time-of-flight method of a 200-fs optical pulse [11, 12]. In Fig. 10.2 is shown the band used for this measurement, togther with examples of the spectra in the time domain. From such data, we can get valuable information about pulse-propagation characteristics. First, the shape of the 200-fs pulse transmitted through a PC is more or less distorted, so, we must be careful in using short pulse propagation in a PC such as the present one. Furthermore, special care must be taken to deduce a v_g-value from the time delay, since the pulse shape is no longer the same as that of the original pulse.

Second, we have observed small v_g-values in the range between $0.05 \times c$ and $0.07 \times c$, c being the light velocity in vacuum. This observation was made for a relatively flat band (the second-lowest one in the Γ-K direction for the TM-like mode) of the guided mode in a PC slab sample with number of periods $N = 26$; see Fig. 10.2. Third, the result observed for a different band of another sample shown in Fig. 10.3 reveals that the value of v_g becomes smaller, as the band edge is approached [12], as also shown in Fig. 10.3. Moreover, the present experiments reveal that the observed values of v_g do not necessarily agree with the estimated values $(\mathrm{d}\omega/\mathrm{d}k)$ for the band calculated for an infinitely long sample.

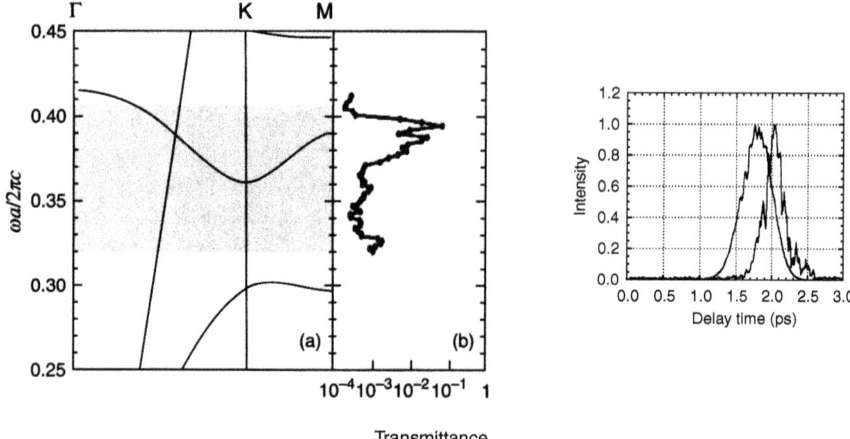

Fig. 10.2. *Left:* the relatively flat second-lowest band for TM-like modes of a 2D PC slab of air-bridge type, together with the observed transmittance spectrum; the straight line represents the air light-line. *Right:* an example of the observed 200-ps pulse that has passed through the sample with $N = 26$; the pulse on the right- and left-hand side passes through the sample with and without the PC, respectively. The former is seen to be delayed relative to the latter; quoted from [12]

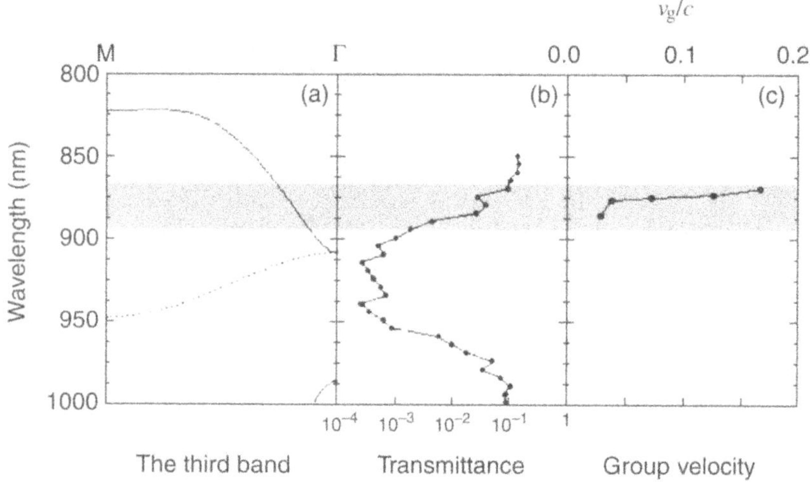

Fig. 10.3. Variation of the group velocity v_g at the band edge of the third-lowest band in the Γ-M direction of a SC sample with $N = 10$; quoted from [12]

In this connection, we would like to provide evidence indicating clearly the discrepancy, although the result was observed in the submillimeter region [13]. Figure 10.4 shows the variation of the calculated phase shift in the Γ-Z direction for a 3D PC sample with $N = 15$; the sample is of quasi-simple-cubic structure with a on the order of 0.4 mm [13]. For the second-lowest

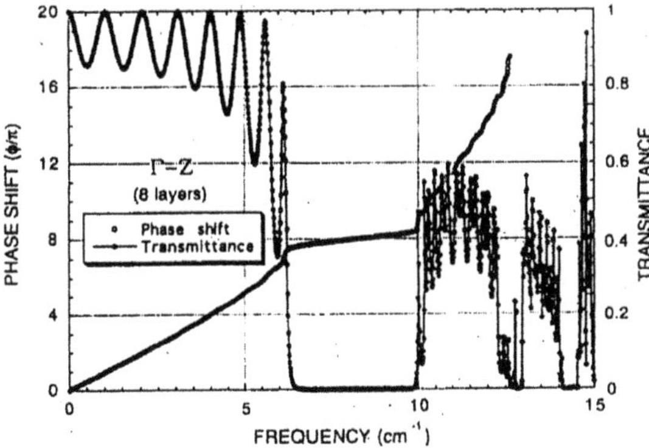

Fig. 10.4. Variation of the phase shift calculated for a 3D PC sample with finite periods of 15; quoted from [13]

band, the step-like variation of the phase shift is clearly seen there; in other words, the phase shift changes markedly around the wavelengths where T is large. This calculated result is supported by the observed one (not shown); the spectrum is presented in Fig. 13.3. From this result it follows that v_g-values there are much smaller than, e.g., those at the wavelengths of the transmission minima, since the derivative of the phase-shift is inversely proportional to v_g. Interestingly, it is noted that the maximum value is more than two orders of magnitude smaller than the minimum.

10.4 Laser Action: Experimental Aspects

The so-called photonic crystal lasers may be classified into two categories: one utilizing a particular point in the BZ such as a high-symmetry point where v_g is small, and the other a PC microcavity. Alternatively, those may be classified into one using a bulk PC or the other utilizing a defect mode.

10.4.1 Band-Edge Laser

Laser action can be easily achieved utilizing a band position where v_g is very small. In this case, the threshold pump-fluence or electric current of laser action becomes very low for the reason that has already been explained in Sect. 10.3; see also the next section. This kind of laser was first reported in 1997 by K. Inoue *etal.* [14], and is described below; at almost the same time a similar phenomenon was independently reported for a 2D PC slab [15].

The original 2D PC used for this purpose is one that we have described in Sect. 5.1, i.e., a parallel array of 1.0-mm-long air cylinders with circular cross-section (diameter $R = 0.67$ μm) etched into PbO glass. A specimen with air cylinders filled with a laser dye solution was optically pumped, i.e., irradiated

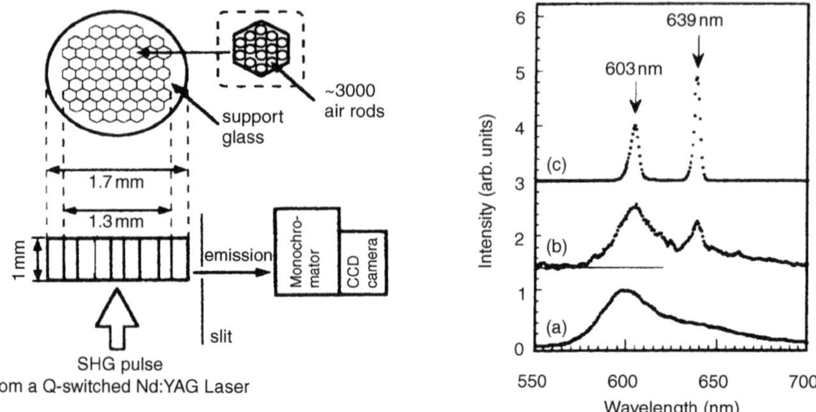

Fig. 10.5. *Left:* schematic showing the experimental setup for observing 2D PC laser action; *right:* variation of the emission spectrum with pumping fluence, from [14]

homogeneously from the bottom plane and parallel to the rod axes, by using a 4 ns pulse of second-harmonic light of a Q-switched Nd:YAG laser.

Emission from the specimen was observed in the 2D PC lattice plane in the way shown in Fig. 10.5. Examples of the emission spectrum taken for sulforhodamine-B dyes in dimethylsulfo-oxide (DESO) are also shown there. The results show variation of the spectral shape with increasing pumping-fluence; (a) a smooth spectrum dominated by spontaneous emission under sufficiently weak pumping-fluence, (b) a spectrum with a distinct line emerg-

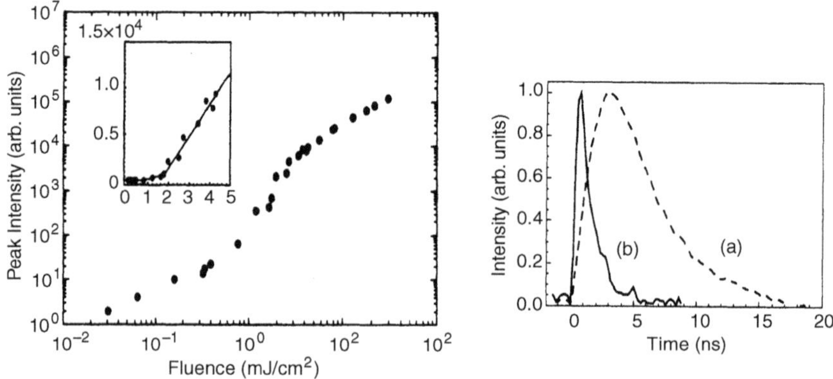

Fig. 10.6. *Left:* dependence of the emission intensity on the pumping fluence; *right:* variation of the output emission pulse before (*dotted line*) and after (*solid line*) the threshold pumping fluence; quoted from [14]

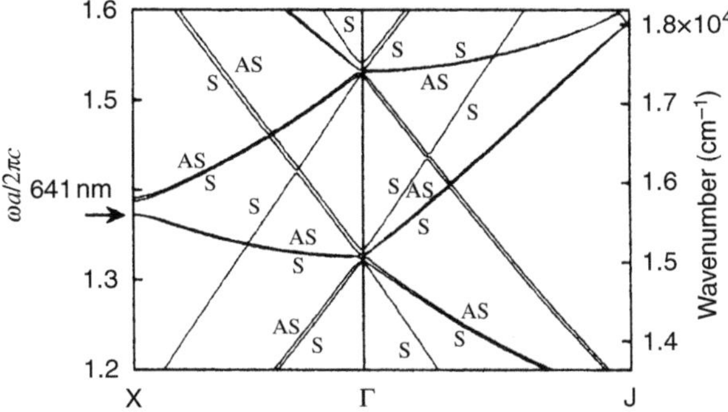

Fig. 10.7. The band structure showing a relatively flat band, the M (X) point of which is used for observing laser action; S and AS refer to symmetric and antisymmetric bands, respectively; quoted from [14]

ing with increasing fluence, and (c) a spectrum reduced to two narrow lines on further increasing the fluence.

Variation of the relative output power around 639 nm with pump fluence shows nonlinear behavior characteristic of laser action with a sudden change of slope at the threshold pump fluence; see Fig. 10.6 (*left*). Importantly, laser action at 639 nm takes place at a wavelength where the relative gain is supposed to be much smaller than that at the peak, i.e., 603 nm. The involvement of laser action is more directly seen by observing the decay time of the emission, which becomes faster due to stimulated emission above the threshold, as seen in the Fig. 10.6 (*right*). Lasing at the longer wavelength of 639 nm corresponds in energy to the $X(M)$ point of the PBS shown in Fig. 10.7, where v_g is very low. The emission pattern observed outside the specimen showed that strong emission was confined to the 2D lattice plane; unexpectedly, however, no significant directionality of the emission within the plane was observed. Anyway, the present lasing action is considered to be a 2D version, in a sense, of the so-called 1 D DFB (distributed feedback) laser.

10.4.2 Vertically-Emitting Laser

A novel type of photonic crystal laser was invented by S. Noda and his coworkers on the basis of the new concept shown in Fig. 10.8, where mode–mode coupling is schematically illustrated for a 2D PC with triangular lattice structure. Let us consider first a mode at the $X(M)$ point, i.e., with k having just the zone-edge value in the Γ-X direction. In this case the light wave is coupled by Bragg diffraction to the counterpart wave propagating in the opposite direction. As a result, laser action should easily occur independently in the three equivalent Γ-X directions, each in a way similar to the DFB laser for a 1D PC slab. The PC laser explained in Sect. 10.4.1 corresponds to this case.

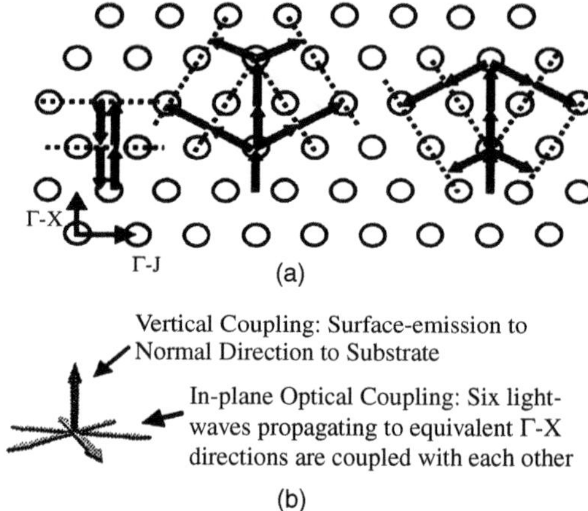

Fig. 10.8. Schematic showing that the laser beam emerges in different directions as a result of coupling of the equivalent modes, depending on the high symmetric point of the 2D PC band; quoted from [17]

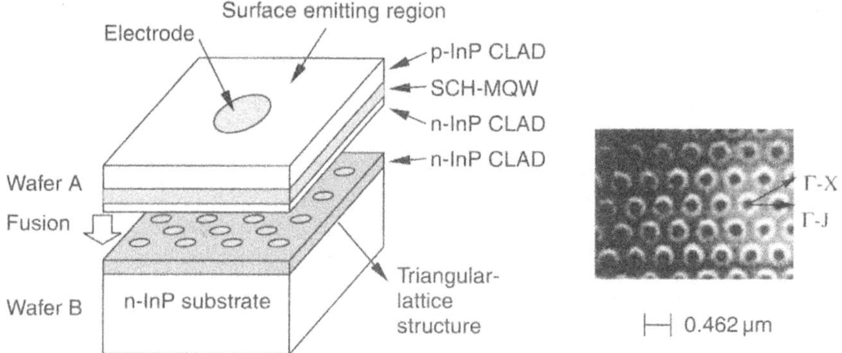

Fig. 10.9. A PC slab sample for laser action by electric current pumping; the 2D PC with triangular lattice structure was embedded by the wafer fusion technique; quoted from [16]

On the other hand, the wave with k twice this value (i.e., at the Γ point) can couple to the other five equivalent waves propagating in the different directions through Bragg diffraction, causing a unique 2D cavity. Note that in this case laser emission can be extracted in the direction perpendicular to the 2D plane, as shown in Fig. 10.8(b).

Laser action using this coupled modes was demonstrated for the first time in the following way [16]. The sample fabricated for this purpose is of an InAs-based multilayer structure, as shown in Fig. 10.9. In this structure the layer for amplifying light is separated, just like a 1D corrugated structure, from the layer for light propagation where air holes are introduced two-dimensionally.

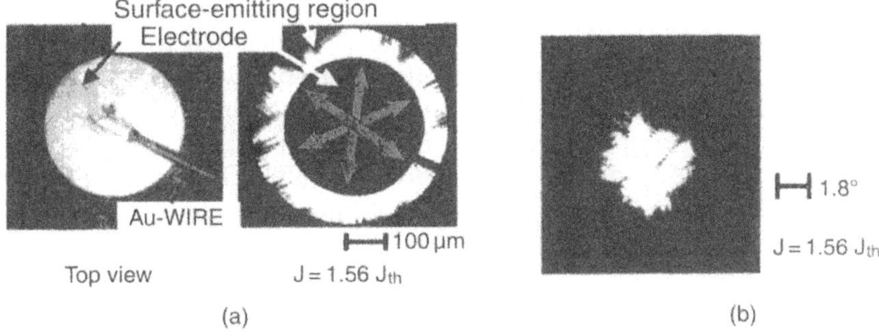

Fig. 10.10. (**a**) Top view of the laser shown in Fig. 10.9 and the near-field pattern of laser emission normal to the top surface; electrode diameter is 350 nm. The six arrows show the six equivalent Γ-X directions. (**b**) The corresponding far-field pattern; quoted from [17]

As the active (amplifying) element, multiple quantum wells composed of InGaAsP/InP were adopted. Pumping was supplied by an electrical current (pulse current) at room temperature through two electrodes located on the upper-most and the lowest-lying layers shown in Fig. 10.9.

Above the threshold a strong stimulated emission leaves the slab sample in the direction perpendicular to the 2D plane. In Fig. 10.10 are shown both the near-field (a) and far-field patterns (b) observed from the above electrode side. Importantly, the laser has a surprisingly narrow directionality of solid angle 1.8°. Comparison of the observed behavior including the emission wavelength with the corresponding band structure shown in Fig. 10.11 nicely

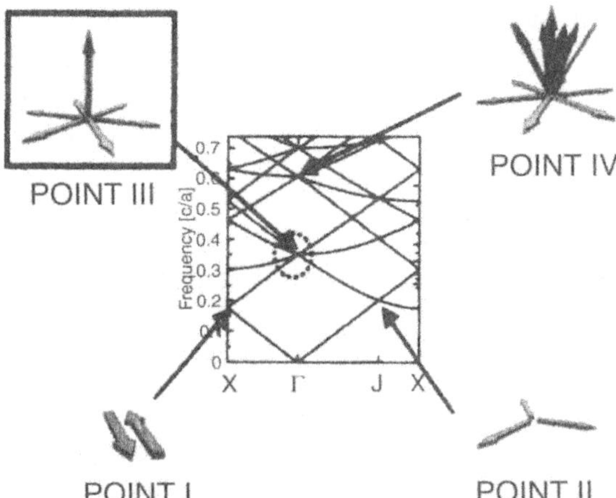

Fig. 10.11. The band structure corresponding to the 2D PC sample shown in Fig. 10.9 and coupling of the six equivalent modes at the Γ point shown there causes the present laser emission in the direction perpendicular to the 2D plane; quoted from [17]

explains that the laser light goes into free space in the perpendicular direction as a result of coupling among six equivalent modes at the Γ point, which has already been explained in Fig. 10.8(b). Furthermore, the wavelengths observed at several different near-field points are the same, indicating that the single-mode laser action takes place over a large area in the present case. This laser is considered as a new type of vertical or VCSEL laser of high quality; it is remarked that the present laser is not intended to have a low threshold.

10.4.3 Defect Laser

As already explained in Sects. 2.6 and 2.7.2, the point (or 0D) defect mode serves as a small optical cavity. Such a small cavity, or a microcavity can be easily produced on the basis of a 2D PC or a 2D PC slab. The most typical example is such that the size of the air cylinder at a particular lattice point is changed. Light can be trapped or confined in this cavity, so localization of light is easily achieved. By utilizing this type of microcavity made of a semiconductor-based PC slab of the air-bridge type and of air-hole type, lasing action is expected to occur when the Q-value is high enough. Scherer and his collaborators succeeded for the first time in achieving a laser operating at low temperatures by optical pumping [18]. Although the Q-value can be made very high, laser action at room temperature has been difficult to observe until very recently.

10.4.4 Other PC Lasers

Very recently, laser action utilizing a line-defect mode in a 2D PC slab, instead of the point-defect mode, has also been demonstrated by optical pumping, where a small v_g of the local mode is responsible for the laser action [19]. Furthermore, it has also been demonstrated that in a combination of point- and line-defect modes laser action easily takes place [20]. For details, see Chap. 11.

10.5 Laser Action: Numerical Aspects

10.5.1 Treatment of the Onset of Lasing

In this section we present a numerical study of the PB effect on the onset of lasing. We consider a periodic array of rods of circular cross-section in free space. The rods are made of a substance which has a population inversion, i.e., a substance with a greater number of excited atoms than that of ground-state atoms. To express the population inversion, we add a *negative* imaginary part to ε of the cylinders. To discard the unnecessary complexities, we put

$$\varepsilon = \varepsilon_1 + \mathrm{i}\gamma \tag{10.5}$$

for ε of the cylinders with real part ε_1 and imaginary part γ, both assumed to be ω-independent. We array the cylinders in a square lattice of lattice constant a. The resulting PC is of typical 2D type. In what follows we use the symbol r for the radius of cylinders and ω for the frequency normalized by $2\pi c/a$, c being the light velocity. The parameter N stands for the number of layers in the thickness direction of the slab PC. We take the x and y axes along the two sides of the square with the y axis taken in the thickness direction. The z axis is along the axis of the cylinders.

By this model the onset of the laser oscillation can be identified with the appearance of the divergence of the transmittance T in the parameter space of (ω, γ, N). We examine $T(\omega, \gamma)$ of plane-wave light incident in the y direction on the system with fixed N. The position of the divergence gives the frequency and the threshold value of γ for the onset of lasing in the PC of N layers of cylinders. The laser light leaves in the y direction, i.e., normal to the external surfaces. The first application of this method to catch the onset of laser action was made by Yariv and collaborators in examining lasing in DFB (distributed feedback) lasers [21].

The BS is calculated by the vector KKR equation for a periodic array of cylinders (see Sect. 4.1.1) and the transmitted amplitude for the slab of a finite N of PC is obtained by applying the layer-doubling method (see Sect. 4.4) [22]. Assigning a complex value to ε of cylinders, with a negative imaginary part to cover the gain, adds nothing serious to the numerical treatment. In this chapter we restrict ourselves to the case of $\varepsilon_1 = 2.0$ and $r/a = 0.3$, in the frequency range of $\omega < 1.0$. Using this very modest periodic variation of ε, the characteristic features of lasing in PCs is understood in a qualitatively reliable manner.

For the real ε-value of 2.0, we give in Fig. 10.12 the BS for $\boldsymbol{k} = (0, k_y, 0)$ and $\omega < 1.0$ as a function of k_y. Of course this is the result for $N = \infty$ obtained by the PB calculation. Two bands are marked by thick curves. The first is the lowest band which terminates at the BZ boundary at $\omega = 0.409$. Its eigen-electric-field is polarized in the z direction, i.e., it is the TM band or E band in the terminology of Sect. 3.2 (the band immediately next to it is x-polarized (TE or H band)). So, the frequency $\omega = 0.409$ is the lower edge of the first BG of the z-polarized PB. We call the range near this band edge range I. The second range, range II, is a higher frequency region near the very flat band shown by the solid curve. The electric field of this band is polarized in the x direction. Besides being very flat, the band is characterized by its band minimum located at $k_y = 0.12$ and $\omega = 0.880$. The nearby band which crosses this band in Fig. 10.12 is z-polarized and plays no part in the transmission of the x-polarized light. These two ranges are selected so that each of them contains within it one band-edge with large band-bending, which is expected to bring about a remarkable effect in the onset of lasing.

Figure 10.12 shows T for range I of z-polarized plane-wave light propagating in the y direction (normal incidence). It shows a bird's eye view of T

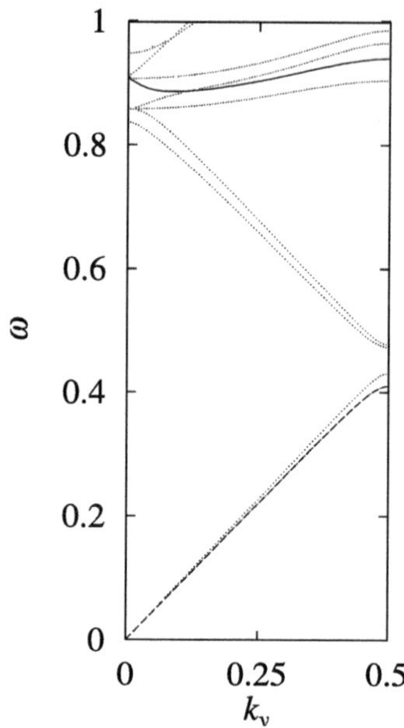

Fig. 10.12. Band structure of arrayed cylindrical rods with $r/a = 0.3$ and $\varepsilon = 2.0$ for the wavevector $\boldsymbol{k} = (0, k_y, 0)$. The bands shown by bold dashed and solid curves define range I and II of the frequency, respectively

in the (ω, γ) plane for the case of $N = 32$. We calculate T in the range $0 > \gamma > -0.03$. By gradually increasing N from the case of $N = 1$, we confirmed that nothing special happens in the calculated region of (ω, γ) until $N \leq 8$. For $N = 16$, a rapid increase of T is seen near the band edge with increasing $|\gamma|$. However, there is not yet a conspicuous development of any peak. The case of $N = 32$ of Fig. 10.13 is the first one where a T-peak indeed appears. Note that the layer-doubling method deals only with the cases $N = 1, 2, 4, \cdots 2^n, \cdots$, and accordingly the thickness of $N = 32$ is a value sufficient for the onset of lasing in range I. We confirmed that the height of the peak of Fig. 10.13 is as high as 10^5. Also we checked that the reflectance has a divergent peak simultaneously with T at the same (ω, γ, N). Note that the divergence appears at $\omega \simeq 0.409$, just at the lower edge of the first band gap of the z-polarized light. This reflects the fact that the slab with $N = 32$ is thick enough to realize the first BG of $N = \infty$ and that the slow v_g at the band edge is the cause of the first appearance of lasing peak there.

10.5.2 Lasing Involving Group-Velocity Anomaly

A more dramatic feature of the T-map is observed in range II. We calculate T for the x-polarized light in range II of $0.87 < \omega < 0.93$ (for light of normal incidence as above). As for the N dependence, we have confirmed

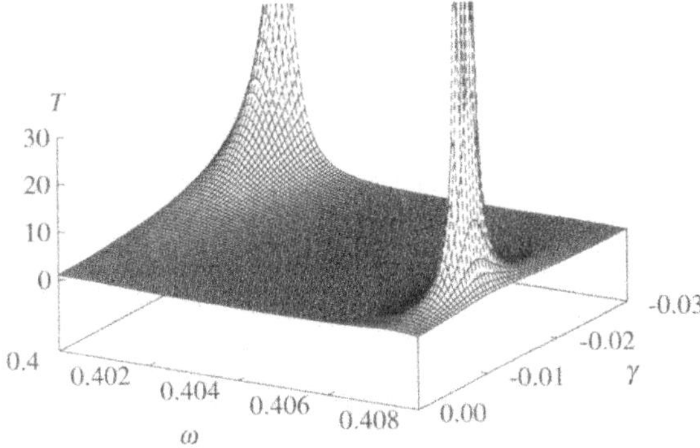

Fig. 10.13. Transmittance in the (ω, γ) plane of z-polarized light for the case of $N = 32$ in range I. The height of the peak is over 10^5. The frequency region shown is just below the first band gap of the z-polarized light of Fig. 10.12

that a remarkable enhancement of T occurs even in the case of $N = 1$. In the cases of $N = 2, 4$, two divergent peaks appear in the calculated region of (ω, γ) with $0 > \gamma > -0.03$. In these cases the slab is too thin to realize the PBS shown in Fig. 10.12. Indeed, the peak frequencies have no special relationship with the PBS of range II. However, the appearance of the peaks at these earlier stages obviously shows that laser action is more easily induced on the EM modes positioned in the higher-frequency range. As compared to range I, an enhanced confinement effect and hence higher Q-values of the EM modes set up inside are responsible for this feature. The results of $N = 1, 2, 4$ of range II imply that if we can use a PB of such large angular momenta as employed in the cavity QED, an enormous advantage will result, as far as the onset of lasing is concerned.

We show in Fig. 10.14 the T-map of $N = 32$, which allows the analysis based on the band scheme shown in Fig. 10.12.

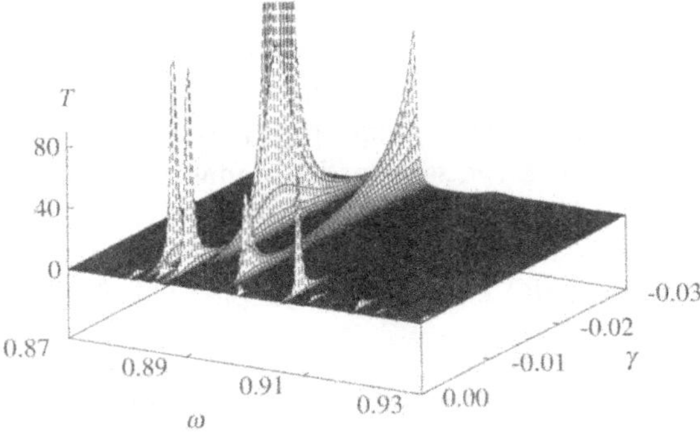

Fig. 10.14. Transmittance of range II for x-polarized light in the case of $N = 32$

It shows the simultaneous appearance of a number of peaks. One reason for the complex intensity map of range II is that two band states of different k_y get excited by a single incident light wave just above the band minimum. Another reason is that the high finesse of the interference fringe gives rise to many divergent peaks. The coexistence of several peaks suggests the possibility of cross-saturation effects between them in the case of multimode lasing and the possibility of bistability via the nonlinear coupling. Figure 10.14 thus suggests the diversity of the lasing phenomena involving PCs. Note that the magnitudes of $|\gamma|$ of the peak positions are much smaller than that of Fig. 10.13, showing that the threshold value of $|\gamma|$ for the onset of lasing is made smaller, if we use a flat PB [23, 24]. The analogous decrease of $|\gamma|$ was observed for the T-map calculated with increasing values of r and ε_1; a larger r leads to an increase of the region of active substance and a larger value of ε_1 induces an enhanced confinement effect or an enhanced Q-value of the EM normal modes.

It is interesting to compare the present results with those of a 1D multilayer stack of thin films [25]. It is found that in the 1D case any frequency range has a T-map analogous to that of range I shown in Fig. 10.13. Since the lowest BG of a 2D PC arises due to the diffraction associated with the reciprocal lattice point of $\boldsymbol{h} = (0, \pm 1)$, the appearance of the first BG in Fig. 10.12 is essentially of 1D nature and the analogy stated above is natural. The conclusion is thus that the advantage of a 2D PC over the 1D PC appears when a higher band participates in the lasing. In other words, we do not take full advantage of the EM confinement effect in the distributed feedback effect of the multilayered films. As a typical 3D effect we examined T from the slab PC of spheres arrayed in a cubic lattice of the same volume fraction. We find that the lasing involving the lowest PB is not very different from the 1D and 2D systems and that higher bands have qualitatively the same characteristics as the 2D ones obtained here.

10.6 Nonlinear Optical Phenomena

By utilizing the unique PBS in either a 2D or 3D PC, the phase matching (PM) can be easily fulfilled at several wavelengths, as already described in Chap. 2. In addition, the umklapp process can also be utilized, since the reciprocal vector $\boldsymbol{q}$ is on the same order in magnitude as the relevant wavevectors $\boldsymbol{k}_i$ involved in the phenomenon under study; it has already been proved that this process is useful in SHG for a 1D periodic system (1D PC). We describe below the SHG case as an example. The SHG signal intensity $I^{2\omega_1}$ is expressed as,

$$I^{2\omega_1} = A(c/v_g)^2 \left(I^{\omega_1}\right)^2 \left| \frac{\sin(|\Delta \boldsymbol{k}|Na/2)}{\sin(|\Delta \boldsymbol{k}|a/2)} \right|^2, \tag{10.6}$$

where N is the number of layers, and the phase mismatch $\Delta \boldsymbol{k}$ is given by

$$\Delta \boldsymbol{k} = 2\boldsymbol{k}_1 - \boldsymbol{k}_2 \pm \boldsymbol{q} \tag{10.7}$$

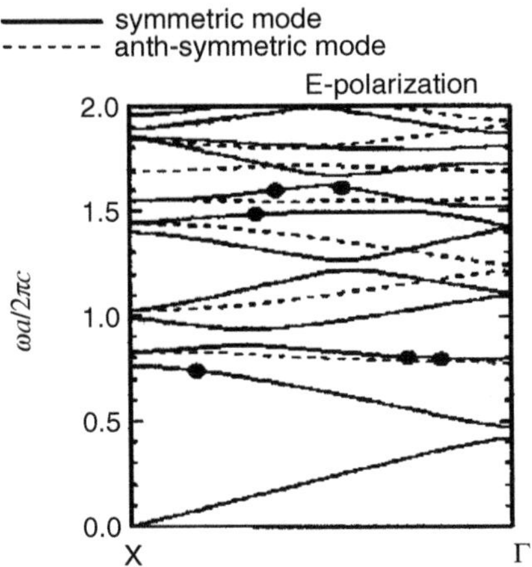

Fig. 10.15. An example showing that the PM condition holds at three wavelengths in a rather narrow wavelength region of a 2D PC

where $\boldsymbol{k}_1$ and $\boldsymbol{k}_2$, and ω_1 and $\omega_2 = 2\omega_1$ are the wavevectors, and angular frequencies of incident (fundamental) and second-harmonic light, respectively; I^{ω_1} is the fundamental light intensity, v_g is the group velocity of 2ω wave, and A is a constant. It has already been proved experimentally that this kind of umklapp process holds in the nonlinear phenomena in the X-ray region. It is noted that, as is well known, the wavelength of X-rays is comparable to the size of the BZ, or on the order of the inverse of the atomic distance comprising the conventional lattice.

We present below an example of the collinear SHG case for a 2D PC of air-hole type made of arrayed fibers, described in Sect. 5.1. The calculated band structure for E-polarization shown in Fig. 10.15 indicates that the PM condition is fulfilled at three wavelengths between the fourth lowest band and the higher bands; the coupled and uncoupled bands are denoted by solid and dotted lines, respectively. Importantly, the SHG signal is generated between the coupled bands, which is required from the fact that the overlap integral of the relevant electric fields must not vanish. Experimentally, a small number of noncentrosymmetric organic particles was randomly introduced into the air holes; the amount was such that it did not affect the BS. Then, variation of the SHG signal intensity with the wavelength of incident light was observed in the forward direction.

The result is shown in Fig. 10.16, where the signal enhancement is found to occur at three wavelengths. These wavelengths turn out to be consistent with those calculated in terms of the BS. From this example, one can readily conjecture that the PM condition should be satisfied at many wavelengths for the other configurations of polarization, such as combinations of H-H and E-H, where the first and the second E or H refer to the polarization of the incident and the SHG light, respectively.

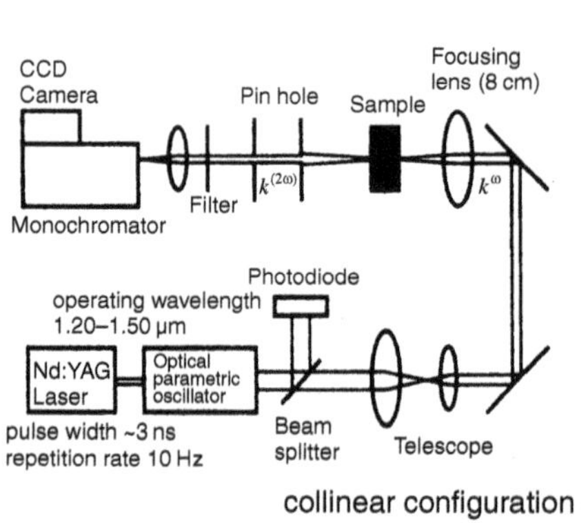

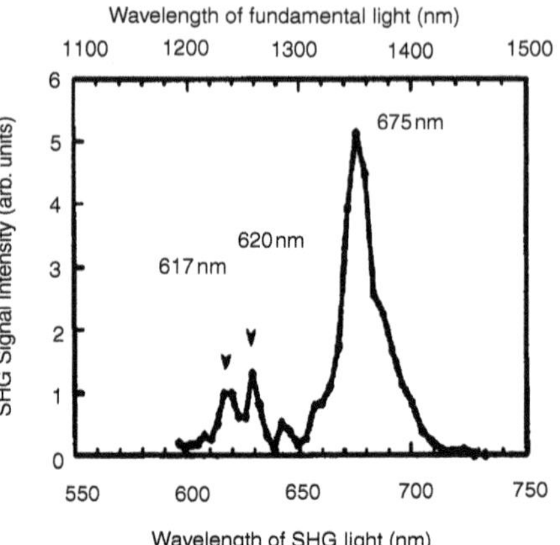

Fig. 10.16. The experimental result showing that an enhanced SHG signal was observed at a few wavelengths corresponding to the band structure shown in Fig. 10.15

Next, (10.4) also includes concretely the fact that use of a small v_g allows the signal to be generated very effectively, as is the case with laser action. As has been described thus far in this chapter, this is generally true with interaction between light and matter in a PC.

Part II. Indirect Interaction: Smith–Purcell Radiation

The light emission concerned here is not from an electron in an atom but from a free electron traveling with high velocity near the surface of a PC. We present the theoretical spectrum and discuss the PB effects in it in comparison with the spectrum obtained by using a conventional diffraction grating.

10.7 Emission of Light from a Traveling Electron

A traveling electron is known to emit light which is appreciable only in the vicinity of its trajectory. The prohibition of emitting a propagating photon is a direct result of the difference of the dispersion relations of a free electron and a free photon, which prevents energy and momentum conservation from being satisfied simultaneously in the emission process. Near a periodic array of scatterers of light, the situation changes drastically. The evanescent light from the electron can acquire an umklapp momentum shift by being scattered by the periodic scatterers and has a finite possibility of being converted to observable plane-wave light. Since the pioneering discovery by Smith and

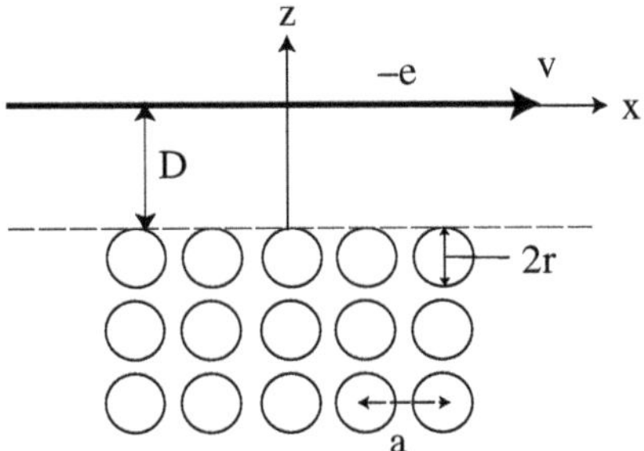

Fig. 10.17. Electron beam and a slab photonic crystal of arrayed spheres. The electron of velocity v is running in the x direction at distance D from the top of the first layer of spheres. Dielectric spheres, each of radius r, are arrayed in a simple cubic array of lattice constant a

Purcell (SP) [26], diffraction gratings have been used as a converter to make use of periodic 1D light-scattering by surface corrugations.

The well-definedness of the PBS naturally leads us to expect that an emission of photons might be achieved with higher efficiency if a PC is used instead of a diffraction grating.

We consider an electron of charge $-e$ ($e > 0$) traveling with velocity v, parallel to the surface of a PC of finite thickness. As shown in Fig. 10.17, the PC we are examining is made of dielectric spheres of radius r arrayed in a simple cubic lattice with lattice constant a. On the xy plane ($z = 0$), we place the centers of the spheres of the first layer, with the x axis taken in the (1,0) direction of the square lattice and place the whole system consisting of N stacked layers in the region $z < 0$. The lateral extension in the x and y directions of the PC is assumed to be infinite. Let the electron travel in the x direction at distance $D + r$ above the xy plane, where D is the distance between the electron trajectory and the top of the first-layer spheres.

Since we are interested in how a photon is influenced by Bragg reflection, the frequency ω of the emitted photon is assumed to be in the visible or longer wavelength range. With $\hbar\omega$ much smaller than the beam energy, which is assumed to be hundreds of kilovolts or higher in the relativistic range, the recoil of the electron due to the photon emission can be neglected. The emission process is then treated simply by solving Maxwell's equations using the beam current as the source term. This is an assumption usually employed in the treatment of the Smith–Purcell radiation (SPR) and works very well [27–30]. We treat the emission of light by dividing the whole process into two stages, the emission of evanescent light from the beam and the subsequent Bragg reflection of it by a PC.

With the origin of time t fixed at the instant at which the electron passes the point $x = 0$, the current density of the electron with velocity v is expressed by

$$\boldsymbol{J}(\boldsymbol{r}, t) = -e\hat{x}v\delta(x - vt)\delta(y)\delta(z - D - r), \tag{10.8}$$

where $\hat{x}$, the unit vector in the x direction, specifies the current direction. The Fourier transform of $\boldsymbol{J}(\boldsymbol{r}, t)$ with respect to t, $\boldsymbol{J}(\boldsymbol{r}, \omega)$, provides a source term in Maxwell's equations for the emission of light of frequency ω. From (10.8) we obtain

$$\boldsymbol{J}(\boldsymbol{r}, \omega) = -e\hat{x}\mathrm{e}^{\mathrm{i}k_x x}\delta(y)\delta(z - D - r), \tag{10.9}$$

with the key relation

$$k_x = \frac{\omega}{v}. \tag{10.10}$$

The source term $\boldsymbol{J}(\boldsymbol{r}, \omega)$ has a $\mathrm{e}^{\mathrm{i}k_x x}$ dependence and so does the light irradiated in the first-stage process. When the wavevector of the first-stage photon is denoted as $\boldsymbol{k}_{\mathrm{i}}$ (the suffix i stands for the initial photon) defined by

$$\boldsymbol{k}_{\mathrm{i}} = (k_x, q_y, \Gamma), \tag{10.11}$$

the free-space dispersion relation of light leads to

$$\Gamma = \sqrt{\left(\frac{\omega}{c}\right)^2 - \left(\frac{\omega}{v}\right)^2 - q_y^2} = \sqrt{\kappa^2 - k_x^2 - q_y^2}, \tag{10.12}$$

κ being ω/c. Since $v < c$, Γ is always a pure imaginary number and hence the direct light from the running charge is evanescent away from the trajectory. From $\boldsymbol{J}(\boldsymbol{r}, \omega)$ given above, the electric field is then obtained as

$$\boldsymbol{E}(\boldsymbol{r}, \omega) = \frac{e}{2c^2\varepsilon_0}\omega \int \frac{\mathrm{d}q_y}{2\pi}\left(-1 + \frac{1}{\beta^2}, \frac{q_y}{\kappa\beta}, \frac{\Gamma}{\kappa\beta}\right)\frac{\mathrm{e}^{\mathrm{i}\boldsymbol{k}_i\cdot\boldsymbol{r}}}{\Gamma}, \tag{10.13}$$

with $\beta = v/c$, where ε_0 is the dielectric constant of the vacuum. The detailed derivation is given in [31, 32]. Note that q_y of the wavevector $\boldsymbol{k}_{\mathrm{i}}$ is arbitrary and that the integral over q_y is needed in (10.13).

In the second stage, we show how the first-stage evanescent light given by (10.13) is influenced by periodic scattering.

We illustrate the first and second stage in Fig. 10.18, taking the case of $q_y = 0$, for simplicity. In the $q_x\omega$ plane, two lines are drawn to show the dispersion relations $\omega = cq_x$ and $\omega = vq_x$. They will be referred to as the c line and v line, respectively. The former specifies the edge of the light cone and the latter shows the dispersion curve of the emitted evanescent light, because at $\omega = \omega_0$ the c and v lines give $q_x = \omega_0/c$ and $q_x = \omega_0/v$, respectively, the latter being just k_x given by (10.10). In the second-stage calculation, the evanescent light given by (10.13) is treated as an incident wave to the PC. The role of the photonic crystal in the SP mechanism is in essence to induce umklapp scattering of the incident light of wavevector $\boldsymbol{k}_{\mathrm{i}}$.

On the upper side of the PC, a reflected wave is observed. Since the momentum shift due to a 2D reciprocal lattice vector is expressed as

$$(\Delta q_x, \Delta q_y) = \frac{2\pi}{a}(n_x, n_y) \tag{10.14}$$

with integers n_x and n_y, the wavevector of the scattered light is given by

$$\boldsymbol{k}_{\mathrm{s}}(n_x, n_y) = \left(k_x - n_x \frac{2\pi}{a}, q_y - n_y \frac{2\pi}{a}, \Gamma_{n_x n_y}\right), \tag{10.15}$$

with

$$\Gamma_{n_x n_y} = \sqrt{\kappa^2 - \left(k_x - n_x \frac{2\pi}{a}\right)^2 - \left(q_y - n_y \frac{2\pi}{a}\right)^2} \tag{10.16}$$

by energy conservation. The (n_x, n_y) waves with real $\Gamma_{n_x n_y}$, i.e., the waves with $\boldsymbol{k}_{\mathrm{s}}(n_x, n_y)$ lying inside the light cone of Fig. 10.18, can propagate as a plane wave and are detectable at a distant observation point. Figure 10.18 shows the appearance of these propagating waves for the case of $n_y = 0$. We have a series of v lines (infinite in number) shifted horizontally by the quantity $-(2\pi n_x/a)$. In the frequency region within the light cone, i.e., along a shifted v line bounded by the two c lines $\omega = \pm c q_x$, we have one SPR emission band. Thus, the entire spectrum is composed of $h_{10}, h_{20}, \cdots$ emission bands, where the subscript 10, for example, specifies the momentum shift $(n_x, n_y) = (1, 0)$. The angle of the emitted photon relative to the z axis (surface normal) is determined by the direction of $\boldsymbol{k}_{\mathrm{s}}$ given by (10.15).

The point of the calculation of the SPR spectrum is how to treat multiple light-scattering by a PC in umklapp scattering. We carry out this part by the layer-doubling method treated in Sect. 4.4.

In this way, we can obtain the amplitude transmission and reflection coefficients of the incident evanescent wave of $\boldsymbol{k}_{\mathrm{i}}$. Since the light reflected back is our concern, the 3×3 amplitude reflection tensor $\boldsymbol{r}(\boldsymbol{k}_{\mathrm{s}}; \boldsymbol{k}_{\mathrm{i}})$ of an N-layer PC must be obtained for the conversion process from $\boldsymbol{k}_{\mathrm{i}}$ to $\boldsymbol{k}_{\mathrm{s}}$. Recovering the

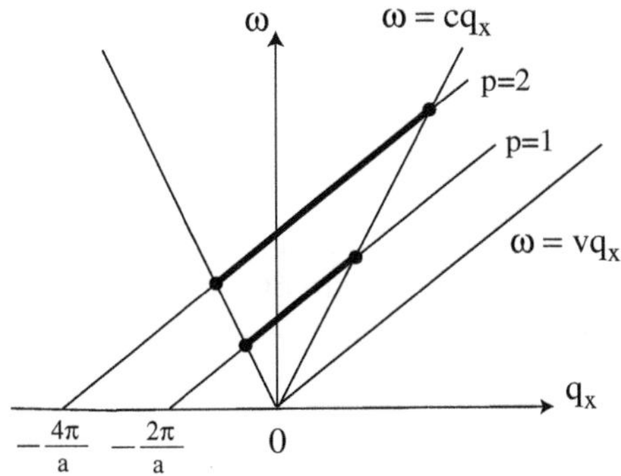

Fig. 10.18. Light conversion using an umklapp momentum shift. The line $\omega = v q_x$ shows the dispersion relation of the evanescent wave (called the v line in the text). Two umklapp-shifted v lines are drawn (h_{10} and h_{20} lines). The lattice constant of the reciprocal lattice is $2\pi/a$. The light lines $\omega = \pm c q_x$ define the light cone. The part of the shifted v line bounded by them (shown by the thick solid lines) defines the SPR band

q_y integral of (10.13), we obtain the umklapp reflected wave in the following form:

$$\boldsymbol{E}_{\mathrm{s}}(\boldsymbol{r},\omega) = \frac{e}{4\pi\varepsilon_0 c}\sum_{n_x n_y}\int dq_y \exp(\mathrm{i}\boldsymbol{k}_s\cdot\boldsymbol{r})\boldsymbol{E}_1^{(n_x n_y)}(k_x,\, q_y,\, \Gamma), \tag{10.17}$$

where

$$\boldsymbol{E}_1^{(n_x n_y)}(k_x,\, q_y,\, \Gamma) = \boldsymbol{r}(\boldsymbol{k}_{\mathrm{s}};\boldsymbol{k}_{\mathrm{i}})\boldsymbol{E}_0(k_x,\, q_y,\, \Gamma), \tag{10.18}$$

with

$$\boldsymbol{E}_0(k_x,\, q_y,\, \Gamma) = \omega\left(-1+\frac{1}{\beta^2},\, \frac{q_y}{\kappa\beta},\, \frac{\Gamma}{\kappa\beta}\right)\frac{\mathrm{e}^{-|\Gamma|(D+a)}}{c\Gamma}. \tag{10.19}$$

We define the SPR intensity in the following form in terms of the integrand of (10.17):

$$I(n_x,\, n_y)_{q_y} = |\boldsymbol{E}_1^{(n_x n_y)}(k_x,\, q_y,\, \Gamma)|^2. \tag{10.20}$$

The set (n_x, n_y) defines one SPR band. When q_y and ω are both fixed, a sharp spot is obtained in the (n_x, n_y) band because $\boldsymbol{k}_{\mathrm{s}}$ is uniquely determined by (10.15). The spot that has the highest intensity of all will be that of $(n_x, n_y) = (1, 0)$, the band with the shortest umklapp shift, which appears along the h_{10} line of Fig. 10.18. If we sweep ω in the (1,0) band, the spot moves along the shifted v line. If we sweep q_y in addition, the direction of $\boldsymbol{k}_s$ varies in the y direction. If, in particular, we restrict the direction of observation to within the xz plane, the observed intensity is given by the contribution from $q_y = 0$. To describe the SPR intensity of this case, we consider

$$I(1,\, 0)_0 = |\boldsymbol{E}_1^{(10)}(k_x,\, 0,\, \Gamma)|^2 \tag{10.21}$$

by putting $n_y = q_y = 0$ in (10.20) and call this quantity simply the SPR intensity of the h_{10} band.

10.7.1 SPR Spectrum and PBS

We show in Fig. 10.19 the 2D PBS of the slab of $N = 4$ and h_{10} emission band [32]. See Sect. 3.4 for the origin of the PBS of a slab of finite thickness, which is obtained by the plot of the DOS peaks. In Fig. 10.19(a), the BS is shown for the 2D wavevector $\boldsymbol{q} = (q_x, 0)$ as a function of q_x. Only part of the PBS is presented (see [32] for more about the PBS). The parameters of the PC used in the calculation are $\varepsilon = 3.2^2$ and $r/a = 0.42$. The calculated h_{10} emission spectrum is shown in (b). The emission intensity I is calculated for $\beta = 0.9$ and $D = 0.5a$.

Comparison of the two panels shows that there is a very good correspondence between the resonant peaks of the emission spectrum and the position of the dispersion curves cut by the h_{10} line. We see that the width of the resonances depends rather strongly on the band index. Since the PBs in question

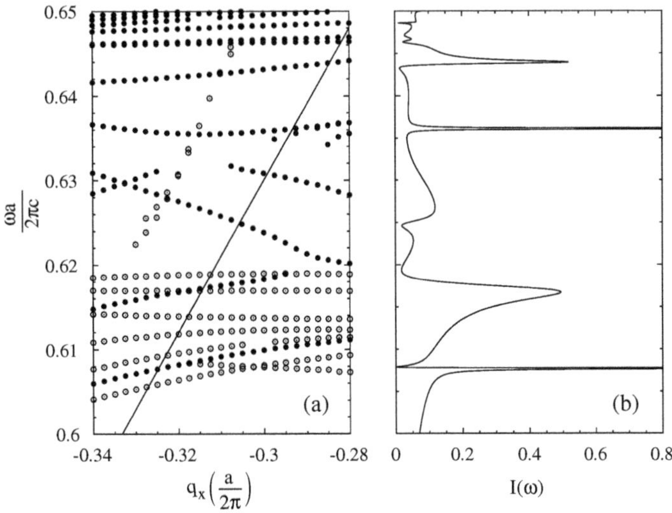

Fig. 10.19. Band structure and SPR spectrum on the h_{10} line. Panel (**a**) shows the band structure of the slab with $N = 4$ for a narrow region of $\boldsymbol{q} = (q_x, 0)$ as a function of q_x, and (**b**) shows the plot of $I(1,0)_0$ defined by (10.21) for $v/c = 0.9$. The straight line in (**a**) is the shifted v line (h_{10} line of Fig. 10.18). The filled (open) circles of the band structure show the dispersion relations of the p-active (s-active) bands

lie within the light cone, they are leaky bands which have finite lifetime. It is therefore concluded that the magnitudes of their lifetime determine the peak values of the emission peaks. This is one of the common features of the optical response of PCs involving leaky PBs (see, e.g., the calculation of transmittance given in Sect. 4.3.4).

Also, it is remarkable in Fig. 10.19 that only the PCs which are active to the p-polarized band can react in the photon emission. The s-polarized PCs (polarization vector in the y direction) leave no trace in the emission spectrum. Their appearance is symmetry forbidden in the spectrum of $(m, n) = (-1, 0)$ shown here, as discussed in [32].

As N increases, the peaks become sharper reflecting the enhanced Q-values of the normal modes caused by the improvement of the mode confinement. Also, the emission band becomes more and more complicated with increasing N due to the enhanced band population. Furthermore, with increasing m and n in the umklapp momentum transfer in (10.15), the emission intensity becomes small very quickly, which indeed supports the neglect of the umklapp shifts other than the largest contribution $n = 1$. These features are discussed in detail in the work cited above [31, 32].

10.7.2 Photon Yield of PC versus Diffraction Grating

To compare PCs with diffraction gratings in the emission efficiency, let us take as an example of the latter a perfectly conducting grating with a rectangular groove profile. This type of grating has often been employed in the grating SPR both theoretically and experimentally. To make the geometry of the two systems as similar as possible, we compare a PC of $N = 1$ and $r/a = 0.25$ with a grating having groove width and height relative to the grating period a both equal to 0.5 [32]. A vertical view of the two systems, when superimposed, is given in Fig. 10.20(a). The distance D of the beam trajectory from the grating is measured from the top surface of the grating.

For the emission cross-section of the h_{10} band, Fig. 10.20(b) compares the two systems for $\beta = 0.9$ and $D = 0.5a$, in a logarithmic scale. We have calculated the absolute magnitudes of the cross-section of the photon emission. See [32] for the definition of the cross-section. The solid curve is for the PC and the dashed curve is for the grating. We can see a marked difference between the two curves both in magnitude and line shape. Numerous resonant peaks in the PC arise due to photonic band excitations. It is notable that the heights of the sharp peaks are larger than those of the grating, generally by about one order of magnitude. Fine structure arises in the curves of the grating, too. In Fig. 10.20(b) we find the sharp dips and kinks appear in the grating curve at $(\omega a/2\pi c) \simeq 0.9, 1.4, 1.9, \cdots$, together with broad peaks located between them. These are typical features observed experimentally in SPR spectra of gratings [27, 28]. It should, however, be noted that the peaks present in the grating case are too broad to compare their FWHM with those

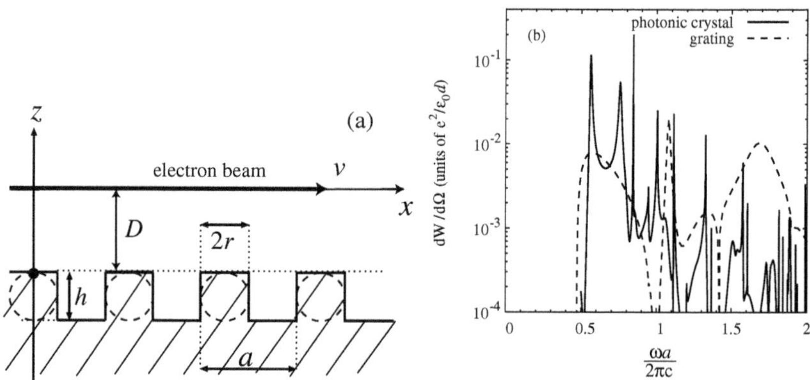

Fig. 10.20. Comparison with a diffraction grating. The profiles are given in (**a**) for the diffraction grating with a rectangular groove profile and a free-standing PC of $N = 1$ used in the comparison. The cross-section of emission of light is compared in (**b**). The results for the h_{10} SPR band are given for $\beta = 0.9$ and $D/a = 0.5$. The solid and dashed lines represent the spectra of the PC and of the grating, respectively

of the PB excitation. Because of this, the peak heights of the grating SPR are much lower than those of the PC. The fine structure positioned at those frequencies noted above appears when the h_{10} line intersects the equally spaced c lines associated with umklapp shifts of integer multiples of $2\pi/a$. This shows that the BS of the perfectly conducting grating is obtained by folding the light lines to the first BZ and that the quality factors of the excited bands are much smaller than the PBs of the arrayed dielectric spheres. This is another way of saying that the spectral difference between the two system is caused by the difference of the degree of the electromagnetic confinement effect.

Although we can obtain almost monochromatic and enhanced emission lines by using the excitation of PBs, it should be pointed out that the integrated value of the spectral intensity of PC does not look very different from that of the grating.

To summarize, the most peciliar effect of the SPR spectrum from a PC slab is that the enhanced emission peaks are caused by the resonant excitation of the PBs and that their sharpness is determined by the lifetime of the excited PBs. The peak values are one order of magnitudes larger than the SPR intensity from a metal grating.

We have given a numerical calculation for a PC made out of arrayed spheres. Considering that the measured T of the $N = 1$ PC of arrayed Si_3N_4 spheres of mm range agrees remarkably well with the calculated prediction [33], we expect the comparison in SPR spectrum to also be favorable. We hope to report the related experiment in the near future.

References

1. L. I. Schiff, *Quantum Mechanics*, 2nd ed. (McGRAW-HILL, New York, 1955) chap. XIII
2. M. E. Purcell, Phys. Rev. **69**, 681 (1946)
3. Z.-Y. Li, L.-L. Lin, and Z.-Q. Zhang, Phys. Rev. Lett. **84**, 4341 (2000)
4. V. Lousse, J.-P. Vigneron, X. B. Bouju, and J.-M. Vigoureux, Phys. Rev. B**64**, 201104 (R) (2001)
5. A. Yamanaka and K. Inoue, unplublished data
6. M. D. Tocci, M. Scalora, M. J. Bloemer, J. P. Dowling, and C. M. Bowden, Phys. Rev. A**53**, 2799 (1996)
7. E. P. Petrov, V. N. Bogomolov, I. I. Kalosha, and S. V. Gaponenko, Phys. Rev. Lett. **81**, 77 (1998)
8. S. Noda, M. Imada, M. Okano, S. Ogawa, M. Mochizuki, and A. Chutinan, IEEE J. Quantum Electronics, **38**, 726 (2002)
9. S. Fan, P. R. Villeneuve, and J. D. Joannopoulos, Phys. Rev. Lett. **78**, 3294 (1997)
10. T. Baba , K. Inoshita, H. Tanaka, J. Yonekura, M. Ariga, A. Matsutani, T. Miyamoto, F. Koyama and K. Iga, IEEE/OSA J. Lightwave Thechnol. **17**, 2113 (1999)

11. K. Inoue, N. Kawai, Y. Sugimoto, N. Carlsson, N. Ikeda, and K. Asakawa, *Microphotonics-Materials, Physics and Applications*, eds. K. Wada *etal.* (MRS, Pennsylvannia, 2001) pp. E.3.2.1
12. K. Inoue, N. Kawai, Y. Sugimoto, N. Carlsson, N. Ikeda, and K. Asakawa, Phys. Rev. B**65**, 121308 (2002)
13. T. Aoki, M. Wada, J. W. Haus, Z. Y. Yuan, M. Tani, K. Sakai, N. Kawai, and K. Inoue, Phys. Rev. B**64**, 045106 (2001)
14. K. Inoue, M. Sasada, J. Kawamata, K. Sakoda, and J. W. Haus, Jpn. J. Appl. Phys. **38-2**, 2B, L157 (1999)
15. M. Meier, A. Dodabalapur, A. Mekis, R. E. Slusher, J. Rogers, and J. Joannopoulos, Appl. Phys. Lett., **74**, 7 (1999)
16. M. Imada, S. Noda, A. Chutinan, T. Tokuda, M. Murata, and G. Sasaki, Appl. Phys. Lett. **75**, 316 (1999)
17. S. Noda and M. Imada, IEICE TRANS. ELECTRON. **E85-C**, 45 (2002)
18. O. Painter, R. K. Lee, A. Yariv, A. Scherer, J. D. O'Brian, P. D. Dapkus, and I. Kim, Science, **284**, 1819
19. A. Sugitatsu and S. Noda, Electron. Lett. **39**, 213 (2003)
20. K. Inoshita and T. Baba, Electron Lett. **39**, 844 (2003)
21. A. Yariv, *Quantum Electronics*, 3rd edn. (John Wiley & Sons, New York, 1987) pp 600–650
22. K. Ohtaka, T. Ueta and K. Amemiya, Phys. Rev. B **57**, 2550 (1998)
23. K. Sakoda, Opt. Express **4**, 167 (1999)
24. K. Sakoda, K. Ohtaka and T. Ueta, Opt. Express **4**, 481 (1999)
25. J. P. Dowling, M. Scalora, M. J., Bloemer and C. W. Bowden, J. App. Phys. **75**, 1896 (1994)
26. S. J. Smith and E. M. Purcell, Phys. Rev. **92**, 1069 (1953)
27. O. Haeberlé, P. Rullhusen, J.-M. Salomé, and N. Maene, Phys. Rev. E **49**, 3340 (1994)
28. O. Haeberlé, P. Rullhusen, J.-M. Salomé, and N. Maene, Phys. Rev. E **55**, 4675 (1997)
29. F. J. Garcia de Abajo, Phys. Rev. Lett. **82**, 2776 (1999);
Phys. Rev. E **61**, 5743 (2000)
30. P. M. van den Berg, J. Opt. Soc. Am. **63**, 689 (1973);
J. Opt. Soc. Am. **63**, 1588 (1973);
J. Opt. Soc. Am. **64**, 325 (1974)
31. K. Ohtaka and S. Yamaguti, Optics and Quantum Electronics, **34**, 235 (2002); Optics and Spectroscopy, **91**, 506 (2001)
32. S. Yamaguti, J. Inoue, O. Haeberlé and K. Ohtaka, Phys. Rev. B **66**, 195202 (2002)
33. T. Kondo, M. Hangyo, S. Yamaguti, S. Yano, Y. Segawa, and K. Ohtaka, Phys. Rev. B **66**, 33111 (2002)

11 Photonic Crystal Devices

T. Baba

At the early stage of research, photonic crystals (PCs) were mainly discussed out of physical interest. However, studies of its device applications started and have expanded worldwide since the middle of the 1990s. Presently, PCs are expected to be one of the key technologies in the next era of optoelectronics. The first part of this chapter discusses the principles and future prospects of various PC devices.

11.1 How to Use Photonic Crystal Properties

Figure 11.1 summarizes the relation of PCs with other optics and various applications now being developed. Considering that PCs exhibit their functionalities by multidiffraction and multiscattering, a PC can be said to be a kind of hologram in a broad sense. A novelty of PCs, compared with conventional holograms, is the precise design method, namely the photonic band

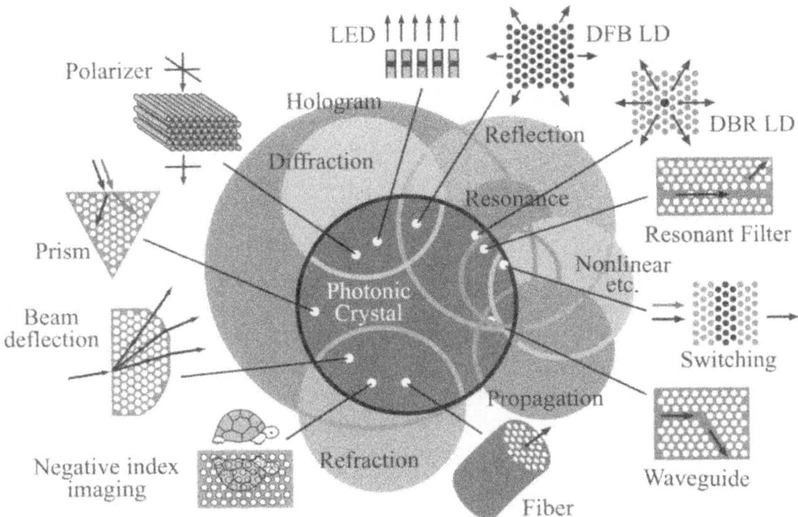

Fig. 11.1. Photonic crystal optics and various applications

(PB) calculation, which is very effective for estimating their functionalities and performance.

The PB diagram indicates that there are three different frequency ranges for light, which can be utilized for real applications, as shown in Fig. 11.2. The first one is the lowest frequency range below the first zone folding of the PB. The gradient of the lowest straight PB is determined by the effective refractive index n of the PC and it is different for different polarizations (Figure 11.2 is drawn for one polarization). Classically, this characteristic is called form birefringence. Since the PB calculation precisely predicts the effective index for each polarization, the index is artificially controlled by the PC structure. The second one is the photonic band gap (PBG), which means the omni-directional stopband. It is one of the most unique properties of PCs, and actually it was the main topic at the early stage of PC research. The PBG can be used as a reflector for light that is to enter the PC from arbitrary directions. It is applied to reflection-type devices, e.g., lasers and waveguides. The third one is the higher frequency range above the PBG where complex PBs exist. The slope of a PB is proportional to the group velocity v_g of light. Therefore, the horizontal band at a band-edge means zero v_g and the localization of light energy. In 2D and 3D PCs, such a zero v_g or a small v_g appears not only at a band-edge but in many bands. They will be effective for the enhancement of various interactions of light with materials of the PC. In addition, the 2D or 3D distribution of bands, the so-called dispersion surface, provides unique light propagation in PCs. Thus, this frequency range can be used in transmission-type functional devices.

Most devices presently studied use 2D PCs because of the relatively easy fabrication process compatible with LSI planar technology. In addition, most of the 2D PC devices exhibit their functions for light propagating in the 2D plane. Therefore, how to confine light in the third direction and how to

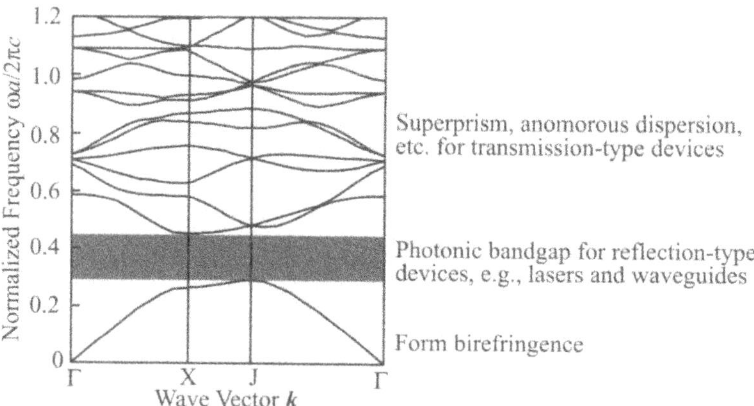

Fig. 11.2. Photonic band diagram for a 2D PC of holes in a triangular lattice and three frequency ranges for different application targets

couple external light to the 2D plane of a PC are sometimes discussed as important issues to realize high efficiency devices. This is the reason why the PC slab, which has a thin membrane structure with high refractive index contrast boundaries and confines light inside the 2D plane by total internal reflection, is widely studied. 3D PCs have a higher potential for the control of light emission and propagation. The fabrication process is still the main issue for 3D crystals. However, the detailed discussion of their applications will increase with the development of the fabrication process.

11.2 Light Emitters

PC light emitters so far studied are categorized into four types, as shown in Fig. 11.3, i.e., (a) ultralow threshold microlaser having a point-defect active region in a uniform PC, (b) high-power distributed feed back (DFB) laser utilizing the whole area of a uniform PC, (c) high-power and stable single-mode vertical cavity surface emitting laser (VCSEL) with an air-hole array, and (d) light-emitting diode (LED) with high extraction efficiency by PCs.

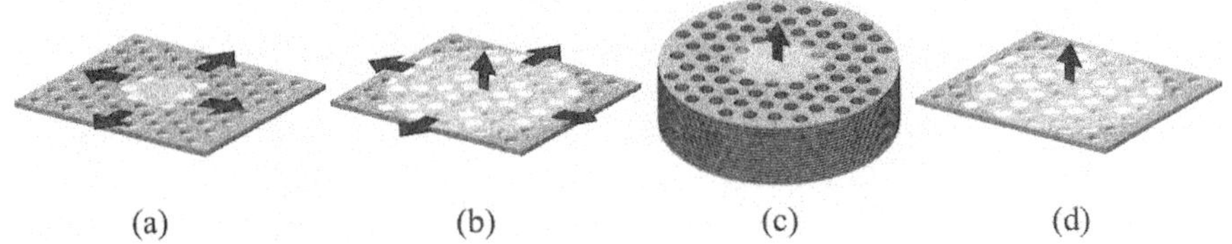

Fig. 11.3. Four photonic crystal light emitters: (**a**) point-defect laser; (**b**) band-edge laser; (**c**) VCSEL; (**d**) light-emitting diode

11.2.1 Point-Defect Laser

This is the first application that was discussed for PCs in 1987 [1]. The basic concept arises from impurity doping in semiconductors; the electron energy level appears in the electronic band gap, which spatially localizes the electron wave near the impurity atom. In the field of PCs, such impurities are commonly called defects. Similarly to the electronic case, light is strongly localized at a defect induced into a PC exhibiting a PBG. Thus it acts as an ultrasmall laser cavity. Such a cavity has the potential to exhibit an ideal cavity quantum electrodynamic effect including ultimate control of spontaneous emission. We can expect the development of an ultralow threshold laser or so-called thresholdless laser, which is usable as an internal light source in a high-density functional photonic integrated circuit and in quantum communication and computation systems operated by a single photon.

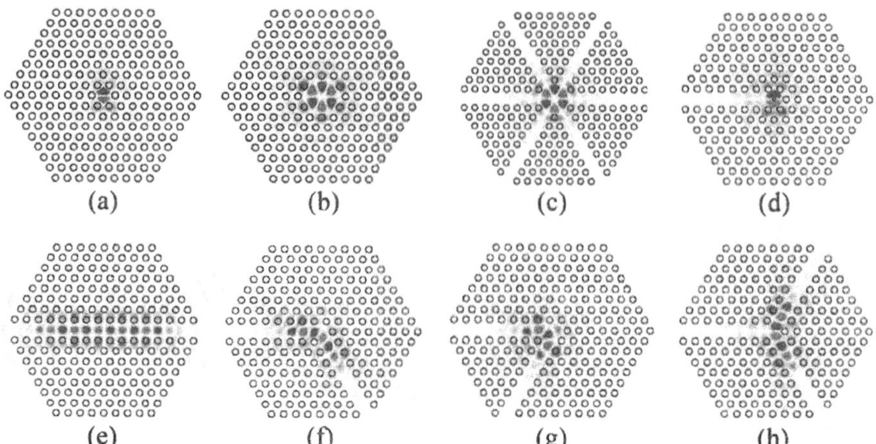

Fig. 11.4. Localized modes at various defects in a 2D photonic crystal of triangular lattice holes. The modes were calculated by the finite difference time domain method

The simplest defect is the deformation of one unit cell, e.g., one missing element from a uniform crystal [2]. But various defects also cause strong localization of light. Figure 11.4 shows magnetic field profiles of such localized modes calculated by the finite-difference time-domain (FDTD) computer simulation [3]. Here, the 2D PC of holes in a triangular lattice is considered, which exhibits a PBG for the polarization parallel to the 2D plane. Not only the smallest point defect, but also larger defects, line and point composite defects, modified line defects, and so on, maintain similar localized modes. For the simple line defect (e), an expanded mode appears. Such an expanded mode comes from the PB edge of a waveguide mode [4] (the details of the waveguide mode will be discussed in Sect. 11.3). Therefore, light localization in the composite defects and in the modified line defects are provided by the cutoff condition of the waveguide mode [5].

As a real device, the semiconductor PC slab with a defect has been studied, because of its multidimensional optical confinement by the PBG and total internal reflection. The first lasing operation was demonstrated by photopumping at low temperature in 1999 [2], and since then, lasing modes in various defects have been investigated. For three reasons, the GaInAsP/InP compound system has been widely used as a basic material [6]. The first reason is the emission wavelength longer than 1 μm, which is suitable for silica fiber-optical communication; the second reason is the mature epitaxial technology; and the third reason is the small surface recombination velocity v_s, which minimizes the nonradiative recombination at deeply processed structures of the photonic crystal slab having a large surface area for the unit cell volume. The typical v_s measured for this system is $1\text{–}2\times10^4$ cm/s. This value is 1–2 orders of magnitude lower than that of GaAs/AlGaAs. Figure 11.5

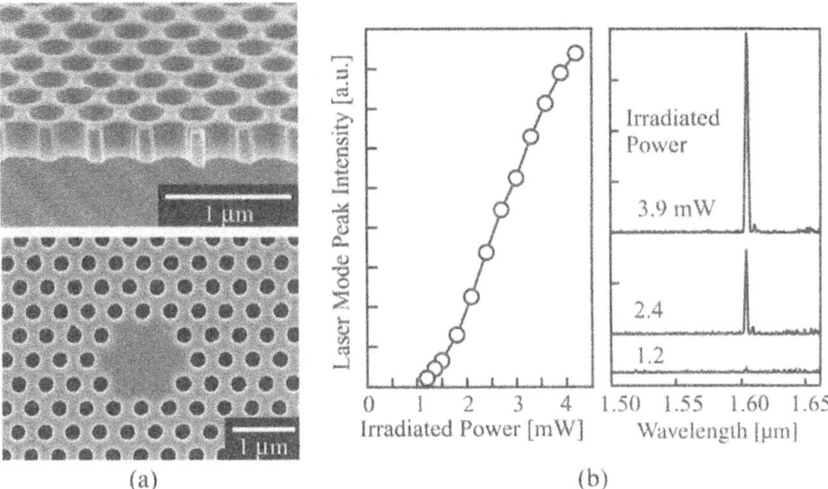

Fig. 11.5. Fabricated photonic crystal laser and lasing characteristics: (**a**) scanning electron micrographs (SEM) of a GaInAsP photonic crystal slab and point-defect cavity; (**b**) room temperature lasing characteristics observed by pulsed photopumping

shows a point defect cavity in a GaInAsP PC slab and room temperature lasing characteristic observed for pulsed photopumping. Here, the threshold power of total irradiated light was 1.4 mW [3]. Thus far, a modified single point defect cavity recorded the lowest threshold of ~ 0.2 mW [7].

In such lasers, however, the internal quantum efficiency is still restricted by surface recombination to $< 30\%$ in a typical PC slab. The threshold pump power at low temperature is of order 100 μW–1 mW, which is nearly 10 times higher than that of other microcavity lasers such as VCSELs and microdisks at room temperature. Therefore, an effective surface treatment that suppresses the surface recombination is crucial in PC lasers. For example, methane plasma irradiation reduces v_s of GaInAsP to $\sim 7 \times 10^3$ cm/s. Figure 11.6 shows the change of carrier lifetime τ_c of 1.55 μm compressively strained quantum wells measured with various diameters $2r$ of micropillars at room temperature by phase resolved spectroscopy [8]. The slope of the $\tau_c^{-1} - (2r)^{-1}$ characteristic gives half of v_s. The reduction in v_s by methane plasma irradiation corresponds well to the increase in photoluminescence intensity. Further optimization of the material, the treatment and the structure are desired to obtain a sufficiently low v_s of less than 7×10^3 cm/s. For this purpose, the investigation of other materials such as the GaN system, which may have a low v_s, is also important.

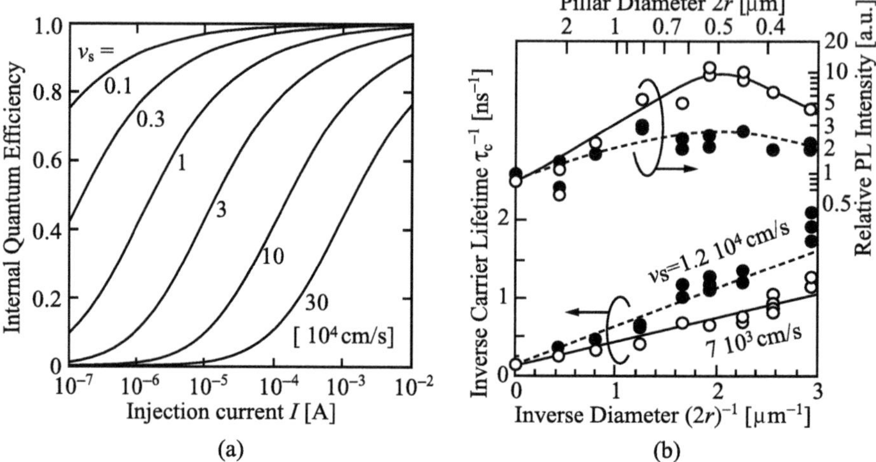

Fig. 11.6. Influence of surface recombination on light emission characteristics: (a) theoretical estimation of internal quantum efficiency with injection current for a pillar-type semiconductor photonic crystal, which exhibits the lowest-order photonic band gap for the polarization parallel to the pillars; (b) experimental results of carrier lifetime and photoluminescence intensity in a GaInAsP/InP pillar-type 2D photonic crystal. Closed circles and dotted lines indicate as-formed structures, while open circles and solid lines indicate those after CH_4 plasma irradiation

11.2.2 Band-Edge Laser

This is the expansion of a conventional DFB laser toward a 2D periodic structure. The lasing mode can be understood in two ways; one is the zero v_g at a band edge, and the other is the standing wave of coupled modes. A unique

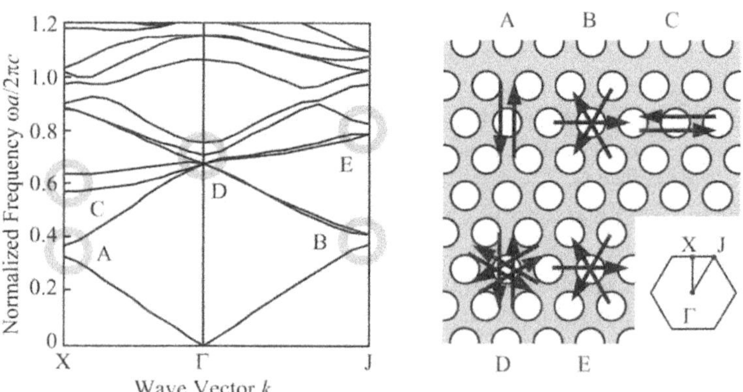

Fig. 11.7. Photonic band diagram for a 2D photonic crystal and schematic of optical feedback at various band edges

advantage of this laser is the coherent lasing operation over the wide 2D area of a uniform PC. This is confirmed from unique far-field patterns (FFPs), which can be explained by PBs [9, 10]. Figure 11.7 shows an example of a PB diagram for a triangular lattice 2D PC. Each band edge generates a different 2D feedback, as illustrated in the figure. In particular, a diffracted beam in the vertical direction with a single-lobe FFP, a single frequency and a single polarization was demonstrated by using the Γ point of the second band [11]. This laser is expected to be a high-power single-mode surface-emitting laser. It will be at a practical level by improving two general conditions: one is the threshold current or power due to the wide area cavity, and the other is the low light extraction efficiency due to the different directions of the resonance and light extraction.

11.2.3 VCSEL

This is a relatively new topic since 2001 [12]. The VCSEL is a microcavity laser that achieves optical feedback in a 1D PC, i.e., a semiconductor multilayer stack. It is now commercially available and used as a low-cost light source in local area networks. Presently, however, the single-mode light output is limited to < 8 mW due to unstable lateral modes for the large-aperture of the device [13]. The use of a PC, as illustrated in Fig. 11.8, is expected to suppress higher lateral modes in a large-aperture VCSEL by the mechanism of single-mode propagation in PC holy fibers, as will be discussed in Sect. 11.4.

11.2.4 High Extraction Efficiency LED

This is a subject studied by several groups since 1997 [14]. In normal LEDs, the light extraction efficiency strongly restricts the total efficiency to $< 10\%$. Therefore, its improvement is an important issue for future LED displays and white light illuminators. At the first stage of research, the use of a PC slab and

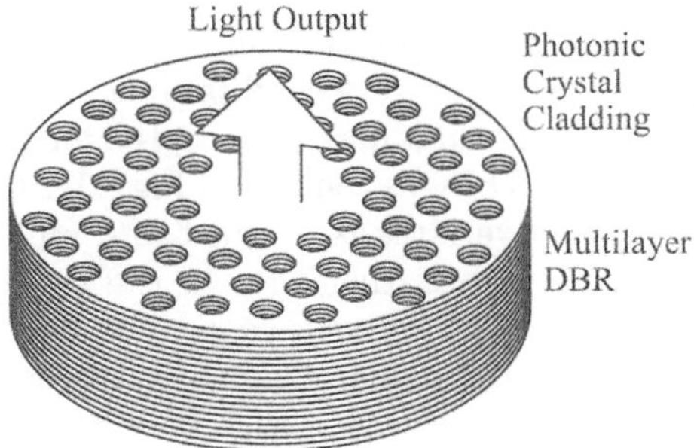

Fig. 11.8. Schematic illustration achieving a VCSEL by using a fiber

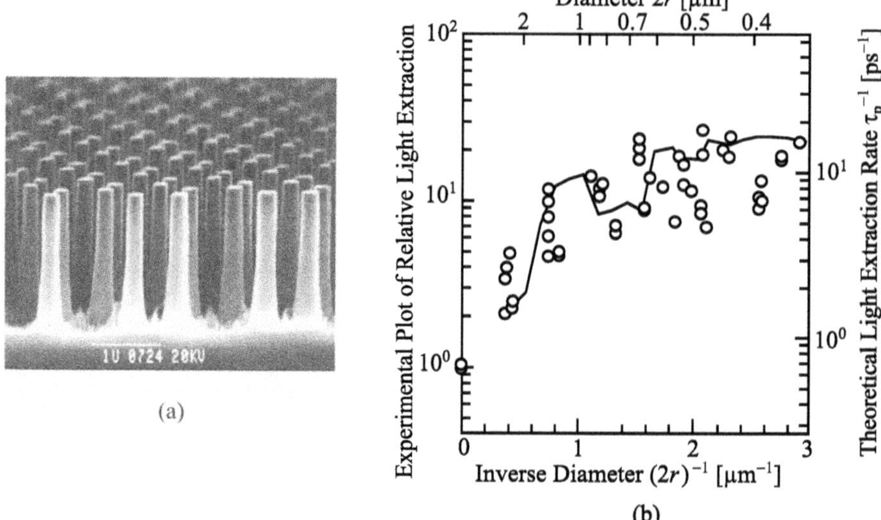

Fig. 11.9. Enhancement of light extraction efficiency measured and calculated for a GaInAsP/InP 2D pillar-type photonic crystal

a micropillar array was investigated, since the 2D periodicity of these structures shifts the tangential component of the **k** vector so that the internal light satisfies a light extraction condition called the light cone. Theoretically, they improve the extraction efficiency to > 80%. Such a high extraction efficiency was experimentally demonstrated in micropillars, as shown in Fig. 11.9 [15]. Here, the maximum enhancement of over 20 times was evaluated through photoluminescence and carrier lifetime measurements. However, the evaluation of carrier lifetime indicated a large surface recombination, as mentioned above. It reduced the internal quantum efficiency to < 20% and disturbed the improvement of the total efficiency. In addition, the formation of metal electrodes for current injection is essentially difficult in these structures. To avoid the first problem, a structure with a PC pattern separated from the light-emitting area was studied [16]. But they still need a thin slab structure with low n claddings, for which the second problem still remains.

Another simpler structure that may be accepted for low-cost LED production is the surface grating 2D PC, as shown in Fig. 11.10(a) [17]. The internal light emitted from the active layer is coupled into guided light around the active layer and the free propagating light in the semiconductor. The free propagating light not extracted from the semiconductor by the total internal reflection has the largest solid angle. Therefore, the extraction efficiency is improved by applying the reciprocal lattice vector of the grating to the k vector of the free propagating wave and changing the propagating angle to that extracted to free space. Figure 11.10(b) compares the FDTD simula-

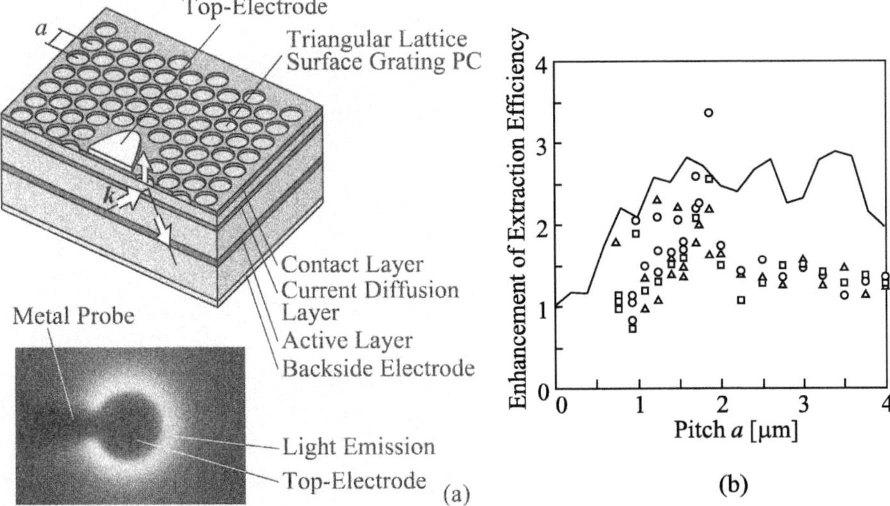

Fig. 11.10. GaInAsP/InP LED structure with surface grating 2D photonic crystal: (**a**) schematic structure and observed near-field pattern; (**b**) measured and calculated enhancement in total efficiency of LED

tion and experimental results obtained for GaInAsP/InP LEDs with the 2D surface grating PC. The improvement of the efficiency by a factor 2–3 was confirmed; this was determined by the solid angle effect and the diffraction efficiency. This value is attractive when considering the shallow etching process of ~ 0.5 μm depth for the surface grating, a relatively large lattice constant of several μm, low wavelength sensitivity, robustness against structural imperfections and adaptability to arbitrary materials. Therefore, this structure is not only effective for semiconductor LEDs but also for any spontaneous-emission-based light emitters such as an organic electroluminescence (EL) device.

11.3 Optical Waveguides

Figure 11.11 summarizes various waveguides based on the PC concept. All of them utilize line defects, a series of defects in a PC. Most studies are concentrated on PC slabs. This section first describes the detail of this waveguide, and then outlines other types.

11.3.1 Line-Defect Waveguide in a Photonic Crystal Slab

The line defect induced into a uniform PC acts as an optical waveguide. This is the so-called PC line-defect waveguide. First, it was demonstrated in a numerical simulation for square lattice and triangular lattice 2D PCs composed

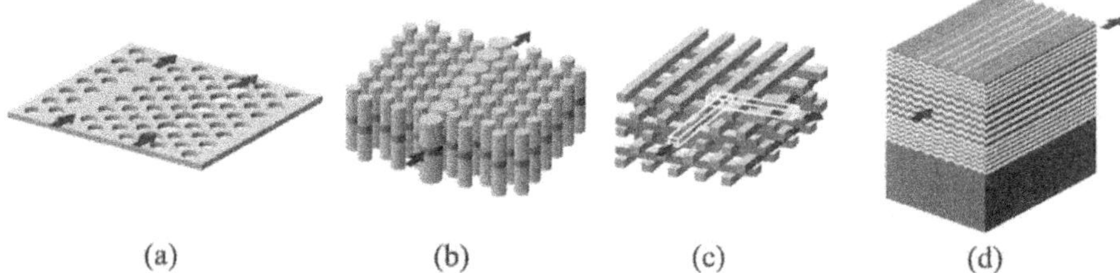

Fig. 11.11. Various waveguide structures based on photonic crystal line defects: (**a**) photonic crystal slab type; (**b**) pillar type; (**c**) 3D wood-pile type; (**d**) auto-cloning type

of dielectric pillars with infinite height [18, 19]. The polarization of light was limited to that of TM mode having the electric field parallel to the pillars and exhibiting the photonic bandgap inside the 2D plane. However, the channel is air in this case, so it is difficult to confine light inside the 2D plane. Therefore,

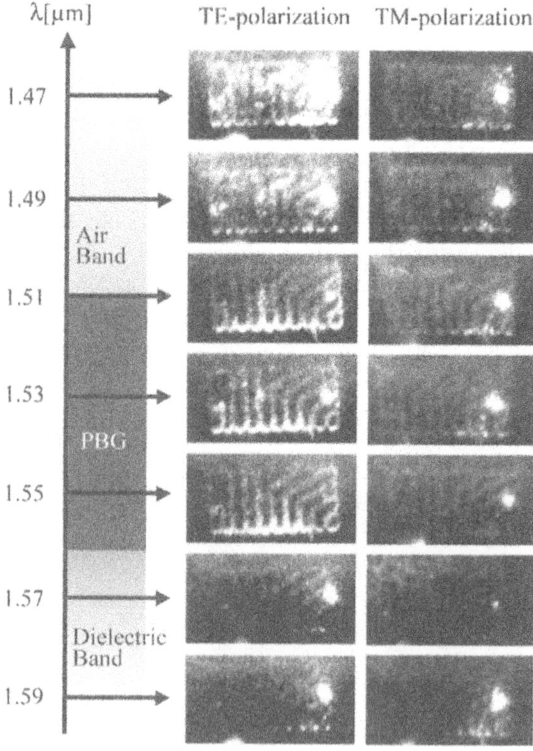

Fig. 11.12. Near-field patterns observed from the top of the photonic crystal slab type waveguide fabricated by bonding GaInAsP/InP film on top of the InP/SiO_2 host substrate

in the experiment, a PC slab composed of holes was employed [20]. The 2D PC of holes exhibits a wide PBG inside the 2D plane for polarization parallel to the 2D plane (TE polarization). In the PC slab, the band gap also occurs for the TE-like polarization. In this case, the line defect is a dielectric channel, and the light is confined in the vertical direction by total internal reflection. In the first experiment, a GaInAsP semiconductor film with holes was bonded on a SiO_2 film. The light propagation in the line defect was observed at fiber communication wavelengths, as shown in Fig. 11.12. Here, wavelength and polarization dependences were observed, which were related to PB properties.

At the first stage of the experiment, however, the pure guided mode condition was not clear. The observation of light propagation from the top suggested that it was not a pure guided mode but a leaky mode. This leaky mode condition is the same as the light extraction condition for PC LEDs, i.e., the light cone. At the second stage of the research, the PB diagram for the waveguide was investigated in detail, as shown in Fig. 11.13. Here, the light cone lies at the higher-frequency range than the pure guided mode region. Two requirements were clarified for the pure guided mode condition overlapping with the PBG [21]. One is the relatively small diameter of holes, so that PBs of waveguide modes are pulled down to below the light line, which is the boundary of the guided mode condition and the light cone. The other is the air cladding sandwiching the slab, which allows a wide frequency range and a large v_g. These conditions were confirmed in a waveguide fabricated by the well-improved process. An important improvement was the use of a silicon-on-insulator (SOI) wafer [22–24] and a GaAs/AlGaAs epitaxial film [25]. This is a high-quality commercially available substrate having a Si slab, and is transparent at fiber communication wavelengths. The PC wave-

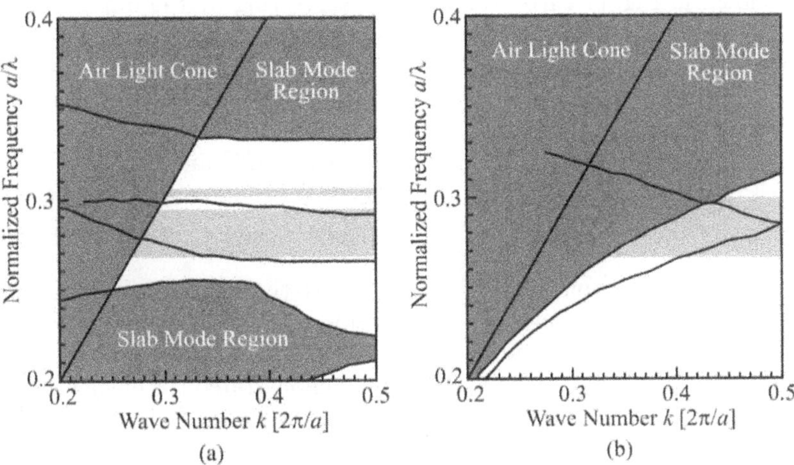

Fig. 11.13. Photonic bands of a single line defect in a photonic crystal slab: (**a**) TE-like polarization; (**b**) TM-like polarization

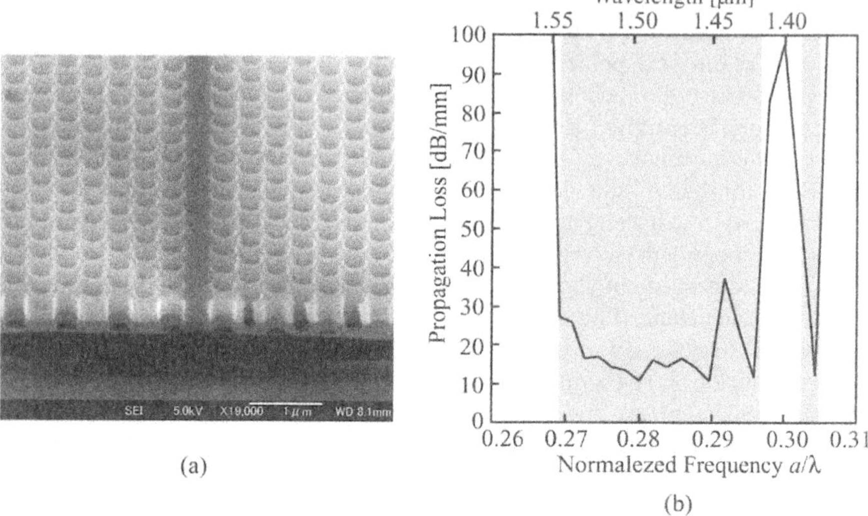

Fig. 11.14. Fabricated single line defect waveguide and propagation loss: (**a**) picture of waveguide fabricated on an SOI wafer; (**b**) measured propagation loss spectrum

guide was simply fabricated only by opening holes, as shown in Fig. 11.14(a), and forming the air-bridge structure to satisfy the second issue. As a result, the pure guided mode, which corresponds well to theoretical expectation, has been observed with no light leakage along the waveguide. At present, the propagation loss of this waveguide is evaluated to be as large as 1–10 dB/mm, as shown in Fig. 11.14(b), due to light scattering by the imperfection of the structure. Therefore, it will be reduced by the development of process techniques. In addition, this waveguide has the potential of lower scattering loss than that in a high n contrast waveguide, since the PB essentially restricts the existence of radiation modes. Simple perturbation theory of the scattering loss for an index-confinement-type waveguide and the assumption that the PC waveguide is equivalent to the index-confinement with no radiation modes indicates that the loss would be reduced by one order of magnitude [24].

The biggest topic of this waveguide is the feasibility of ultra-compact waveguide elements such as sharp bends, small branches, and short directional couplers [19]. These elements were first numerically calculated with pillar-type PC models. The results well demonstrated the efficient light transmission through these elements. However, the modal behavior is much more complex and the design window for efficient transmission is much narrower for the PC model. To avoid these problems, various modified structures have been investigated [26, 27]. Figure 11.15 shows some examples of modified bends, for which the 2D FDTD calculation predicted a high transmission

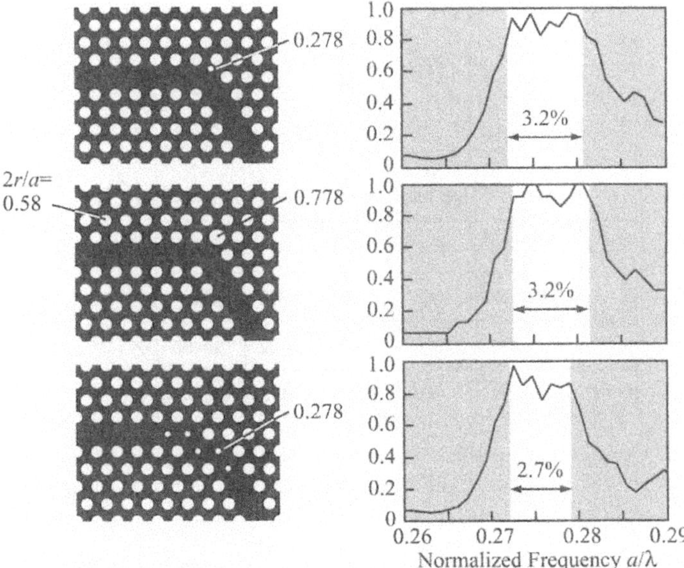

Fig. 11.15. Modified bend structures in a photonic crystal waveguide and their transmission spectra calculated by the FDTD method

efficiency over 95% in a relatively wide spectral range (i.e., nearly 6% of the center frequency). However, the condition that simultaneously satisfies a higher efficiency and a wider transmission range in a 3D structure is now still under investigation.

Presently, many groups are studying this waveguide as a platform for the next era of high-density photonic integrated circuits. However, one should pay attention to the fact that similar sharp bends, branches, etc., can also be realized by a high index contrast waveguide [28–30]. The combination of PC waveguides and high index contrast waveguides will be a smart choice. As unique applications of the PC waveguide, an ultralow group velocity [31] at a band edge and a point defect cavity filter [32] are attractive. Another important issue is the optical coupling between the waveguide and the single-mode fiber. Two methods are being studied; one is the mode size converter [33] and the other is a grating coupler [34]. Recently, the first one achieved a low loss of 0.8 dB/port. Another issue is the fragility of the air-bridge structure. It strongly restricts the flexibility of the optical wiring and the stacking of additional layers. Modified width and modified shape of line defects are being studied to engineer the PB of the waveguide so that the restriction of the air-bridge structure is avoided [23].

Figure 11.16 summarizes the comparison of waveguide loss, device size, integration scale, etc. among a conventional silica waveguide, a high index contrast waveguide called a Si photonic wire and a PC waveguide. Here, the

	Roughness σ [nm]	Relative Index Diff. Δ [%]	Scattering Loss α_s [dB/mm]	$\propto \sigma^2 \Delta^{2.5}$ S [mm²]	Chip Size r [mm]	Bend Radius N [piece] (Device Number)
Silica	50	0.3	<0.01	50^2	>2	50
Si PW	20	45%	10	0.05^2	0.002	50
	4	45%	0.4	$>1^2$	0.002	20000
Si PC	4	45%	<0.1	$>4^2$	0.002?	>100000?

Fig. 11.16. Comparison of silica waveguide, Si photonic wire (PW) waveguide and Si photonic crystal (PC) waveguide for high-density photonic integration

waveguide loss is assumed to be dominated by the scattering loss and the scattering loss of the PC waveguide is assumed to be 10 times smaller than the photonic wire due to the suppression of radiation modes by the PBG. In addition, the device size is assumed to be simply determined by the bend radius. As shown in Fig. 11.16, the photonic wire achieves a drastic reduction in device size and an ultra-high density integration. However, with the PC waveguide, larger-scale integration will be possible by reducing the process roughness to less than 10 nm and reducing the scattering loss to 1 dB/cm order.

11.3.2 Other Types of Waveguide

In the line-defect waveguide, point defects are directly connected with each other to become a line defect. On the other hand, another type of waveguide is also studied, in which each point-defect is separated by some period of a PC [35]. It is called the coupled cavity waveguide, since neighboring defects are weakly coupled with each other. Light of a certain frequency range propagates through the waveguide by repeating the power transition from one defect to another defect. Theoretically, it achieves high efficiency light transmission through any sharp bends. However, the frequency range is limited due to its principle of operation and the transmission efficiency is not robust against the fluctuation of the defect size. It is now studied not only as a photonic circuit element but also as a dispersion compensation device.

Similarly to the waveguide in the PC slab, the pillar-type waveguide was proposed, as illustrated in Fig. 11.11(b) [36]. Each pillar has a three-layer waveguide structure, and larger pillars are used as line defects. The PB calculation showed the existence of a PBG and pure guided modes for the TM-like polarization.

An index confinement type line defect waveguide was also studied. It was demonstrated in a corrugated layer structure fabricated by bias sputtering on a corrugated substrate and named an autocloned PC [37]. Since it does not have a PBG, it is not suitable for high-density integration. The advantage of this type is the simple fabrication process achieved by only one time etching and one time deposition. Some functional devices utilizing its unique dispersion characteristics are expected.

Thus far, 2D PC waveguides have been discussed. However, almost the same principle can be considered for 3D crystals. A line defect waveguide composed of one missing bar in a woodpile 3D PC and its bend were calculated and fabricated. Calculation predicted light propagation with smaller dispersion and a wider range of high transmission efficiency at a sharp bend than in a 2D PC [38].

11.4 Optical Fibers

Figure 11.17 illustrates three kinds of PC fibers. The first two are those pioneered by Russell, et al. [39, 40] and partly commercialized by a few fiber companies. They are fabricated by drawing a pure silica glass preform with many air holes. Therefore, the fibers have periodic structures in their cross-sections and uniform structures along the optical axis. The first one is called a holey (or microstructured) fiber and the second one is called a photonic band-gap fiber. The former has a dielectric defect and the latter has an air defect at the center of the PC as the fiber core. The third fiber is different from the other two types; it has axially symmetric multilayers [41]. It has also been partly commercialized.

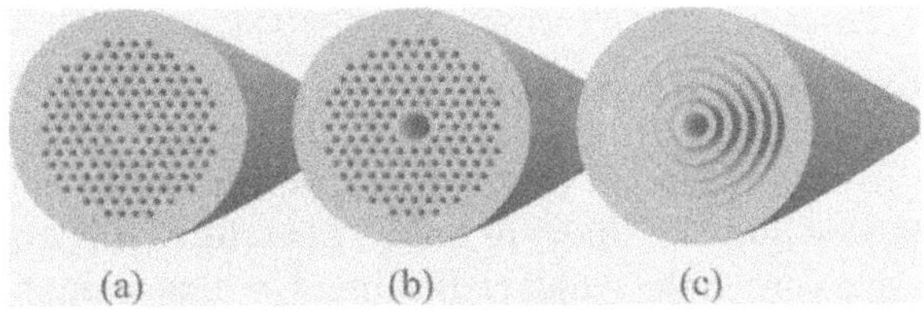

Fig. 11.17. Cross-sectional structures of photonic crystal fibers: (**a**) holey fiber; (**b**) photonic band-gap fiber; (**c**) Bragg fiber

11.4.1 Holey Fiber

The principle of light propagation in the holey fiber is not a PBG but total internal reflection. Light is confined around the center silica core by the difference between the index of the core and the effective index of the PC cladding. The propagation loss was reduced to $\sim$ 0.5 dB/km by improving the uniformity of the holes and additional losses coming from the absorption and contamination. This fiber has unique features for the single-mode condition and related strong optical confinement, the dispersion characteristic, and the polarization characteristic.

Regarding the first point, longer wavelength light penetrates into air holes in the PC cladding, so the effective index of the cladding decreases and the optical confinement into the core is strengthened. Shorter wavelength light is well confined in the silica in the PC cladding, so the effective index approaches the core index and the optical confinement is moderately weakened. As a result, the single-mode condition is maintained in a very wide frequency range from visible to near-infrared. This feature is applied to a small (large) core size single-mode propagation that enhances (reduces) the optical power density and the nonlinearity. Rare-earth-metal-doped amplifications, Raman amplification, four-wave mixing, super-continuum radiation, etc., were experimentally demonstrated [42].

Regarding the second point, a very large absolute value of positive or negative dispersion and a zero dispersion wavelength are arbitrarily designed by changing the lattice and the diameter of holes. For example, a negative dispersion of −2000 ps/nm/km was theoretically predicted at a wavelength λ of 1.55 μm [43]. In the experiment, zero dispersion was realized at a wavelength of 0.8 μm, and a positive dispersion of 200 ps/nm/km was achieved in the wavelength range of 0.6–1.0 μm [44]. A high performance dispersion compensation fiber and a wide spectral range zero dispersion fiber are expected.

Regarding the third point, unique birefringent characteristics and polarization maintaining functions were also investigated theoretically and experimentally. For this purpose, a large number of air holes are not necessarily required. Various fibers with fewer holes are being investigated.

11.4.2 Photonic Band-Gap Fiber

This fiber is based on reflection by the PBG, which can be controlled by the design of the lattice and the shape of holes. Therefore, single-mode propagation in a large size core and a small radius bend, which cannot be achieved in a holey fiber, are expected. These features are promising for high power transmission with small nonlinearity. Recently, the propagation loss was reduced to dB/km order. The remaining issue is the relatively narrow transmission range dominated by the range of the PBG.

11.4.3 Bragg Fiber

This fiber was invented based on the discovery of the omni-directional reflection characteristic of 1D PCs, i.e., dielectric multilayer films [45]. In classical optics theory, it is well known that the stop band of an alternating stack of two different dielectric media has dependences on the incident angle and polarization. Actually, however, an appropriate choice of refractive indexes of the two media allows a fixed stop band for any directions and polarizations. This characteristic was used as a cladding of a hollow core fiber, which is sometimes called a Bragg fiber. It is not only expected to be a high-power transmission fiber but also an optical communication fiber, which is free from loss mechanisms in silica fibers.

11.5 Wavelength Filters

A high-performance wavelength filter is the key component in wavelength division multiplexing (WDM) systems. PCs are expected to offer a compact filter for this purpose. Fundamentally, there are three different types, i.e., resonant type, directional coupler type and diffraction type, as shown in Fig. 11.18.

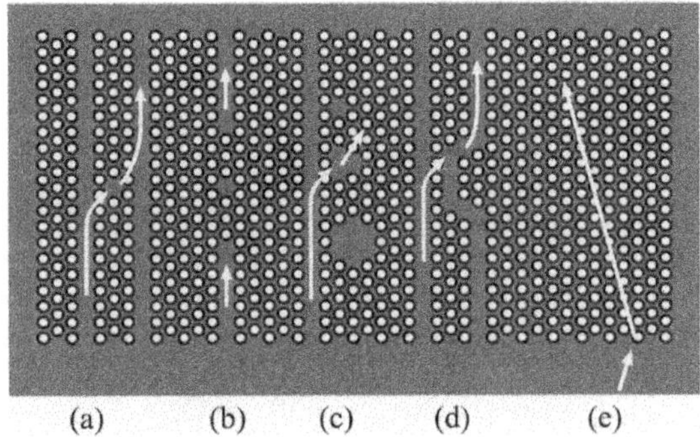

Fig. 11.18. Various types of wavelength filter in a 2D photonic crystal: (**a**), (**b**) resonant filters with parallel waveguides and series waveguides, respectively; (**c**) resonant filter coupled with free space; (**d**) directional coupler; (**e**) diffraction filter based on superprism effect

11.5.1 Resonant Type Filter

A point defect in a PC can be an ultimately small cavity for sinusoidally oscillating light waves. Therefore, it can be a resonant filter with the largest possible free spectral range (FSR). Although the FSR in an actual point defect cavity is restricted by the width of the PBG, it will cover, for example, the C band and L band fiber communication ranges of 1.53–1.61 μm wavelengths. By improving the Q factor to 1000–100 000, an ultimately large finesse of over 10 000 will be possible.

So far, the combination of such a point defect (or a slightly larger defect) and line-defect waveguides, as shown in Fig. 11.18(a)–(c), has been investigated. For the extraction of resonant light, the coupling to another waveguide [46] and the direct coupling to free space [32] were proposed and the latter was demonstrated as an add/drop filter. The wide range tunability of the resonant frequency is obtained by controlling the defect size. However, this tunability also becomes a disadvantage in any resonant type filters. To get a target resonant frequency, the defect size must be controlled with nanometer order precision. For the production of a practical device, a post process such as trimming will be indispensable. Current interest for researchers is the improvement of the Q-factor and the efficiency. But in a real WDM system, the control of the filter function is necessary to obtain high spectral efficiency. The simple Lorentzian function of a resonant filter must be changed to a box-like shape by using a multiresonance. For this purpose, the design of coupled defects will be an important issue in future.

11.5.2 Diffraction Type Filter

Peculiar dispersion characteristics in the frequency range higher than the photonic bandgap, as shown in Fig. 11.2, can be used as a diffraction type filter. Due to its angular deflection of light, it was called a superprism filter [47].

In a standard 1D diffraction grating, the simple zone folding of the dispersion characteristic generates wavelength-dependent diffraction waves and this characteristic is used as a filter. In a higher dimensional PC, more complex zone folding occurs, which enhances the wavelength sensitivity (angular dispersion) and modifies the beam propagation. This characteristic is understood from the dispersion surface analysis, i.e., the drawing of contour plots of band curves over the BZ, as shown in Fig. 11.19. In a PC, light propagates in the gradient direction of the dispersion surface. Therefore, depending on the deformation of the dispersion surface, light propagation exhibits collimated-beam-like propagation, and concavity- and convexity-lens-like propagation. Since this characteristic also has a strong wavelength dependence, a large angular dispersion is achieved. However, if this characteristic is utilized for a filter, the above-mentioned light behavior must be considered to estimate the wavelength resolution of the filter. Figure 11.20 shows mappings of three important parameters showing the degree of beam collimation, the wavelength sensitivity and the wavelength resolution for the second photonic band in a triangular lattice 2D PC [48]. Here, the gray curves denote the locus of dispersion characteristics against a certain incident angle of light. Since the incident angle is generally fixed in a filter, the dispersion characteristic along one such equi-incident-angle curve should be used. At the early stage of research, a high resolution was expected at an abruptly changing dispersion characteristic. However, such a characteristic does not continue along an incident-angle curve, so the usable wavelength range is limited. In addition, the simultaneous increase in angular dispersion and the divergence

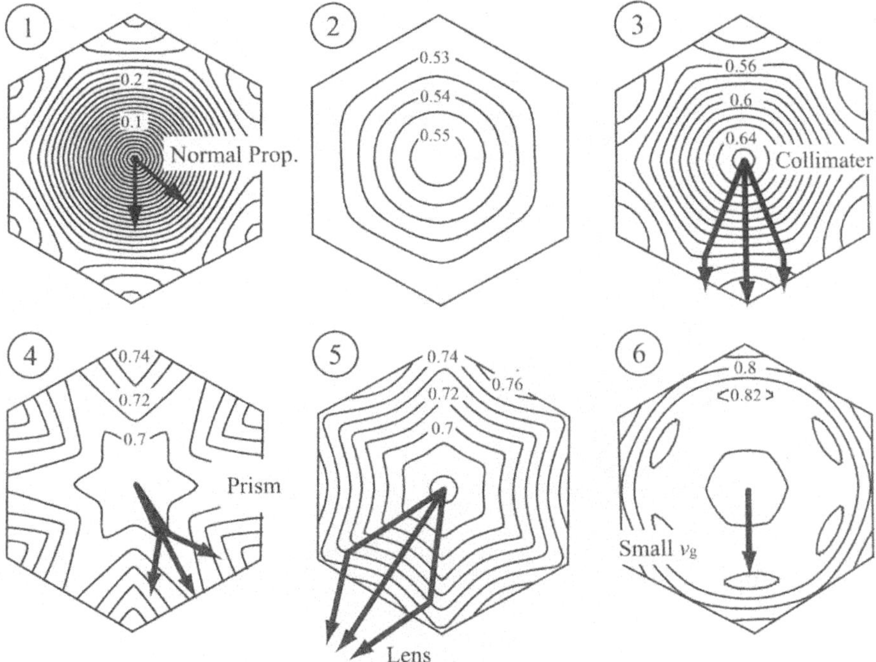

Fig. 11.19. Dispersion surface calculated for each band of a 2D photonic crystal

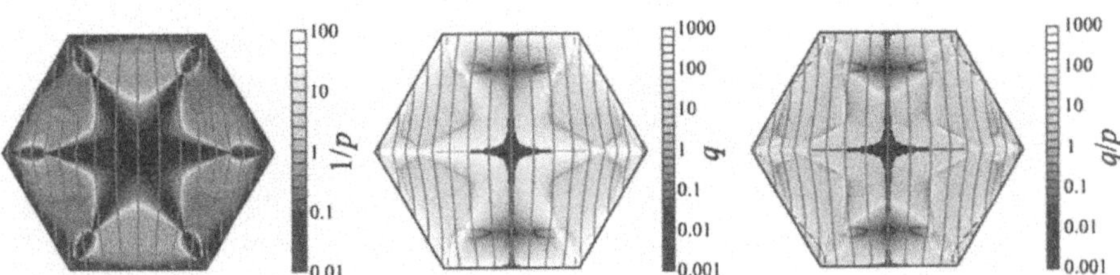

Fig. 11.20. Beam collimation parameter $1/p$, wavelength sensitivity parameter q and wavelength resolution parameter q/p, which are calculated for a 2D photonic crystal

of the light beam rather degrades the resolution, and seriously reduces the available number of resolution points. As seen in Fig. 11.20, a high-resolution condition along an incident-angle curve is rather obtained by a condition slightly apart from the abruptly changing dispersion characteristic. Under this condition, the resolution parameter can be higher than that for a normal diffractional grating. However, a total length of cm order is required for spatially separating two different wavelength beams. This length is similar to that of a silica-based array waveguide grating (AWG) filter. The reason of this long length is the complex zone-folding itself. In normal diffractional gratings and AWGs, the resolution is improved by increasing the diffraction order. On the other hand, the superprism uses a lower order (second to fourth) band, which corresponds to lower-order diffraction, since higher-order bands have complex overlap between bands resulting in multibeam outputs. Now, size reduction and high resolution are being investigated by modifications of the structure and the principle of light beam separation.

11.6 Polarization Filters

Photonic crystals have many boundaries with a high n contrast. In general, such a structure exhibits strong polarization dependence. In a photonic band diagram for a 2D PC, band curves are very different for two orthogonal polarizations. This feature is easily used for a polarization selective filter. The simplest way to realize such a filter is to use the photonic band gap for one polarization [49]. This allows a reflection type filter. A 2D PC composed of multilayers on a corrugated substrate was used as a vertical input type filter and a high performance of < 0.5 dB transmission loss and -50 dB extinction ratio was demonstrated. They are sufficient for practical use, so it is now at the stage of discussing production costs.

11.7 Dispersion Compensators

By optimum design of the dispersion characteristic of a bulk photonic crystal or a PC waveguide, the dispersion compensation factor will be much larger than that in a dispersion compensation fiber. In particular, the characteristic of a coupled cavity waveguide is expected to allow wide band compensation due to the symmetric band folding at the band edge [50]. It is expected to allow miniaturization and tunability in the dispersion compensation device. However, the estimated device size is still of the order of mm to cm. Therefore, the propagation loss of the waveguide should be less than 1 dB/cm. In addition, the normalized frequency, which is used for dispersion compensation, is relatively high (> 0.5). This leads to the problem of large leakage loss of light due to the light cone, when the waveguide is fabricated into a PC slab. The suppression of this loss will be the primary consideration of this kind of device.

11.8 Light Control Devices

In the next era of fiber communication networks, the development of a photonic switch and a wavelength tuner will be two of the most important devices. In conventional devices, the carrier plasma effect, the electro-absorption

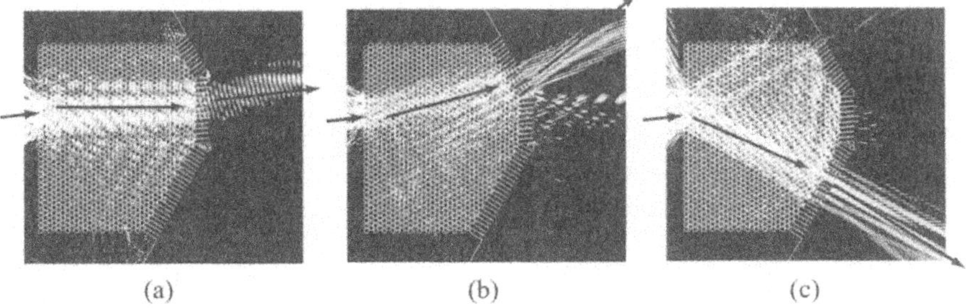

Fig. 11.21. Numerical demonstration of light deflection device by the FDTD method, where the superprism phenomenon in a 2D photonic crystal of holes is utilized: (**a**) normalized frequency of 0.55; (**b**) 0.61; (**c**) 0.70

effect, the Pockels effect, the Kerr effect, the cross-gain modulation effect, the four wave mixing effect, etc., have been used. The semiconductor amplifier is also increasing in importance due to its variety of functions including the simple on/off switching effect by carrier injection. Up to now, discussions on these effects with PCs are limited. However, the point defect mode and the small v_g band can enhance the internal optical intensity, which results in the enhancement of these effects. A similar effect is seen in a Fabry–Perot etalon. There is a trade-off relation between the enhancement and the narrowing of the transmission band. In addition, high transmission is obtained only at the top of the Lorentzian function in a simple Fabry–Perot etalon. Therefore, a short optical pulse with a wide spectrum cannot gain this enhancement as it is. This restriction can be relaxed by a coupled cavity design, which flattens the transmittance around the center of the resonant frequency. In a higher dimensional PC, localized light at a point defect realizes a similar effect. It will be possible to couple light to the defect by using line defect waveguides, as discussed in Sect. 11.5. On the other hand, a higher-dimensional photonic crystal exhibits a photonic band whose group velocity is smaller than 1/10 of that in free space. Such bands close to band edges have a large dispersion, so they are not suitable for the control of a short pulse. But those in the middle of the Brillouin zone realize zero dispersion with small group velocity.

Another light control device expected for PCs is the deflection device of an optical beam, which will be used in scanners, displays and spatial optical switches. For this device, the superprism effect is also effective [51]. As shown in Fig. 11.21, light deflection is possible for different wavelengths of the incident beam in a PC superprism with angled output ends.

All the devices discussed here are light transmission type devices. Therefore, high transmission efficiency is an important issue. Since the higher frequency bands have complex field profiles of Bloch functions, the coupling of external light (e.g. a plane wave) is generally low. However, the coupling can be improved by using an appropriate interface of the PC, which extinguishes the field mismatch. Gradual changes of the size and the shape of the PC,

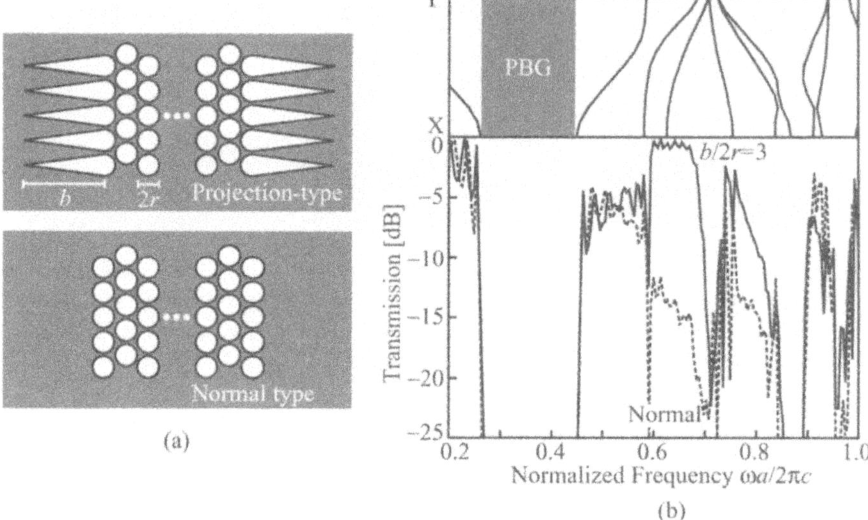

Fig. 11.22. High-efficiency input/output interface for light transmission through a 2D photonic crystal of holes in a triangular lattice: (**a**) schematics of projection-type and normal interfaces; (**b**) photonic band diagram and transmission spectra calculated by the FDTD method

as shown in Fig. 11.22, were proposed and a high transmission efficiency of $> 93\%$ was theoretically demonstrated [52].

11.9 Harmonic Generation

PCs have two effects for the improvement of harmonic generation efficiency. One is multidimensional phase matching and the other is the enhancement of internal light by the small v_g band. As an example of the first one, a $LiNbO_3$ 2D PC was fabricated with a lattice constant a of several μm to 10 μm, and multibeam harmonic generation was experimentally demonstrated [53]. This experiment also showed an easier phase matching condition by the PC. On the other hand, the latter effect is similar to those discussed in Sect. 10.6. In this case, the lattice constant will be less than 1 μm. If the combination of these effects allows phase matching, strong enhancement and a wide tolerance, it will be a unique key technology for harmonic generation.

References

1. E. Yablonovitch, Phys. Rev. Lett. **58**, 2059 (1987)
2. O. Painter, R. K. Lee, A. Scherer, A. Yariv, J. D. O'Brien, P. D. Dapkus, and I. Kim, Science **284**, 1819 (1999)

3. T. Baba, K. Inoshita, D. Sano, A. Nakagawa and K. Nozaki, SPIE West, 5000-05 (2003)
4. A. Sugitatsu and S. Noda, Electron. Lett. **39**, 213 (2003)
5. K. Inoshita and T. Baba, Electron. Lett. **39**, 844 (2003)
6. T. Baba, IEEE J. Sel. Top. Quantum Electron. **3**, 808 (1997)
7. M. Loncar, T. Yoshie, A. Scherer, et al., Appl. Phys. Lett. **81**, 2680 (2002)
8. H. Ichikawa, K. Inoshita and T. Baba, Appl. Phys. Lett. **78**, 2119 (2001)
9. M. Meier, A. Mekis, A. Dodabolapur, A. Timko, R. E. Slusher, J. D. Joannopoulos, and O. Nalamasu, Appl. Phys. Lett. **74**, 7 (1999)
10. M. Notomi, H. Suzuki and T. Tamamura, Appl. Phys. Lett. **78**, 1325 (2001)
11. M. Imada, S. Noda, A. Chutinan, T. Tokuda, M. Murata, and G. Sasaki, Appl. Phys. Lett. **75**, 316 (1999)
12. H. J. Unold, M. Golling, R. Michalzik, D. Supper, and K. J. Ebeling, European Conf. Opt. Commun., Th.A.1.4 (2001)
13. D. Zhou and L. J. Mawst, IEEE J. Quantum Electron. **38**, 12 (2002)
14. S. Fan, P. R. Villeneuve, J. D. Joannopoulos, and E. F. Schubert, Phys. Rev. Lett. **78**, 3294 (1997)
15. T. Baba, K. Inoshita, H. Tanaka, J. Yonekura, M. Ariga, A. Matsutani, T. Miyamoto, F. Koyama and K. Iga, J. Lightwave Technol. **17**, 2113 (1999)
16. M. Boroditsky, T. F. Krauss, R. Coccioli, R. Vrijen, R. Bhat and E. Yablonovitch, Appl. Phys. Lett. **75**, 1036 (1999)
17. T. Baba and H. Ichikawa, Optoelectronic and Commun. Conf., 9C2-1 (2002)
18. A. Mekis, J. C. Chen, I. Kurland, S. Fan, P. R. Villeneuve, and J. D. Joannopoulos, Phys. Rev. Lett. **77**, 3787 (1996)
19. J. Yonekura, M. Ikeda, and T. Baba, J. Lightwave Technol. **17**, 1500 (1999)
20. T. Baba, N. Fukaya, and J. Yonekura, Electron. Lett. **27**, 654 (1999)
21. A. Chutinan and S. Noda, Phys. Rev. B **57**, R2006 (2000)
22. M. Loncar, D. Nedeljkovic, T. Doll, J. Vuckovic, A. Scherer, and T. P. Pearsall, Appl. Phys. Lett. **77**, 1937 (2000)
23. M. Notomi, K. Yamada, A. Shinya, J. Takahashi, C. Takahashi, and I. Yokohama, Electron. Lett. **37**, 293 (2001)
24. T. Baba, A. Motegi, T. Iwai, N. Fukaya, Y. Watanabe and A. Sakai, IEEE J. Quantum Electron. **38**, 743 (2002)
25. N. Kawai, K. Inoue, N. Ikeda, N. Carlsson, Y. Sugimoto, K. Asakawa, and T. Takemori, Phys. Rev. Lett. **86**, 2289 (2001)
26. A. Chutinan A, M. Okano and S. Noda, Appl. Phys. Lett. **80**, 1698 (2002)
27. T. Baba, Proc. SPIE **4870**, 306 (2002)
28. A. Sakai, G. Hara and T. Baba, Jpn. J. Appl. Phys. **40**, L383 (2001)
29. A. Sakai, T. Fukazawa and T. Baba, IEICE Trans. Electron. E85-C, 1033 (2002)
30. T. Fukazawa, A. Sakai and T. Baba, Jpn. J. Appl. Phys. **41**, L1461 (2002)
31. M. Notomi, K. Yamada, A. Shinya, J. Takahashi, C. Takahashi, and I. Yokohama, Phys. Rev. Lett. **87**, 253902 (2001)
32. S. Noda, A. Chutinan, and M. Imada, Nature **407**, 608 (2000)
33. K. Yamada, M. Notomi, A. Shinya, I. Yokohama, T. Shoji, T. Tsuchizawa, T. Watanabe, J. Takahashi, E. Tamechika and H. Morita, Proc. SPIE **4870**, 324 (2002)
34. D. Taillaert, W. Bogaerts, P. Bienstman, T. F. Krauss, P. Van Daele, I. Moerman, S. Verstuyft, K. De Mesel and R. Baets, IEEE J. Quantum Electron. **38**, 949 (2002)

35. A. Yariv, Y. Xu, and R. K. Lee, A. Scherer, Opt. Lett. **24**, 711 (1999)
36. S. G. Johnson, P. R. Villeneuve, S. Fan, and J. D. Joannopoulos, Phys. Rev. B **62**, 8212 (2000)
37. T. Sato, Y. Ohtera, N. Ishino, K. Miura and S. Kawakami, IEEE J. Quantum Electron. **38**, 904 (2002)
38. S. Noda, K. Tomoda, N. Yamamoto, and A. Chutinan, Science **289**, 604 (2000)
39. J. C. Knight, T. A. Birks, P. St. J. Russell, and J. P. de Sandro, J. Opt. Soc. Am. A **15**, 748 (1998)
40. R. F. Cregan, B. J. Mangan, J. C. Knight, T. A. Birks, P. S. Russell, P. J. Roberts, and D. C. Allan, Science **285**, 1537 (1999)
41. S. D. Hart, G. R. Maskaly, B. Temelkuran, et al., Science **296**, 510 (2002)
42. T. A. Birks, J. C. Knight, B. J. Mangan, and P. St. J. Russell, IEICE Trans. Electron. E84-C, 585 (2001)
43. T. A. Birks, D. Mogilevtsev, J. C. Knight, and P. St. J. Russell, IEEE Photon. Technol. Lett. **11**, 674 (1999)
44. J. C. Knight, J. Arriaga, T. A. Birks, A. Ortigosa-blanch, W. J. Wadsworth, and P. St. J. Russell, IEEE Photon. Technol. Lett. **12**, 807 (2000)
45. Y. Fink, J. N. Winn, S. Fan, C. Chen, J. Michel, J. D. Joannopoulos and E. L. Thomas, Science **282**, 1679 (1998)
46. S. Fan, P. R. Villeneuve, J. D. Joannopoulos and H. A. Haus, Phys. Rev. Lett. **80**, 960 (1996)
47. H. Kosaka, T. Kawashima, A. Tomita, M. Notomi, T. Tamamura, T. Sato, and S. Kawakami, Phys. Rev. B **58**, R10096 (1998)
48. T. Baba and T. Matsumoto, Appl. Phys. Lett. **81**, 2325 (2002)
49. Y. Ohtera, T. Sato, T. Kawashima, T. Tamamura, and S. Kawakami, Electron. Lett. **35**, 1271 (1999)
50. K. Hosomi and T. Katsuyama, IEEE J. Quantum Electron. **38**, 825 (2002)
51. T. Baba and M. Nakamura, IEEE J. Quantum Electron. **38**, 909 (2002)
52. T. Baba, and D. Ohsaki, Jpn. J. Appl. Phys. **40**, 5920 (2001)
53. V. Berger, Phys. Rev. Lett. **81**, 4136 (1998)

12 Application to Ultrafast Optical Planar Integrated Circuits

K. Asakawa and K. Inoue

In this chapter, 2D PC slab waveguides including several functional waveguides among various PC devices are taken up in particular. Namely, results on the design, fabrication and characterization of those 2D PC waveguides necessary for application to ultrafast optical planar integrated circuits are described in detail.

12.1 Introduction

Since the 1970s, a photonic integrated circuit (PIC) has long been of great concern among optoelectronic engineers who have devoted themselves to the advancement of conventional discrete optical components toward a highly integrated, resultantly cost-effective and reliable optical circuit. The concept and substance of the PIC have changed from $LiNbO_3$- and glass-based waveguides in the 1970s, via semiconductor-based OEICs (optoelectronic integrated circuits) in the 1980s, to silica-based practical PLCs (planar light wave circuits) in the 1990s. Since the proposal of a PC in 1987, on the other hand, a variety of computer simulations, fabrications and characterizations of the PC have been performed aimed at an advanced PIC. In particular, a semiconductor-based 2D PC slab has been allocated to an extremely miniaturized PIC of practical importance because of the large potential ability to provide ultrasmall light sources, optical switches and waveguide components by utilizing conventional semiconductor planar epitaxial growth and microfabrication technologies [1–11].

Turning to the recent Internet Protocol (IP)-based telecommunication world, on the other hand, it is now intensively urging us to develop novel optoelectronic devices/components for DWDM (dense wavelength-division-multiplex)/OTDM (optical time-division-multiplex)-based terabit optical communications by about 2010. Several optoelectronic technology roadmaps have already predicted the perspective of such future device/system demands based on several network-layer hierarchies.

Figure 12.1 shows a typical optical communication technology roadmap established by the Optoelectronic Industry and Technology Development Organization (OITDA) in 1999 [12]. The figure has been condensed from the

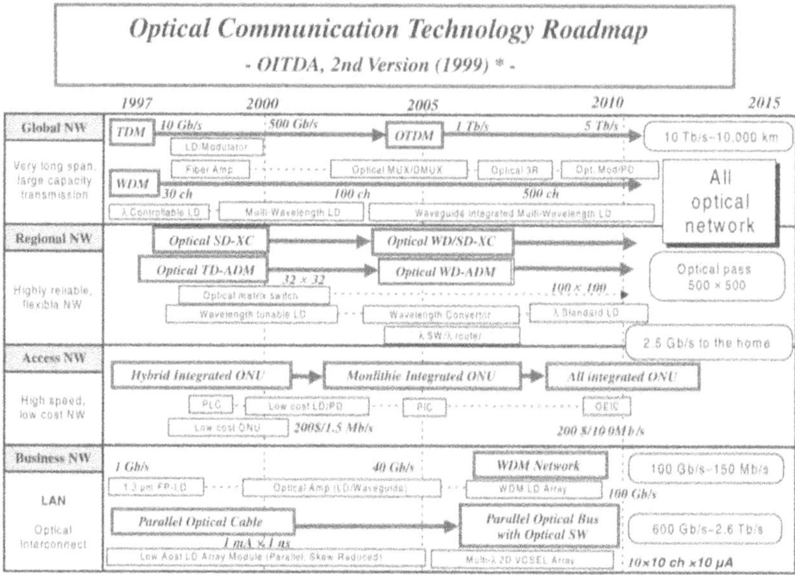

Fig. 12.1. Typical optical communication technology roadmap condensed from the original version which was established by the Optoelectronic Industry and Technology Development Organization (OITDA) in 1999 [12]

original three diagrams into one under the approval of the OITDA. According to this roadmap, novel optical devices to be developed in the next ten years will be divided primarily into two groups. For business and access networks, low cost and integrated/arrayed modules/circuits such as laser diode (LD)/photodetector (PD) arrays and vertical-cavity surface-emitting-laser (VCSEL) arrays are required as low-end devices. For regional and global networks, on the other hand, multiwavelength LD and optical MUX/DEMUX are required as typical high-end devices for the DWDM/OTDM system. Taking into account the appearance of the photonic network imaginable in 2010–2015, all-optical devices based on the advanced PIC will be expected in all the hierarchies in Fig. 12.1. Above all, ultrafast optical signal processing devices are very desirable if they are integrated monolithically in a small chip installed with small waveguides. One reason for this is that otherwise time jitters due to the geometrical fluctuation or anomalous dispersion in large waveguide platforms will degrade the pulse transmission characteristics seriously. Another reason is that a small latency in an ultra-small device, negligible as compared with the pulse width/period, will have the possibility of creating new signal processing devices. A 2D PC-based PIC is desirable for these applications on account of these reasons.

Figure 12.2 shows a schematic diagram showing examples of possible 2D PC-based integrated signal processing PICs. The diagram involves three kinds of optical node switches, that is, an ADD/DROP switch, an optical

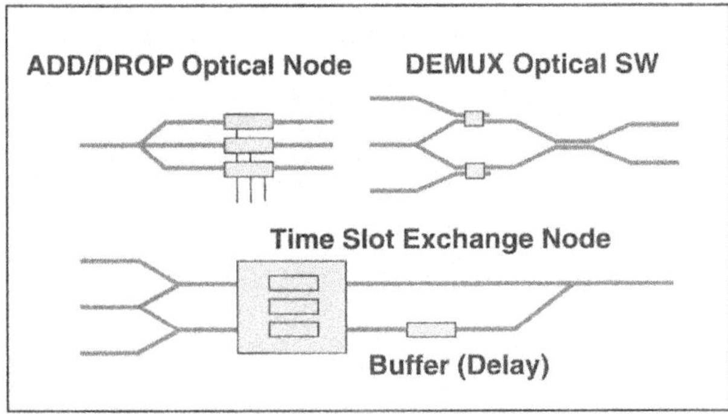

Fig. 12.2. Schematic diagram showing examples of 2D PC-based integrated signal processing PICs involving an ADD/DROP switch, an optical DEMUX switch and a time-slot exchange switch

DEMUX switch and a time-slot exchange switch. The materials responsible for all-optical switching functions are third-order optical-nonlinearity (ONL) (χ^3)-based quantum structures such as quantum wells and quantum dots. Furthermore, a 2D PC-based optical delay element offers the potential for remarkable reduction in size with the assistance of a low v_g specific to the 2D PC slab waveguide. Common to all these miniaturized waveguides, efficient coupling of external light into the 2D PC region can hopefully be achieved, for example, by taking advantage of the continuous vertical guiding of light by extending the interconnecting waveguides to the PC region itself [12].

12.2 Why Photonic Crystal-Based Ultrafast All-Optical Switches, PC-SMZ?

As an application to integrated ultrafast optical signal processing devices, a 2D PC-based ultrasmall symmetrical Mach–Zehnder (SMZ)-type all-optical switch has been proposed so far [13]. Hereinafter, the device is referred to simply as the PC-SMZ. A conventional SMZ-type all-optical switch has already been demonstrated in several switching operations at speeds of ps (picosecond) to sub-ps, the speed not restricted to the carrier lifetime of the semiconductor ONL materials [14]. The PC-SMZ is on a similar principle to the conventional SMZ switch. In addition, the PC-SMZ is aimed at (1) a dramatically down-sized PIC within 500-μm-square, (2) an ultrafast optical switching speed over 40 Gb/s and (3) a reduced optical switching energy (OSE) of less than 500 fJ (femtojoule).

Figure 12.3 shows a schematic picture of the PC-SMZ. Optical waveguides with the SMZ pattern are composed of 2D PC line-defect waveguides, while ONL phase shift arms are selectively embedded with quantum dots (QDs) that exhibit the large ONL. The principle of the PC-SMZ is shown

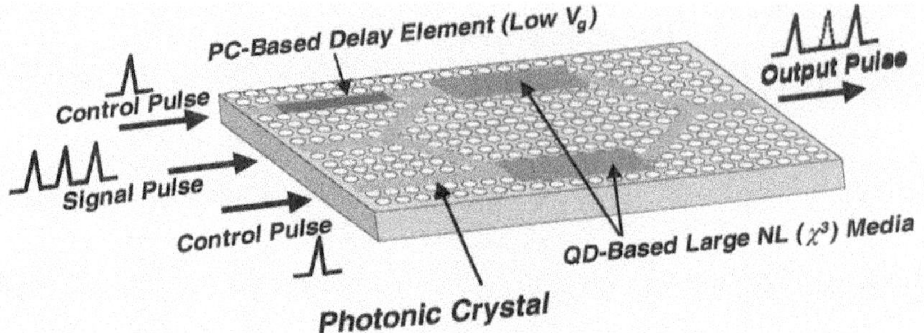

Fig. 12.3. Schematic picture of the integrated SMZ-type all-optical switch (PC-SMZ) based on 2D PC waveguides as an SMZ platform and embedded quantum dots as an optical nonlinear material

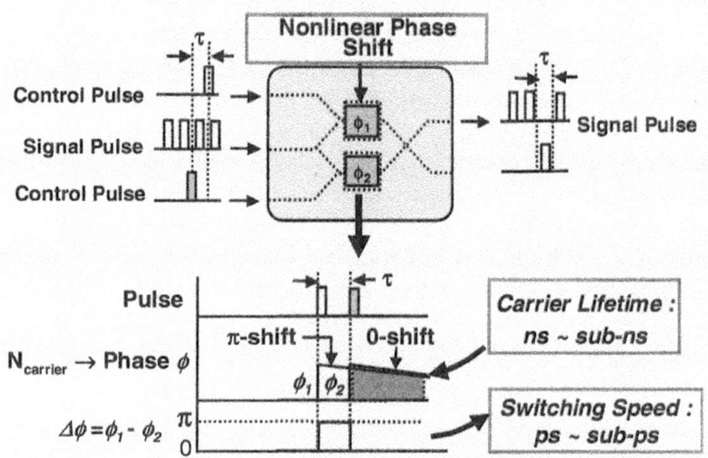

Fig. 12.4. Principle of the PC-SMZ

in Fig. 12.4 based on the conventional SMZ switch [14]. A "switch-on" control pulse incident in the upper ONL arm causes an ONL-induced refractive index change Δn, which leads to a phase shift $\phi_1 = 2\pi(\Delta n/\lambda) \cdot l_{\mathrm{ONL}}$ for a series of signal pulses where λ is the wavelength of the signal pulse and l_{ONL} is the ONL arm length. Similarly, the phase shift ϕ_2 is generated in the lower ONL arm by the "switch-off" control pulse, leading to a phase shift difference $\Delta\phi = \phi_2 - \phi_1$ at the combined Y-junction. Only when $\Delta\phi = \pi$, are the signal pulses switched spatially. Here, the time response of the Δn in the semiconductor is in general rapid in excitation and slow in relaxation, say, on the order of ns to sub-ns as the carrier lifetime in the semiconductor. However, since the Δn's are excited time-differentially by the "switch-on" and "switch-off" control pulses, the Δn's in the tailing periods are cancelled, as shown in Fig. 12.4. Thus, a rapid time-dependent $\Delta\phi = \pi$ on the order of ps to sub-ps can be achieved.

If the structure of the PC-SMZ is fabricated similarly to the conventional SMZ switch, the ultrafast switching operation mentioned above can be achieved as well. However, the conventional SMZ is based on the low refractive-index-contrast waveguide, while the PC-SMZ is composed of an ensemble of high refractive-index-contrast PC lattices. Due to these structural differences in geometry and in size, the PC-SMZ has several key issues to be solved. For reduction of the OSE to an energy level of less than 100 fJ, for example, the conventional SMZ switch adopted the SOA (semiconductor optical amplifier) structure, which is not available in the current PC-SMZ at the moment. Instead, use of a low group velocity (v_g) [15] in the ONL arm in the PC-SMZ is thought to be effective for enhancing the ONL-induced phase shift [16], thus reducing the OSE as well as the ONL arm length. Design, fabrication and characterization of such a low-v_g waveguide are important key issues of the PC-SMZ. Besides this, design of the ONL waveguide which produces $\Delta\phi = \pi$ necessary for SMZ switching with the length as short as possible is another important key issue.

In the following sections, state-of-the-art techniques on design, fabrication and characterization of 2D PC waveguides necessary for implementing the key issues proper to the PC-SMZ are shown in detail. Here, the waveguides include straight, bend, Y-branch, directional coupler (DC) and coupled-cavity (CC) waveguides.

12.3 Fundamental Structures of 2D PC Slab Waveguides

When the 2D PC waveguide is applied to a practical PIC device, several issues common to the vertical and lateral structures have to be carefully addressed for achieving a low-loss, single-mode waveguide. In this section, a 2D PC slab and defect waveguide structure effective for wide applications is described. Computer simulation of the BS for such a waveguide, necessary for proposing, designing and demonstrating a new structure, is also described.

12.3.1 2D PC Slab and Defect Waveguide Structure

Optical beam confinement in the vertical direction is essential for practical application of the 2D PC waveguide to the PIC. Here, core/cladding structures composed of semiconductor hetero-epitaxial layers are referred to simply as the SC (semiconductor clad)-type. Since most of the hetero-epitaxial layers exhibit low refractive index contrast (LIC) in the core/cladding thickness direction, and consequently an optical field penetrates deeply into the cladding layer, the SC-type requires deep 2D PC structures (typically ~ 3 μm in depth in the case of an air-hole array for the near-infrared wavelength range). In

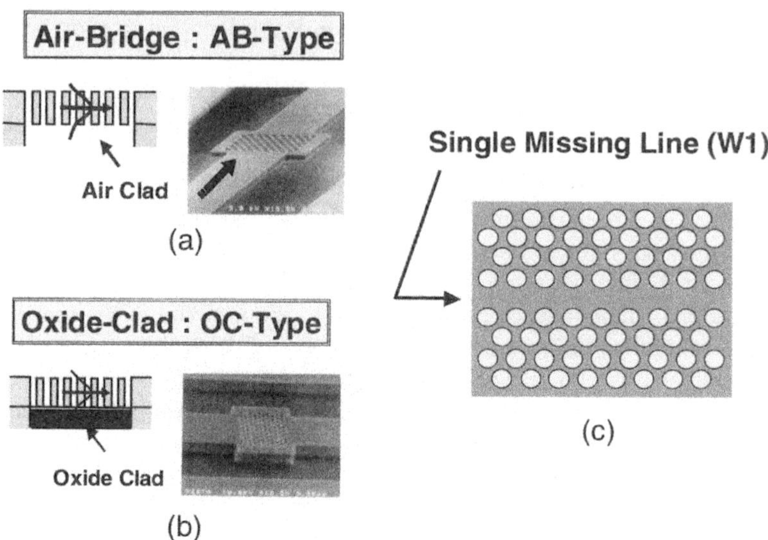

Fig. 12.5. Two kinds of 2D PC slab structures: (**a**) air-bridge (AB-type), (**b**) oxide-cladding (OC-type) structures. (**c**) Top view of a typical 2D PC straight defect waveguide pattern composed of a single missing line (W1)

contrast to this, Figs. 12.5(a) and (b) show two kinds of 2D PC slab structures, that is, air-bridge (AB)-type and oxide-cladding (OC)-type structures, respectively. These structures are categorized in the high refractive index contrast (HIC) system. SEM (scanning electron beam microscope) photographs in Figs. 12.5(a) and (b) show examples of GaAs-based HIC systems with core thicknesses of 200–270 nm. The samples are bulk 2D PC air-rod arrays without defect waveguides.

Defect waveguides are made by introducing missing periodic 2D PC patterns in the waveguide shape. Figure 12.5(c) shows a top view of the typical 2D PC line-defect straight waveguide pattern composed of a single missing line (referred to as W1 according to the number of the missing line) in the triangular-lattice air-rod array. The lattice constant a and air-hole diameter R for demonstrating the PC-SMZ at 1.3–1.55 μm wavelength are 350–420 nm and $\sim$ 200 nm, respectively. As shown later, this AB-type-based waveguide exhibits a single mode (for TE polarization) with bandwidth of $\sim$ 80 nm.

12.3.2 Simulation of Band Structure

The modal behavior of an optical beam propagating in the 2D PC waveguide is investigated by the band structure calculated typically with a plane wave expansion method or 2D-/3D finite-difference time-domain (2D-/3D-FDTD) method. Figure 12.6 shows a BS of the 2D PC slab (AB-type) defect straight waveguide (W1) corresponding to Fig. 12.5(c), calculated by using the 3D-

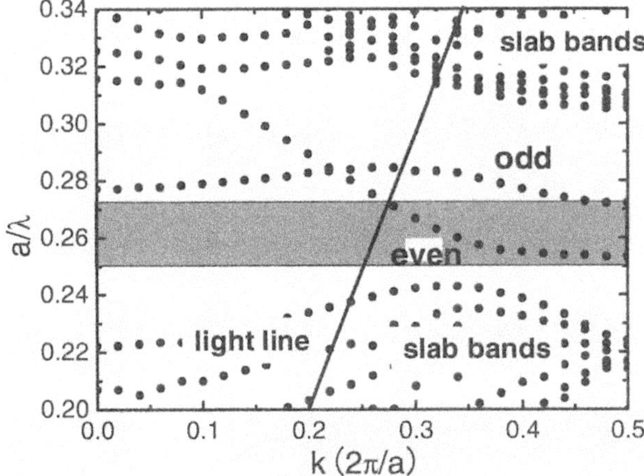

Fig. 12.6. Band structure of the 2D PC slab defect straight waveguide (W1), calculated by using the 3D FDTD method

FDTD method. Two defect-waveguide modes, that is, first-order even and odd modes, appear in the wide band gap lying between the lower slab band (dielectric mode) and the upper slab band (air mode). In the calculation, a and air-hole radius r are 350 and 110 nm, respectively. The light line, indicated by the solid straight line, crosses the even mode line curve at the frequency of $0.273(a/\lambda)$. Accordingly, a single guided mode with a bandwidth of 80–100 nm in wavelength is located in the frequency range $0.250 \leq a/\lambda \leq 0.273$. This frequency range is used for the demonstration of a variety of waveguides, as shown in this chapter.

12.4 Nanofabrication Technologies for 2D PC Slab Waveguides

In this section, two kinds of key technologies, that is, fine EB (electron beam) lithography and dry etching, are reviewed for fabrication of the 2D PC air-rod array used in the near-infrared wavelength range.

12.4.1 Fine EB Lithography

Fine EB dose adjustment is the most important subject in EB lithography for defining a nanometer-scale pattern like the 2D PC air-rod array of 100–200 nm in size. This procedure is inevitable for suppression of the EB irradiation-induced proximity effect, or otherwise the EB-drawn pattern is deformed more or less due to the unnecessary or nonuniform exposure of the back-scattered (or secondary) EB on the designed pattern. In particular, the

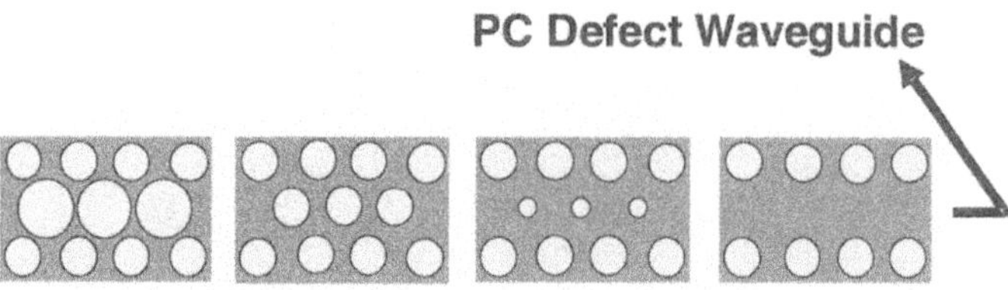

Fig. 12.7. Schematic top-view pattern of the modified straight defect waveguide

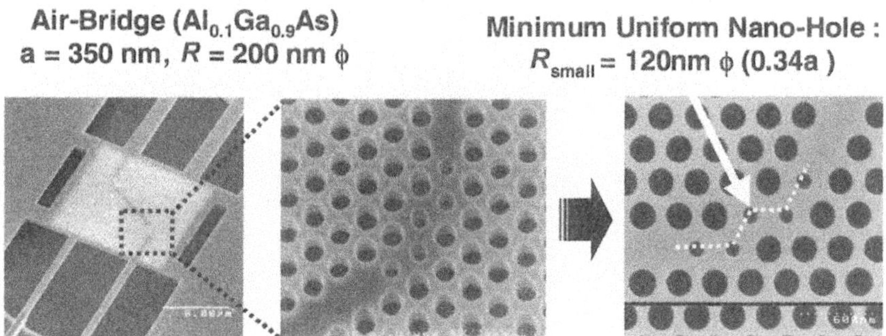

Fig. 12.8. SEM photographs of the fabricated samples including 120-nm-diameter small holes at the corner of the bend waveguides.

precise EB dose adjustment is needed for patterning 2D PC slab waveguides with mixture of a periodic and nonperiodic pattern. Typical examples are as follows: (1) coexistence of a uniform air-rod array and a missing air-hole line as a defect waveguide, (2) insertion of a series of small air-hole arrays in the bend waveguide, and (3) a boundary region between the PC defect waveguide and a non-PC stripe waveguide. Figure 12.7 shows a schematic top-view pattern of the modified straight defect waveguide involved in the group (1), while Fig. 12.8 shows SEM photographs of the fabricated samples including 120-nm-diameter small holes at the corner of the bend waveguide involved in group (2). Actual EB dose adjustment is performed depending on the size of the EB-exposed area to compensate the dose of the overexposed position for making the distortion-free 2D PC pattern. Along with this EB dose adjustment, the side-wall profile of the developed EB resist should also be made vertical enough when the EB resist pattern is used as a dry etching mask. This process leads to the realization of a vertical profile of the dry-etched deep air hole. If these two kinds of adjustments are performed carefully, fluctuations of both the size and position of the air hole array can be reduced to less than 1% (corresponding to the order of $\sim$ 2 nm in scale for the air hole with $\sim$ 200 nm in diameter).

12.4.2 Fine Dry Etching

In the dry etching process, on the other hand, the required etching depth of the air-hole depends on the type of core/cladding structure as mentioned above. The AB- and OC-types with core thickness of 250–300 nm are at most 300–400 nm in etching depth. Nevertheless, precise control of other parameters is necessary for suppressing the fluctuation of the air-hole diameter after etching to less than 2% (corresponding to 4 nm deviation in diameter). Taking notice of the etching with ECR (electron cyclotron resonance)-plasma RIBE (reactive ion beam etching), control of the substrate temperature to $50 \pm 10°C$ is one of the key parameters for refraining from distortion of the side wall profile [13]. When precise control of the etching parameter is established once, the result is available for optimization of the EB-resist-mask process before etching of substrates. Figure 12.9 shows SEM photographs of compared vertical profiles for the EB-resist-mask (upper) and the GaAs etched-hole (lower) at different post-bake temperatures, that is, 130°C (left) and 140°C (right). The upper inset shows an SEM photograph of a top-view air-hole array on the EB resist. After the post-bake at 140°C for 2 min, the EB resist exhibits a desirably vertical profile and accordingly leads to a vertical profile in the GaAs etched air hole, while it leads to a distorted profile on the etched sidewall when the post-bake is performed at 130°C for 2 min. This is simply an example for forming a vertical etched profile but common to all air-hole etching for any kind of 2D PC slab waveguides.

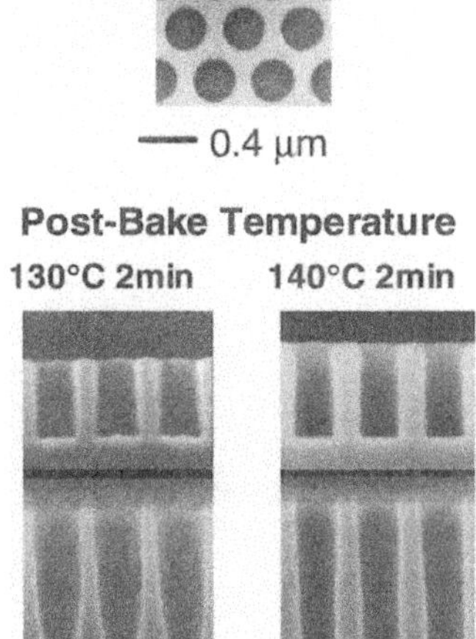

Fig. 12.9. SEM photographs of compared vertical profiles for the EB-resist-mask (upper) and GaAs etched-hole (lower) at different post-bake temperatures

12.5 Fabrication and Characterization of 2D PC Slab Defect Waveguides

In this section, demonstrated results of a variety of 2D PC slab defect waveguides using fine nanofabrication technologies mentioned above are described. The waveguides include straight, bend, Y-branch, directional coupler and coupled-cavity waveguides necessary for constructing the PC-SMZ.

12.5.1 Straight and Bend Waveguides

AB-type straight and bend 2D PC slab waveguides were fabricated by the EB lithography and dry etching process mentioned above. Transmission spectra were measured by using an optical set-up capable of measuring a wide frequency range of the spectrum in a short time by virtue of a halogen white-light source and a high-sensitivity multichannel analyzer [17]. Figure 12.10 shows transmittance spectra for the straight waveguide. The dash-dot line curve shows a measured T-spectrum, while the solid line curve shows a spectrum calculated by using the 3D-FDTD method. The upper SEM photograph shows a top-view pattern of the 2D PC waveguide sample. The total length of the PC is 18 μm and the PC waveguide is sandwiched by the non-PC stripe

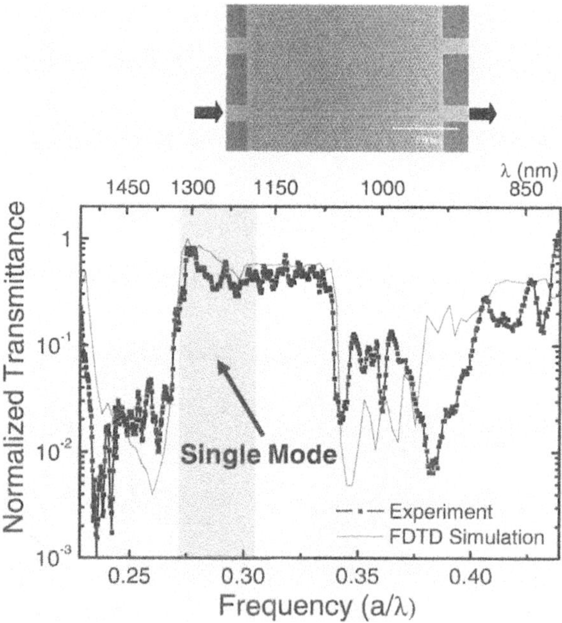

Fig. 12.10. Measured and calculated transmittance spectra for a straight waveguide. The upper SEM photograph shows a top-view pattern of the 2D PC straight waveguide sample

waveguide in the propagation direction. An identical reference sample composed of a non-PC stripe waveguide without the PC waveguide was prepared for calibration of T in the spectrum. In Fig. 12.10, measured and calculated spectra are apparently in good agreement to each other in the wide spectral range from 850 to 1500 nm, in particular, in the sharp band edge and flat band. Moreover, the waveguide exhibits nearly 80% high transmittance in the vicinity of the 1300-nm wavelength. It is also found that the frequency range $0.265 \leq a/\lambda \leq 0.30$ corresponds to the single guided mode. Interestingly, a high T region lies even in the frequency region higher than $0.30(a/\lambda)$, which is proved to be a leaky-mode region judging from the comparison with the BS in Fig. 12.6. This is probably due to the fact that the optical beam energy of the leaky mode is not completely diminished in a waveguide as short as 18 μm in length.

Figure 12.11 shows measured and calculated T-spectra for the double bend 2D PC waveguide. The pattern of the bend waveguide is shown in the upper SEM photographs. Similarly to the result for the straight waveguide in Fig. 12.10, the spectra show good agreement again both in shape and in T itself. As a remarkable difference from the straight waveguide, the bandwidth is significantly reduced to $\sim$ 30 nm. This is due to the bend loss such as mode conversion at the corner. Taking into account application to the PC-SMZ, this

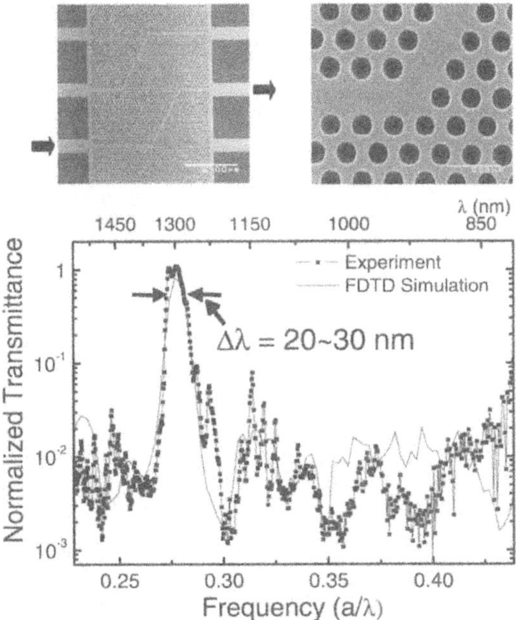

Fig. 12.11. Measured and calculated transmittance spectra for the double bend 2D PC waveguide. The pattern of the bend waveguide is shown in the upper SEM photographs

bandwidth is wide enough for use of control and signal optical pulses detuned, say, by 20 meV (corresponding to ∼ 20 nm in wavelength).

12.5.2 Y-branch Waveguide

A Y-branch 2D PC slab waveguide [18–21] was fabricated and characterized in the same way as the straight/bend waveguides mentioned above. Figures 12.12(a)–(d) show SEM photographs and measured transmission spectra. The total length of the PC portion is 18 μm and the 2D PC waveguide, as shown in Fig. 12.12(a), is sandwiched by the non-PC stripe waveguide. An identical reference similar to the case for the straight/bend sample was prepared for this experiment, too. At the center of the Y-spot, another air hole is added, as shown in Fig. 12.12(b). This structure is aimed at improvement of T, as proposed in [21]. Figure 12.12(c) shows T-spectra at the upper (indicated by C2U in the figure) and lower (indicated by C2L) output ports. It is found that they are in good agreement with each other in the wide spectral range from 850 to 1600 nm. The result suggests that the 2D PC waveguide patterning has been performed identically between the upper and lower branches. It is also noted that the maximum transmittances at the two output ports exhibit nearly 40% (41% and 39%) in the vicinity of 1300-nm wavelength, as indicated by the arrow. In order to investigate the effect of the additional air hole at the Y-spot, another identical Y-branch sample without the additional air hole was fabricated. The measured result is shown by the

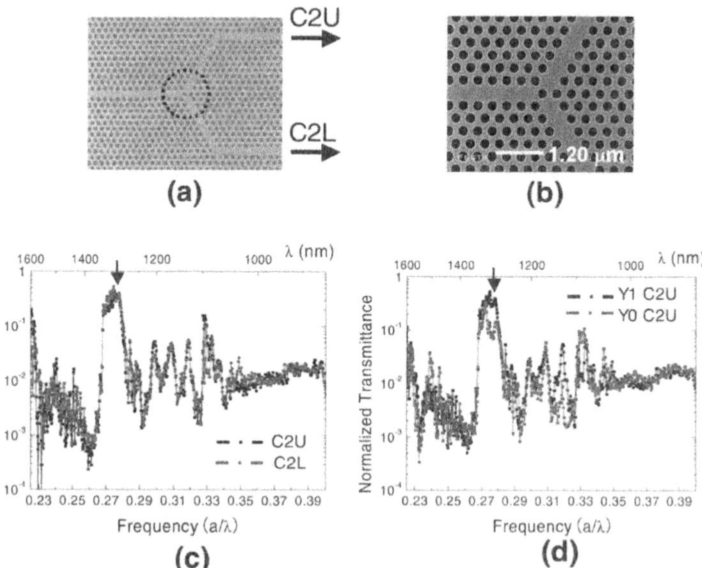

Fig. 12.12. SEM photographs and measured transmission spectra for Y-branch 2D PC slab waveguides with a small hole at the center of the Y-branch

transmission spectrum, as indicated by Y0C2U in Fig. 12.12(d). T of this sample is at most 27% (27% and 26% for the upper and lower output ports) at the wavelength indicated by the arrow. The transmission spectrum for the Y-branch with the additional air hole is indicated by Y1C2U for comparison. From these results, it can be said that the transmittance is increased by 50% by addition of the air hole in the Y-spot.

12.5.3 Directional Coupler

In photonic integrated circuits, a directional coupler (hereinafter referred to simply as a DC) plays an important role as a variety of functional components such as an optical beam splitter and optical switch. In this section, however, the DC is discussed from the viewpoint of the unique application to the output Y-junction in the PC-SMZ.

Specification of Coupler for PC-SMZ. Figure 12.13(a) shows a schematic PC-SMZ defect waveguide pattern. The simplest Y-junction imaginable is a threefold symmetry Y-junction, as shown in Fig. 12.13(b) [18–21]. However, an incident beam propagating through a phase shift arm is equally split into two output beams at the Y-junction, one propagating to an output port and the other returning to another phase shift arm, as shown by the arrows

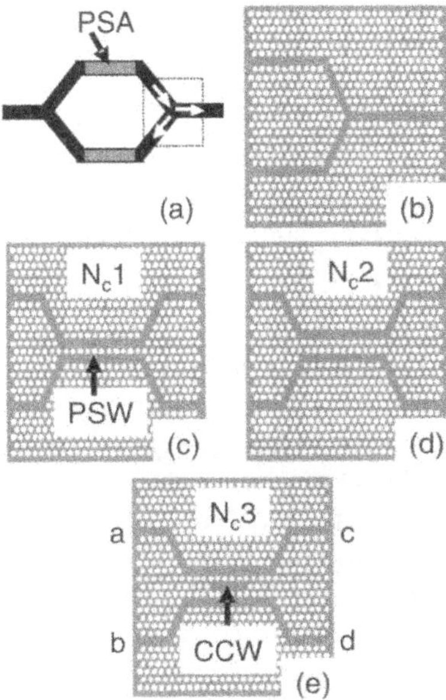

Fig. 12.13. Schematic top view patterns of (**a**) the PC-SMZ switch, (**b**) the 2D PC-based three-fold symmetry Y-junction, (**c**) the N_c1-type DC, (**d**) the N_c2-type DC and (**e**) the CC-DC

in the figure. The latter beam generates a resonant ring mode with a narrow bandwidth, resulting in disturbing the SMZ interference, as noted in [22].

The DC is an alternative structure since it does not generate the loop mode as indicated later. A typical 2D PC-based DC is shown in Fig. 12.13(c), where two parallel-coupled single-defect waveguides (indicated by PSW in the figure) are separated by a single air-hole array in the coupling region. Hereinafter, this type is referred to as type N_c1 according to the number of the air-hole array in the coupling region [23–25]. According to a recent coupled mode analysis on the N_c1-type DC, it was found that the decoupling phenomenon appears in the N_c1-type DC and prevents us from achieving a desirably compact 2D PC DC as long as the 2D PC slab is composed of an air-hole array. This result is not valid in the case of dielectric rod and low n-contrast 2D PC structures. Detailed analyses are reported elsewhere [26].

Coupling-Strength-Controlled Directional Coupler (CC-DC). As shown in Fig. 12.13(d), on the other hand, the N_c2-type DC is preferable to the N_c1-type from the viewpoint of the coupled-mode band structures. However, asymmetric configuration of the air-rod array in the coupling region makes it difficult to a large extent to design the whole PC-SMZ pattern symmetrically. So far, the N_c3-type DC, as shown in Fig. 12.13(e), exhibits the most suitable output Y-junction in the PC-SMZ, as shown later [27]. In this case, simple calculation suggests that insertion of a coupling-strength-controlled defect (hereafter referred to simply as CCD) into the N_c3-type coupling region, as shown in Fig. 12.13(e), enhances the coupling strength of the DC, thus leading to the reduction of the coupling length of the DC. The CCD in this figure is composed of five missing air-holes. The DC with the CCD is referred to as the coupling-strength-controlled-DC (CC-DC) , exhibiting significant reduction of the whole DC length, say, by 1/3 as compared with the N_c3-type DC without the CCD.

Demonstration of the CC-DC. Fabrication and characterization of the CC-DC sample, designed as a 50%-coupler suitable for the Y-junction in the PC-SMZ, are described here. For sample preparation, the lengths of the PSW and the CCD were set to $20a$ (7.2 μm) and $10a$ (3.6 μm), respectively. Figure 12.14(a) shows transmission spectra of the N_c3-type CC-DC calculated by using the 2D FDTD method based on the effective refractive-index approximation. The 2D PC parameters of lattice constant (a), air-hole diameter (R), slab thickness and effective refractive index (N_{eff}) were 360 nm, 216 nm, 250 nm and 2.81, respectively. In the figure, solid (DC_{ac})- and dotted (DC_{ad})-line curves show the respective T at the ports c and d, as indicated in Fig. 12.13(e). For frequencies indicated by the arrows, it is found that nearly 50% coupling is achieved because intensities at the ports c and d are comparable to each other. On the other hand, Fig. 12.14(b) shows calculated transmission spectra of the N_c3-type DC without the CCD. It is noted that, in the frequency range A as indicated in Fig. 12.14(c), coupled intensities DC_{ad} are lower than uncoupled intensities DC_{ac} by 1/2–1/3. This means that the DC without the CCD requires a longer coupling length by a factor of 2–3 for 50% coupling condition. The role of the CCD is briefly explained as follows. Figure 12.14(c) shows band diagrams of the coupled modes, that is, the e-o mode (dotted line) and e-e mode (solid line) for the N$_c$3-type DC without the CCD, calculated by using the 2D FDTD method. The dash-dot line shows a light line. In the frequency range A, the difference Δk_A of the wavenumbers between the e-e and e-o modes is small as compared with that (Δk_B) in the frequency range B defined in the figure. This implies a large coupling strength cannot be expected in the frequency range A. If the CCD is inserted in the coupling region, the coupling strength increases due to the increased effective refractive index in the coupling region. This is the aim of the CCD. Regarding the wavy ripples in the spectra, the CCD plays the role

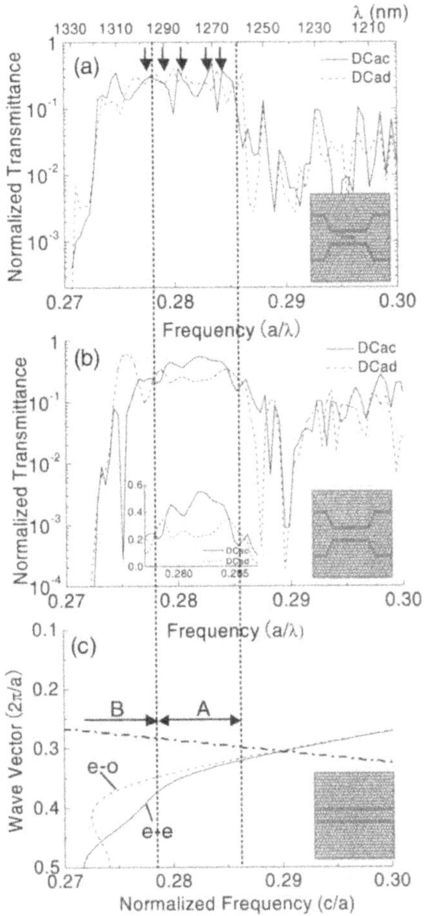

Fig. 12.14. 2D FDTD calculated transmittance spectra for (**a**) the CC-DC, (**b**) the DC without the CCD, and (**c**) band diagrams for the parallel DC

of a frequency-dependent cavity effect in part. A detailed analysis is currently being undertaken.

For the 50% coupling design, calculation on T at the output ports b resulted in the intensity of nearly $\sim 0\%$. This means that the DC used here is suitable for the Y-junction of the PC-SMZ in the sense that it does not generate the resonant ring mode which is undesirable for the SMZ switch operation, as mentioned before. It is also verified that the CC-DC investigated here provides 50%-coupling with a rather small size, say, a PSW length of 7.2 μm. Since the DC without the CCD resulted in a calculated 50%-coupling length of about $60a$ ($\sim$ 20μm), it can be said that the CCD plays the role of reducing the length of the DC by 1/3, as mentioned above.

We have measured transmission spectra over a broad wavelength region from 850 to 1600 nm. Figure 12.15 shows those spectra for the TE-like mode in a region from 1120 to 1380 nm. The spectra are normalized by the reference sample. The solid and dotted lines show spectra for uncoupled intensity DC_{ac} and coupled intensity DC_{ad}, respectively. In Fig. 12.15, the spectrum for a 60° double-bend waveguide is also plotted by a thin-solid line for comparison. In the wavelength range from 1280 to 1325 nm, high-transmission regions for both DC_{ac} and DC_{ad} were obtained and, in particular, a 50% coupling was achieved at a wavelength of 1320 nm. The upper right inset in Fig. 12.15 shows the near-field pattern at two output ports c and d for the

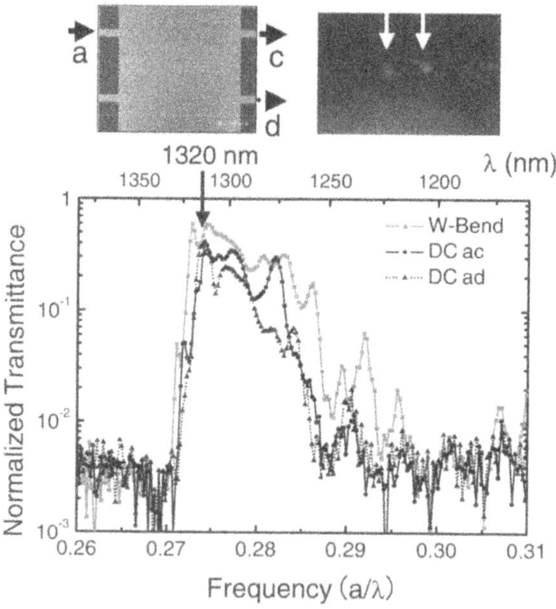

Fig. 12.15. Observed transmittance spectra for the CC-DC. The near-field pattern of output ports observed by the infrared-vidicon camera is shown in the upper right inset

wavelength of 1320 nm, observed with an infrared vidicon camera. As shown by the bright output spots with nearly equal intensities, the CC-DC used in the experiment shows a 50% coupling. The CC-DC shown here is concluded to function for the Y-junction in the PC-SMZ.

12.5.4 Compact and Flat-Band Delay Element

A compact optical delay element is one of the most urgently required components for a fututre ultrafast optical signal processing system. For example, an integrated optical delay element plays an essential role for the PC-SMZ. One possibility for a compact PC-based optical delay element is to use a very small v_g. Recently, a coupled-cavity waveguide (CCW) with the use of an impurity band formed by a coupled defect was proposed and a significant delay for ultrashort pulses was measured [28–33]. It was revealed that quasi-flat impurity bands suitable for the transmission of ultrashort pulses can be achieved by properly designing the configuration of the CCW and carefully choosing the constituting defects. In this section, fabrication and characterization of such an efficient optical-delay line on the 1D CCW are shown.

The CCW is composed of a 1D air-hole array in a silicon-on-silicon-dioxide (SOI) ridge waveguide. The periodically placed defects are formed by intentionally increasing the separation of two neighboring air holes. By properly choosing the lattice constant, quasi-flat impurity bands were achieved at $\sim$ 1.55 μm. The measured impurity bands are in good agreement with those calculated by the FDTD method. A 600-fs delay time was achieved for a 110-fs width optical pulse in fabricated delay lines of only 20 μm [34].

A schematic drawing of the detailed sample structure is shown in Fig. 12.16(a) [35], where various demonstrated parameters are described. The

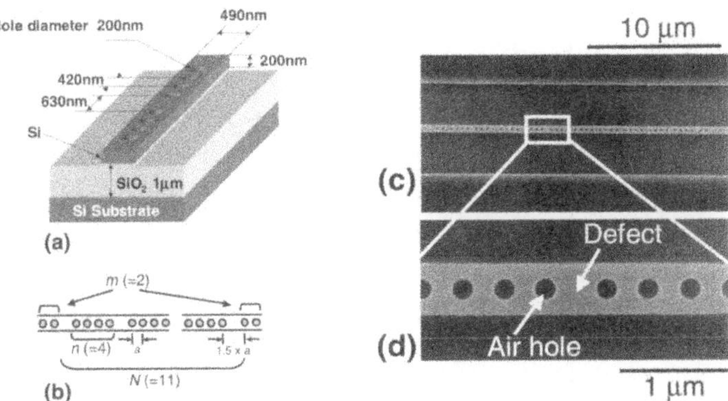

Fig. 12.16. (**a**) Schematic illustration of the 1D PC monorail waveguide. (**b**) Design of the monorail coupled-cavity waveguide. (**c**), (**d**) SEM image of the fabricated PC region of the monorail coupled-cavity waveguide

1D PC structure was fabricated by using EB lithography and RIBE. Figure 12.16(b) shows the configuration in the 1D CCW delay line. There are three primary parameters that control the transmission properties of the impurity bands: the total number of defects (N), the number of normal air holes between two neighboring defects (n), and the number of air holes at both ends (m). Apparently, n determines the coupling strength or bandwidth, while m controls the flatness of the impurity band. The achieved delay is proportional to N. For the experiment, parameters of $N = 11$, $n = 4$, and $m = 2$ were selected as an example. An SEM image of the fabricated structure is shown in Fig. 12.16(c), while an enlarged view is provided in Fig. 12.16(d).

The measured transmission spectrum for the CCW with a designed lattice constant of 420 nm is shown by the solid-line curve in Fig. 12.17. The transmittance was normalized by the reference sample. Apparently, the resulting impurity band is centered at ~ 1.53 µm and has a width of ~ 20 nm. The maximum T was found to be $\sim 20\%$. For comparison, the transmission spectrum for the designed CCW was calculated by using 3D FDTD simulation, as shown by the dashed-line curve in Fig. 12.17. The transmittance peak is located at ~ 1.56 µm and the bandwidth is ~ 50 nm. In order to determine the primary cause of the discrepancy, the position and diameter of each air hole in the sample were measured by using an SEM. The average values for the lattice constant and the diameter of air holes were found to be 408.4 nm and 200.4 nm, respectively. In this way, the transmission spectrum of the fabricated CCW was calculated by using the measured position and diameter of each air hole, as shown by the dotted-line curve in Fig. 12.17. Here, this calculated transmission spectrum was red-shifted from the original ones by 0.59%. By using the actual position and diameter of air holes, the

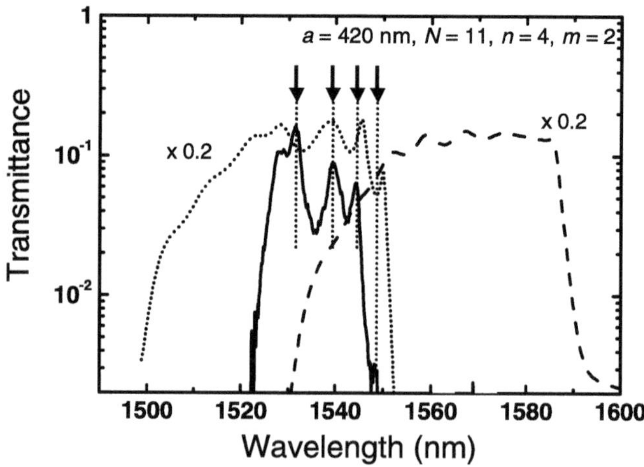

Fig. 12.17. Measured and calculated transmittance spectra for 1D PCs. Dashed line: designed structure; dotted line: fabricated

location of the calculated impurity band was found to be very close to that of the measured impurity band. Even more interesting is the fact that a clear one-to-one correspondence in the transmission peaks can be seen between the two spectra, as indicated by the arrows and dotted-lines. However, the difference in bandwidth remains. For this reason, the effect of the mesh size used in the FDTD simulation is now under investigation.

12.6 Fabrication and Characterization of Long Waveguides

Taking into account the characteristics of several 2D PC slab waveguides mentioned above, the grand design of the total PC-SMZ chip was performed. As shown by the resultant schematic pattern in Fig. 12.18, the total chip size has proved to be within $\sim 500 \times 500\ \mu m^2$. The largest waveguide length of 100–300 μm is allocated to the parallel nonlinear phase shift arms, while the spacing of $\sim 125\ \mu m$ between input channel waveguides is given by the neighboring input optical fibers. The simulated results on the nonlinear phase shift arms are reported elsewhere [15].

For achieving such a hundred-μm-scale PC-SMZ chip, AB-type long 2D PC slab waveguides were fabricated and characterized, as shown here. Figure 12.19 shows several SEM photographs of the cleaved edges viewed from several angles {(a) right angle, (b) inclined-top-view angle, (c) expanded inclined-top-view angle, (d) expanded inclined-bottom-view angle}. The photographs show extremely smooth surfaces. The roughness on the top and bottom waveguide surface is roughly estimated as less than 10 nm. Figure 12.20 shows a plot of measured T powers for the straight waveguide with different lengths from 500 to 2000 μm. It is noted that the propagation loss measured in this way is as small as 1–2 dB/mm. This value is likely to be a record low

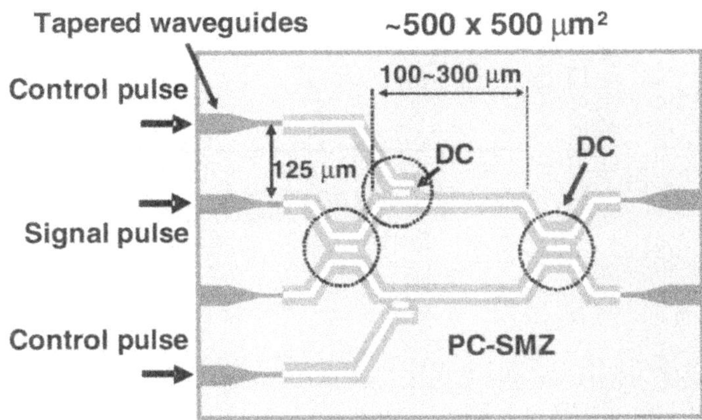

Fig. 12.18. Grand design of PC-SMZ. Total size is around $500 \times 500\ \mu m^2$

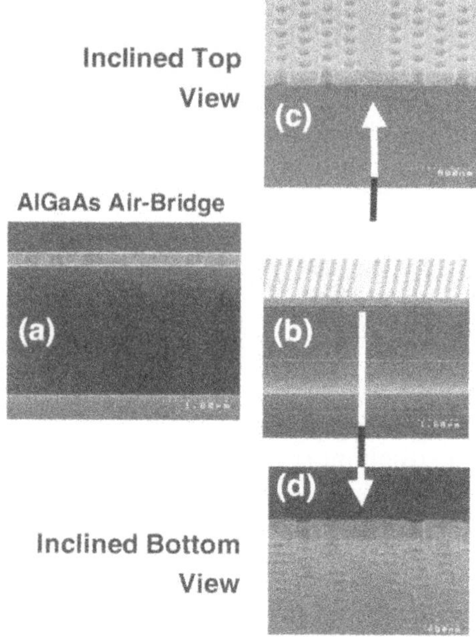

Fig. 12.19. Several SEM photographs of the cleaved edge viewed from several angles

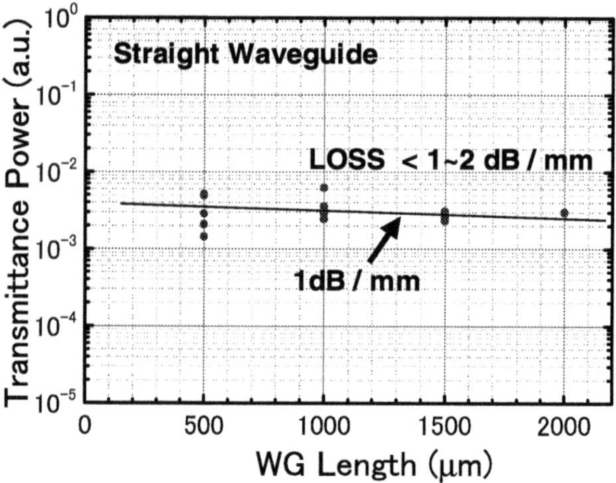

Fig. 12.20. Plot of measured transmittance powers for straight waveguides with different lengths from 500 to 2000 μm

for 2D PC slab waveguide [36] and better than the reported low values for the SOI sample [10, 37]; very recently, a loss as small as 0.7 dB/mm has been attained for a similar sample of 1.0 cm in length [38].

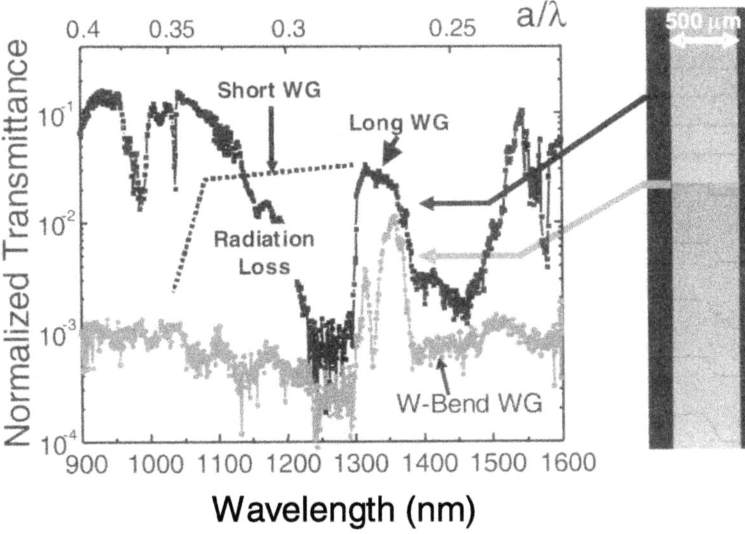

Fig. 12.21. Transmittance spectra for straight and double-bend long waveguides

Figure 12.21 shows T-spectra for the straight and double-bend long waveguides. The top-view SEM photograph of the sample is shown in the right side, where several patterns involving straight, single double-bend, doubled double-bend and tripled double-bend waveguides are seen. High transmission peaks appearing between 1300 and 1380 nm in wavelength are guided modes for the present waveguides.

More detailed analysis of these two kinds of peaks provides us with noticeable results as follows. First, let us compare T of the straight waveguides with the long length (shown in Fig. 12.21) and the short length (shown in Fig. 12.10). For the long waveguide, T at wavelengths shorter than 1300 nm is decreased abruptly. Taking into account that the wavelength of nearly 1300 nm presumably corresponds to the cross-point between the light line and the even guided mode, as shown by the band diagram in Fig. 12.6, the abrupt decrease of the transmission is thought to be due to the leaky mode. For the short waveguide, on the other hand, T corresponding to the leaky mode is high, as indicated schematically by the dotted line in Fig. 12.21. From this result, it is confirmed that the appearance of the high T in the leaky mode region strongly depends on the waveguide length, as predicted theoretically by the concept of the lifetime proper to the leaky mode [39].

Second, the T peak appears at different wavelengths between the long and short straight waveguides. Due to this effect, the maximum peak is separated by about 50 nm in wavelength between the straight and bend long waveguides, the result being different from the case for the short waveguide. These effects are presumably due to the wavelength-dependent propagation loss, namely, due to the group velocity-dependent loss in the current guided mode region. Detailed analysis is necessary for achievement of the PC-SMZ.

12.7 Summary

As an application of the 2D PC slab waveguide to the ultrasmall and ultrafast all-optical switching device, that is, PC-SMZ, several waveguides including straight, bend, Y-branch, directional-coupler, and coupled-cavity waveguides were fabricated by using fine EB lithography and dry etching on a GaAs/AlGaAs substrate. Transmission spectra and near-field patterns were characterized in a wide wavelength range from 850 to 1600 nm with the sample finished to the air-bridge type 2D PC slab.

As a result for the straight and bend waveguides, propagation loss was as small as 1–2 dB/mm, resulting in the high T of more than 80% including reflection at the boundary between PC and non-PC waveguides. Measured T-spectra were in excellent agreement with the spectra calculated by using 2D-/3D-FDTD methods. For the directional coupler, a short coupling length of less than 10 μm for 50% coupling was achieved by virtue of inserting additional 2D PC defects, thus enhancing the coupling strength between the parallel waveguides. On the other hand, the coupled-cavity waveguide was applied to the compact and flat-band delay element by a special configuration of the defect in the 1D air-hole array. This was demonstrated experimentally in the form of the 1D PC monorail on the GaAs substrate. Analysis of the discrepancy between measured and calculated transmission spectra resulted in the necessity of the precise position/size control of the air hole on the order of nearly 2% for realizing the desirable flat-band property.

These waveguides appear to be suitable for achieving the waveguide platform in the PC-SMZ based on the air-bridge type GaAs 2D PC slab.

References

1. J. D. Joannopoulos, R. D. Meade, J. N. Winn: *Photonic Crystals: Molding the Flow of Light*, (Princeton University Press, Princeton, 1995)
2. T. F. Krauss, R. M. De La Rue, and S. Brand: Nature, **383**, 699 (1996)
3. D. Labilloy, H. Benisty, C. Weisbuch, T. F. Krauss, R. M. De La Rue, V. Bardinal, R. Houdré, U. Oesterle, D. Cassagne, and C. Jouanin: Phys. Rev. Lett. **79**, 4147 (1997)
4. T. Baba, N. Fukaya, and J. Yonekura: Electron. Lett. **35**, 654 (1999)
5. S. Noda, A. Chutinan, and M. Imada: Nature **407**, 608 (2000)
6. S. G. Johnson, S. Fan, P. R. Villeneuve, and J. D. Joannopoulos: Phys. Rev. B**60**, 5751 (1999)
7. M. Tokushima, H. Kosaka, K. Tomita and H. Yamada: Appl. Phys. Lett. **76**, 952 (2000)

8. S. Y. Lin, E. Chow, S. G. Johnson, and J. D. Joannopoulos: Opt. Lett. **25**, 1297 (2000)
9. M. Loncar, T. Doll, J. Vuckovic and A. Scherer: J. Lightwave Technol. **18**, 1402 (2000)
10. M. Notomi, A. Shinya, K. Yamada, J. Takahashi, C. Takahashi, and I. Yokohama: IEEE J. Quantum Electron. **38**, 736 (2002)
11. Y. Sugimoto, N. Ikeda, N. Carlsson, K. Asakawa, N. Kawai, and K. Inoue: IEEE J. Quantum Electron. **38**, 760 (2002)
12. See, for example, K. Hirahara, T. Fujii, K. Ishida and S. Ishihara: IEICE Trans. Electron. Vol. E81-C, 1328, (1998)
13. Y. Sugimoto, N. Ikeda, N. Carlsson, K. Asakawa, N. Kawai, and K. Inoue: J. Appl. Phys. **91**, 922 (2002)
14. K. Tajima: Jpn. J. Appl. Phys. **32**, L1746 (1993)
15. Y. Watanabe *et al.*: unpublished
16. H. Nakamura *et al.*: unpublished
17. K. Inoue *et al.*: unpublished
18. S. Fan, S. G. Johnson, J. D. Joannopoulos, C. Manolatou, and H. A. Haus: J. Opt. Soc. Am. B**18**, 162 (2001)
19. Y. Sugimoto, N. Ikeda, N. Carlsson, K. Asakawa, N. Kawai, and K. Inoue: Opt. Lett. **2**, 388 (2002)
20. S. Lin, E. Chow, J. Bur, S. G. Johnson, and J. D. Joannopoulos: Opt. Lett. **27**, 1400 (2002)
21. S. Boscolo, M. Midrio, and T. F. Krauss: Opt. Lett. **27**, 1001 (2002)
22. S. Lan, K. Kanamoto, T. Yang, S. Nishikawa, Y. Sugimoto, N. Ikeda, H. Nakamura, K. Asakawa, and H. Ishikawa: Phys. Rev. B**67**, 115208 (2003)
23. M. Koshiba, Y. Tsuji, and M. Hikari: J. Lightwave Technol. **18**, 102 (2000)
24. S. Boscolo, M. Midrio, and C. G. Someda: IEEE J. Quantum Electron. **38**, 47 (2002)
25. A. S. Sharkawy, S. Shi, D. W. Prather, and R. A. Soref: Opt. Express, **10**, 1048 (2002)
26. Y. Tanaka *et al.*: unpublished
27. Y. Sugimoto, Y. Tanaka, N. Ikeda, T. Yang, H. Nakamura, K. Asakawa, K. Inoue, T. Maruyama, K. Miyashita, and K. Ishida: Appl. Phys. Lett. **83**, 3236 (2003)
28. A. Yariv, Y. Xu, R. K. Lee, and A. Scherer: Opt. Lett. **24**, 711 (1999)
29. M. Bayer, T. Gutbrod, A. Forchel, T. L. Reinecke, P. A. Knipp, R. Werner, and J. P. Reithmaier: Phys. Rev. Lett. **8**, 5374 (1999)
30. M. Bayindir, B. Temelkuran, and E. Ozbay: Phys. Rev. B**61**, R11855 (2000)
31. S. Lan, S. Nishikawa, and O. Wada: J. Appl. Phys. **90**, 4321 (2001)
32. S. Olivier, C. Smith, M. Rattier, H. Benisty, C. Weisbuch, T. Krauss, R. Houdré, and U. Oesterlé: Opt. Lett. **26**, 1019 (2001)
33. S. Lan, S. Nishikawa, Y. Sugimoto, N. Ikeda, K. Asakawa, and H. Ishikawa: Phys. Rev. B**65**, 165208 (2002)
34. S. Nishikawa, S. Lan, N. Ikeda, Y. Sugimoto, H. Ishikawa, and K. Asakawa: Optics Lett. **27**, 2079 (2002)
35. Y. Sugimoto, S. Lan, S. Nishikawa, N. Ikeda, H. Ishikawa, and K. Asakawa: Appl. Phys. Lett. **81**, 1946 (2002)
36. Y. Tanaka, Y. Sugimoto, N. Ikeda, H. Nakamura, K. Asakawa, K. Inoue, and S. G. Johnson: Electron. Lett. **70**, 174 (2004)

37. J. Arentoft, T. Søndergaard, M. Kristensen, A. Boltasseva, M. Thorhauge, and L. Frandsen: Electron. Lett. **38**, 274 (2002)
38. Y. Sugimoto, Y. Tanaka, N. Ikeda, Y. Nakamura, K. Asakawa, and K. Inoue: Optics Express **12**, 1090 (2004)
39. T. Ochiai and K. Sakoda: Phys. Rev. B**63**, 125107 (2001)

13 Photonic Crystals in the Terahertz Region

M. W. Takeda

Introduction

The intermediate region between the infrared and radio waves, comprising the frequency range from about 0.1 THz to 10 THz, (for wavelength: 3 mm–30 μm, and for wavenumber: 3.3 cm^{-1}–333.3 cm^{-1}) is known as the terahertz (THz) region. This frequency range covers the spectrum from millimeter waves to far-infrared waves. The first studies of photonic crystals (PCs) started in the microwave range because of the easy preparation of samples [1]. Developments in the THz region are not just a step in the progression from longer wavelengths to shorter ones; rather, considering the development of practical application techniques, great opportunities lie in the THz region, including millimeter waves.

We discuss the following four subjects in this chapter.

1. Dispersion relation of terahertz wave in PCs revealed by terahertz time-domain spectroscopy.
2. Direct excitation of localized defect modes in a pseudo-simple-cubic 3D PC at terahertz frequencies using a nonlinear crystal defect layer.
3. Dual-periodic PCs using a microstripline.
4. Control of the microwave emitting direction by using diamond lattice structure PC with lattice modification.

In the first section, the details of terahertz time-domain spectroscopy (THz-TDS) and experimental results of the dispersion relation of terahertz waves in PCs revealed by THz-TDS will be discussed. The second section shows the results of pulse THz wave generation by using PC cavities. In the last two sections we will discuss microstriplines and diamond structured PCs as samples for the millimeter and micrometer wave regions.

13.1 Dispersion Relation of Terahertz Waves in Photonic Crystals

The purpose of PC study in the THz region is the direct acquisition of knowledge concerning the propagation characteristics of electromagnetic waves, such as the dispersion relation of electromagnetic waves in a PC [2, 3], which is difficult to measure directly by using any techniques in the infrared or visible regions, as well as device development in this wavelength region. Attention has now turned to the effectiveness with which the knowledge obtained in the THz region can be applied to other wavelength regions, such as the infrared or the visible region, as the scaling law (see Sect.2.1.2) is well established between the lattice constant and the wavelength of an electromagnetic wave in the PC [4, 5]. THz time-domain spectroscopy (THz-TDS) using a pulsed THz electromagnetic wave has also been established after the development of a detection and generation technique for a pulsed THz electromagnetic wave utilizing an ultrashort pulsed laser [6, 7]. This is based on the recent rapid progress made in femtosecond lasers. Studies using THz-TDS regarding the propagation characteristics of THz waves in PCs are introduced in this section.

The detail of the experimental setup for generation and detection of THz radiation is shown in Fig. 13.1 [8]. A mode-locked erbium fiber laser produced 100 fs light pulses with a wavelength of about 780 nm at a repetition rate of 48 MHz. The femtosecond pump pulses were focused by an objective lens on the biased gap of the photoconductive antenna made of GaAs grown at low temperature. The emitted radiation was collimated and focused by a pair of off-axis paraboloidal mirrors onto a specimen. Then the transmitted radiation was again collimated and focused by another pair of off-axis paraboloidal mirrors onto a sampling detector, which was also a photoconductive antenna. The photoconductive detector was gated by femtosecond probe beam pulses that were separated from the pumped pulses by a beam splitter. The dc photocurrent induced by the radiation field of transmission from the specimen on the detector was measured by an amperemeter. By delaying the timing of

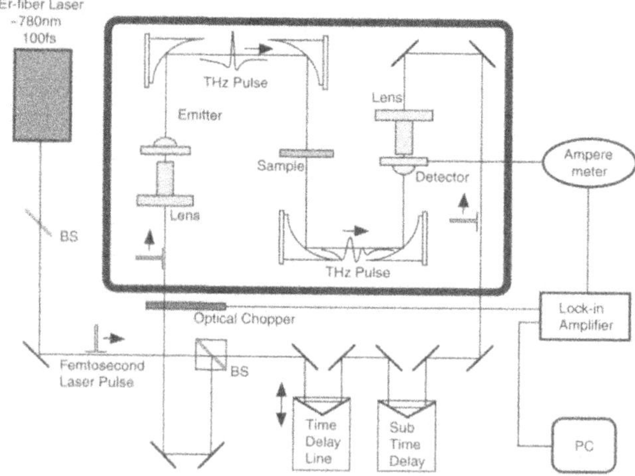

Fig. 13.1. Schematics of terahertz time-domain spectrometer

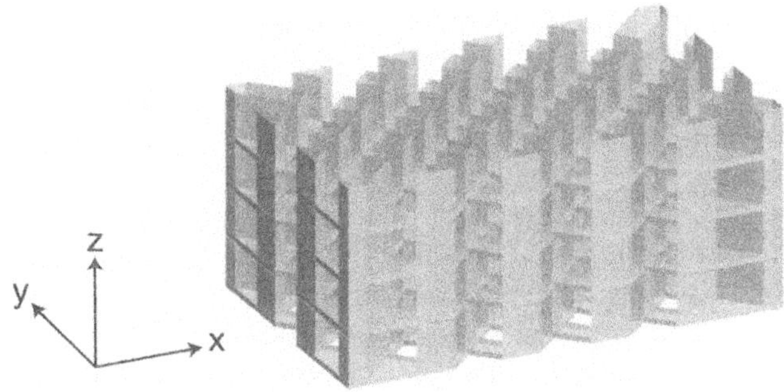

Fig. 13.2. Schematics of simple cubic air-rod lattice photonic crystal

the probe pulse to the pump pulse, the waveform for the electromagnetic pulse was obtained. All the optics for THz radiation were placed in an evacuated box to reduce the THz absorption of water vapor.

A PC of pseudo-simple-cubic lattice composed of square air-rods was fabricated by use of silicon as shown in Fig. 13.2, where a is 0.40 mm, ε is 11.4 and the filling factor is about 0.82 [9].

In Fig. 13.3 the solid line shows the transmission intensity spectrum for the Γ-Z direction. It should be noted that opaque regions exist from 6.5 cm^{-1} to 10.2 cm^{-1} which corresponds to the PBG between the first band and the second band calculated by a plane wave expansion method. The transmission region below 6.5 cm^{-1} corresponds to the first band and that from 10.2 cm^{-1} to 15.9 cm^{-1} corresponds to the second band. The periodic or modulated transmission-features in both regions are due to Fabry–Perot interference effects as discussed below.

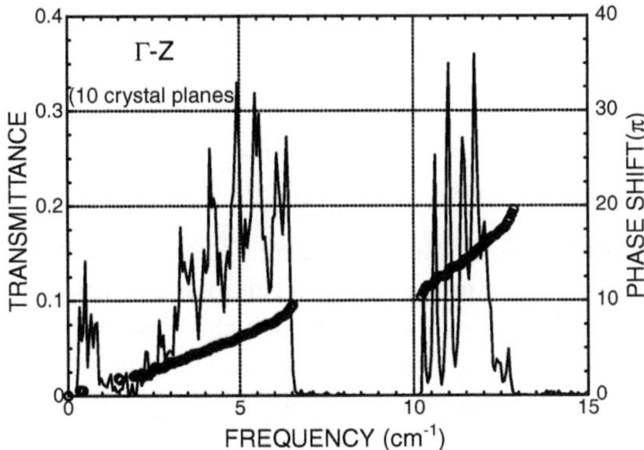

Fig. 13.3. Terahertz transmission spectra of the pseudo-simple cubic air-rod lattice with 10 crystal planes along the z-axis. The solid line denotes the transmission intensity and the open circles the phase shift

The open circles in Fig. 13.3 show the phase shift spectrum for the Γ-Z direction. The row data of phase shift include an ambiguity of $n\pi$ radians. The π radians should be added to the phase of the second band region above 10.2 cm^{-1}, because of the phase shift π between the BG as mentioned in the following section.

From the direct measurement of phase shift by using THz-TDS, we have no phase shift spectra in the BG region between the first and second branches; however, we can infer these from analogy to the electronic band structure of semiconductors, in which the conduction band and the valence band surround the fundamental gap. The gap between the first and second bands occurs at the BZ boundary at $k = \pi/a$. The modes are standing waves with a wavelength of $2a$, twice a for $k = \pi/a$. There are two types of this kind of standing wave. One is that the nodes locate at the center of the high-ε (dielectric constant) layer, and the other is that the nodes locate at the center of the low-ε layer. By the electromagnetic variational theorem the low-frequency modes concentrate their energy in the high-ε region (dielectric band), whereas the high-frequency modes concentrate their energy in the low-ε regions (air band). Therefore the difference between the phase of the mode at the top of the first branch (dielectric band) and that at the bottom of the second branch (air band) must be π.

The transmission intensity shows a maximum at the frequencies corresponding to the standing waves in the specimen (Fabry–Perot effect). In analogy to the standing wave of the mono chord, the difference of phase between the neighboring mode should be π. This feature is seen in the observed spectra shown in Fig. 13.3. In this figure the wavelike oscillation seen in the transmission intensity spectrum is due to the interference between the light reflected by the front and end surfaces of the specimen, the so-called Fabry–Perot mode. The influence of the interference is also observed in the spectrum of the phase shift. That is, the phase shift remains almost constant in the region with low transmission coefficient.

The phase shift, $\phi(\omega)$, is proportional to the wavenumber, $k(\omega)$,

$$\phi(\omega) = \frac{k(\omega)}{d},$$

where d is the thickness of the specimen. The phase shift increases rapidly in the vicinity of the maximum of the transmission intensity and is relatively slow around the minimum as shown in both the observed and calculated phase shift spectra. This fact indicates that the optical density of state (ODOS) is very large around the frequency of each standing wave [10, 11]. Moreover it should be noted that in the vicinity of the band gap edges the ODOS must be very large.

The dispersion relation of an H-polarized electromagnetic wave in a triangular lattice PC that is made of methylpentene polymer (TPX) with air rods [12, 13] is shown in Fig. 13.4. The solid line and open circles indicate

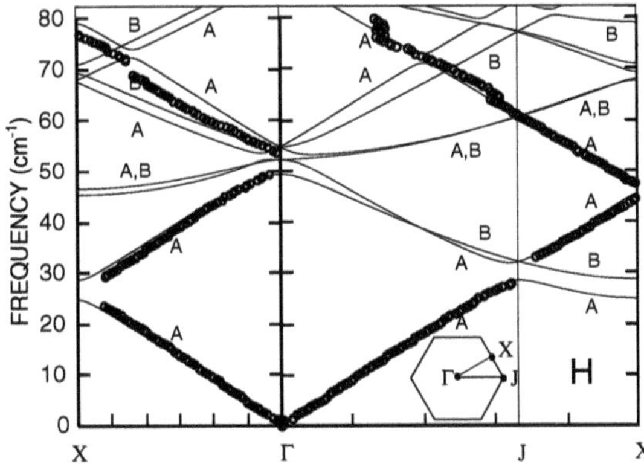

Fig. 13.4. Dispersion relation of THz wave in the triangular air-rod lattice photonic crystal made of TPX polymer

the calculated results from the plane wave expansion method and the measured results estimated from the phase shift spectra, respectively. Both show very good coincidence to each other not only qualitatively but also quantitatively. It should be noted that group velocity v_g of the electromagnetic wave is almost zero at the region where the anticrossing and level repulsion (Browster mode) occurs between the seventh and eighth bands at the Γ-X branch belonging to the A irreducible representation [8].

13.2 Direct Excitation of Localized Planar Defect Modes

In this section, we report on the direct excitation and observation of the defect modes in a PC by exciting the dielectric layer (made with a nonlinear crystal ZnTe) with femtosecond laser pulses. The planar defect modes were introduced by inserting a dielectric layer between the two PC slabs [14]. A nonlinear crystal ((110)-cut ZnTe crystal) substrate with different thickness were used as the dielectric defect layer, thus enabling THz radiation to be generated through difference frequency mixing by excitation with a broadband femtosecond pump laser within the defect layer. There are several advantages in using a nonlinear crystal as the defect layer and hence the direct generation of THz radiation within the defect layer. The first obvious one is that, compared to the simple transmission measurement, we could more clearly see the defect modes generated in the PC because of the higher excitation efficiency of the defect modes in the layer. In contrast, in transmission measurement, most of the radiation power is reflected off due to the high reflectivity of PC at the BG frequency range and is therefore much less strongly coupled to the cavity or the defect modes. Second, we can investigate the effect of a PC on the nonlinear difference frequency mixing through the strongly dispersive property of the PC. Because of the relaxed phase matching condition between the THz wave and optical wave in a PC, we can expect a higher nonlinear efficiency.

Each PC cavity consisted of two four-layered pseudo-simple-cubic air-rod lattices as shown in Fig. 13.2 and a dielectric defect layer with a (110)-cut ZnTe substrate between them. In order to investigate the spectroscopic characteristics of these PCs with such a defect layer, we carried out THz-TDS. The results for the transmission amplitude (not for the power) are shown in Fig. 13.5. The spectra show a transmission stop band from 6.2 cm^{-1} to about 10.5 cm^{-1}, as has already been observed for the bulk PCs without a defect layer (Fig. 13.3). In the middle of the PC BG region, we observed a sharp and distinct transmission peak at 7.9 cm^{-1} (10%) for the sample with a 0.3 mm defect layer.

Next, by exciting directly the defect layer with femtosecond laser pulses, THz radiation was generated through difference frequency mixing among the frequency components within the spectrum bandwidth of the femtosecond laser (or, in other words, by the optical rectification effect). The experimental setup was almost the same as THz-TDS apparatus except that the photoconductive antenna emitter was replaced by the PC cavity. The excitation pulses were delivered from a mode-locked Ti:sapphire laser, with a pulse width, repetition rate, and center wavelength of 150 fs, 76 MHz, and 830 nm, respectively. The laser beam was focused, using an objective lens (×5), onto the defect layer (ZnTe) through an air hole in the PC in the direction of the z axis. The average pump laser power was about 150 mW. The radiation emitted from the opposite side of the sample was collected and focused on to a photoconductive dipole antenna (1 mm long) mounted on a Si hemispherical lens (26 mm diameter) using a pair of off-axis parabolic mirrors. The probe laser pulses, whose average power was about 3 mW, triggered the photoconductive receiver antenna with a variable time delay against the

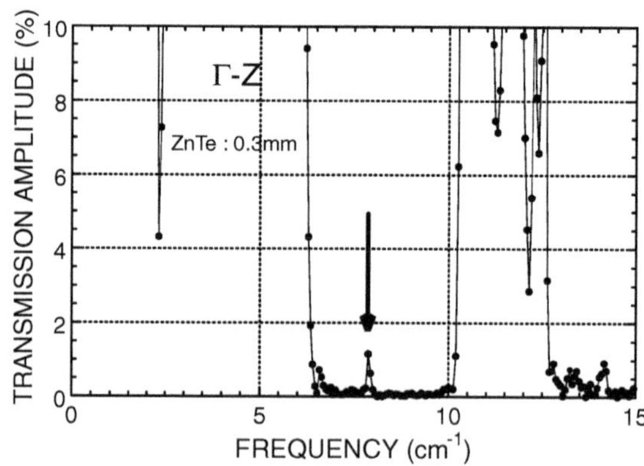

Fig. 13.5. Transmission amplitude spectra of a photonic crystal cavity constructed from simple cubic air-rod lattice and a planar defect of a ZnTe single crystal

pump pulse, enabling the temporal waveform of the emitted THz field to be obtained. The polarization of the pump beam was horizontally polarized and the horizontal polarization component of the THz radiation was detected. The sample was rotated around the beam axis (z axis) to position the $\langle 001 \rangle$ crystal axis of ZnTe at an angle of 26° to the polarization direction of the pump laser, where the THz emission efficiency was maximized. Since our PC was symmetric around the z axis, the rotation of the sample around the beam axis did not affect the transmission properties of our measured spectra. The frequency resolution in these measurements was about 0.12 cm^{-1}, limited by the scanning path difference of the pump and probe beams.

Figure 13.6 shows the THz emission power spectra (the square of the amplitude spectra) obtained by Fourier transforming the measured THz waveforms without normalization by a background spectrum. Therefore, the spectral feature also includes the emission efficiency due to optical rectification in the ZnTe crystal, in addition to the transmission property and defect mode characteristics of the PC. The sample with a defect layer thickness of 0.3 mm showed a spectral peak at 7.9 cm^{-1}, corresponding exactly to that observed in the transmission spectrum in Fig. 13.5. Because the defect layer thickness was not as thick as one monolayer of the crystal structure, it was reasonable to have a single defect mode within the BG region. The spatial distribution of the electric field of the planar defect mode at 7.9 cm^{-1} was calculated by the FDTD method [15]. This mode is asymmetric for the defect plane and localized at the region of the exit side of THz wave propagation. The asymmetric character must be due to the asymmetric cavity.

The Q-factor for the localized defect modes was estimated to be better than 100 from the line width of the peak observed in Fig. 13.5, which was limited by the instrumental frequency resolution ($\sim$ 0.06 cm^{-1}). Considering

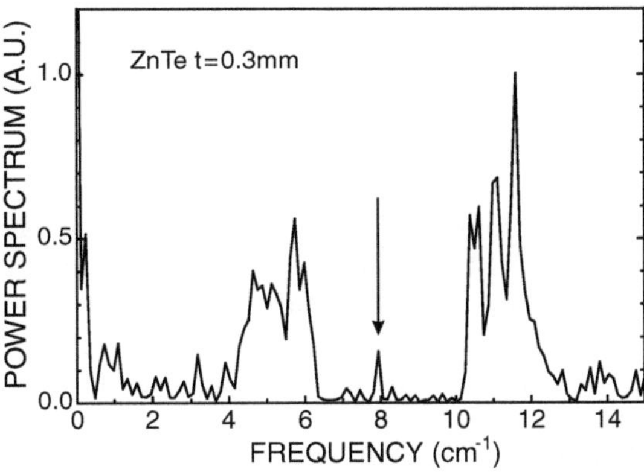

Fig. 13.6. Emission spectra of the photonic crystal cavity constructed from simple cubic lattice and a planar defect of a ZnTe single crystal [14]

the absorption coefficient of ZnTe (< 0.02 cm^{-1} below 300 GHz) in the relevant frequency region, a Q-factor larger than 500 is feasible.

In conclusion, we demonstrated direct excitation of localized defect modes in a 3D (pseudo-simple-cubic) PC using a nonlinear crystal as the defect layer, which was pumped with femtosecond laser pulses. Several defect modes were clearly observed, some of which were difficult to observe using transmission spectroscopy.

13.3 Dual-Periodic Photonic Crystals

Research on a 1D PC has been extended to superlattice structures, such as Harper structures [16] and Fibonacci lattices [17], from simple periodic structures. These superlattices have characteristics such as critical self-similarity and a unique energy spectrum, which is known as a butterfly diagram [18]. Similar research has also been actively performed on optical superlattice structures to investigate photon localization effects and the possible emergence of a fractal structure in the photonic energy spectrum. The dual-periodic structure has the fundamental period λ_1 in the dielectric medium, which is modulated by another periodic function with period λ_2, as discussed in Sect.8.1. To maintain the translation symmetry of the crystal, the value of λ_2 should satisfy the condition $n\lambda_1 = m\lambda_2$, where m/n should be rational. When the ratio of m/n is irrational, the system becomes a quasi-crystal by losing its translation symmetry. When the periodic structure with small fundamental period of λ_1 is modulated by a large period of λ_2 ($p = m/n < 1$), there emerges an optical butterfly diagram and singular Bloch modes with a high Q-factor in the PBG region of the fundamental periodic structure [19, 20]. The dual-periodicity in the dielectric constant ε of a PC is technically very difficult to create. Thus, no report on an experimental study has been made concerning the singular Bloch mode.

Because of this absence of research, we fabricated a dual-periodic PC made with a microstrip line to confirm the existence of the singular Bloch mode. By changing the thickness of the multilayer structure or changing the ε-value of each layer we can control the periodic structure of ε, although changing the layer thickness of a multilayer structure is relatively easier than arbitrarily changing ε. If the periodic structure is made with the microstrip line, the effective ε-value can be controlled by changing the width of the strip line. This feature of a microstrip line is suitable for applying to the fabrication of a PC that consists of a medium with a different ε.

The dual-periodic PC was fabricated by using a dual-periodic microstrip line [22]. The fundamental period of a 1D dual-periodic PC is composed of a repetition of layer A with thickness d_1 and of layer B with thickness d_2. While ε_A of layer A is modulated by a suitable function to introduce a dual-periodicity, ε_B of layer B is kept constant. A position-dependent function is applied as follows (see Fig. 13.7)

$$\varepsilon_A(i) = \frac{\varepsilon_{\max} + \varepsilon_{\min}}{2} + \frac{\varepsilon_{\max} - \varepsilon_{\min}}{2} \cos(2\pi p i),$$

where i is the sequential site number for layer A, and $\varepsilon_{\min}$ and $\varepsilon_{\max}$ are the minimum and maximum values that ε_A can attain. The modulation parameter p must be a rational number $p = m/n$ to preserve the dual-periodicity of the system.

Figures 13.7 and 13.8 show the model of the dual-periodic ε contour and schematics of a real microstrip line with dual-periodicity, respectively.

The solid and open circles in Fig. 13.9 show the transmission intensity and phase shift spectra of the microstrip line with dual-periodicity of $b/a = 1$ and $p = 1/18$ measured by using a network analyzer. The PBG exists in the region from 3.8 GHz to 6.3 GHz, and a sharp peak is seen at 5.15 GHz, which corresponds to the singular Bloch mode due to the dual-periodicity.

The transmission and reflection amplitude and phase shift were calculated for the photonic lattices with the microstrip lines. The method of moments

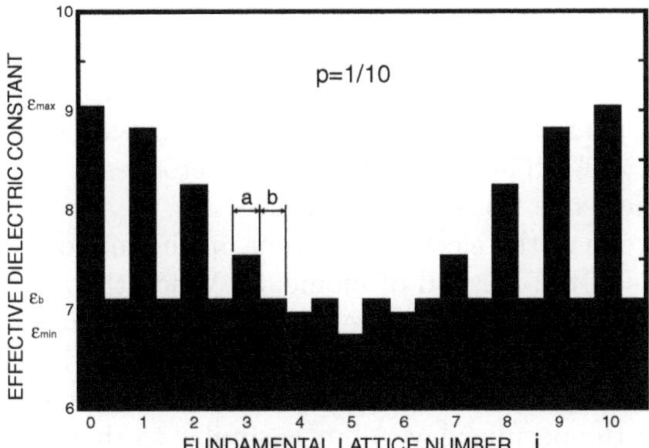

Fig. 13.7. Model of the dual-periodic dielectric constant contour using sinusoidal modulation with $b/a = 1$ and $p = 1/10$

Fig. 13.8. Schematics of microstrip line with dual-periodicity of $b/a = 1$ and $p = 1/18$

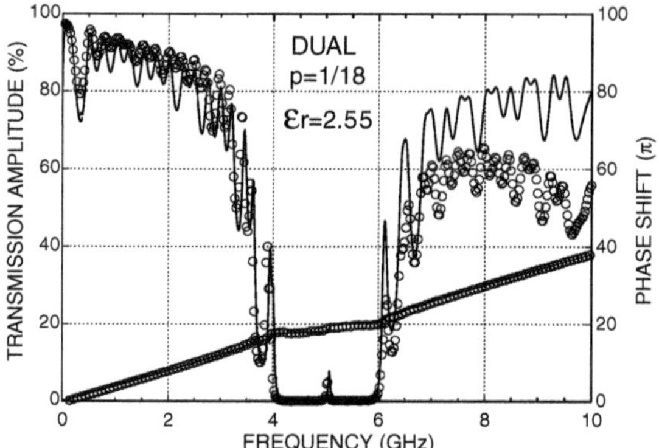

Fig. 13.9. Transmission spectra of microstrip line with dual-periodicity of $b/a = 1$ and $p = 1/18$

was used in the calculations. The solid lines in Fig. 13.9 show the simulation results of the transmission amplitude and phase shift spectra calculated with the same parameters as those of the samples. The simulation results and those of the measurements show very good correspondence in their transmission amplitude and phase shift spectra. In the transmission region, the amplitude does not reach 100%. This is due to the conductor resistance of the microstrip line and the dielectric loss of the substrate.

The amplitude of the electrical current on the microstrip lines was calculated by means of the method of moments. When the impurity line is introduced into a simple periodic microstrip line, the electrical current amplitude in the impurity line is extraordinarily large due to the localization effect. By contrast, the electrical current amplitude of the singular Bloch mode is not emphasized on the peculiar layer, but is similar to that of the standing wave generated by the Fabry–Perot effect. Also, a similar standing wave emerges in the frequency region of the transmission band. This gives us a reason to assume that the singular Bloch mode, which emerges in the dual-periodic structure, is not similar to the impurity mode but is similar to the standing wave mode. This is the mode that shows the maximum amplitude in the transmission band.

13.4 Control of Microwave Emission from Photonic Crystal

The diamond lattice structure is proposed as an ideal structure to achieve a perfect common PBG in all directions. However, such a complex structure is difficult to fabricate. Figure 13.10 shows the PBS of the diamond lattice structure with ε of 7 and the ratio of rod diameter and edge of the cubic lattice is 25/70. The common PBG is seen in the region of normalized frequency

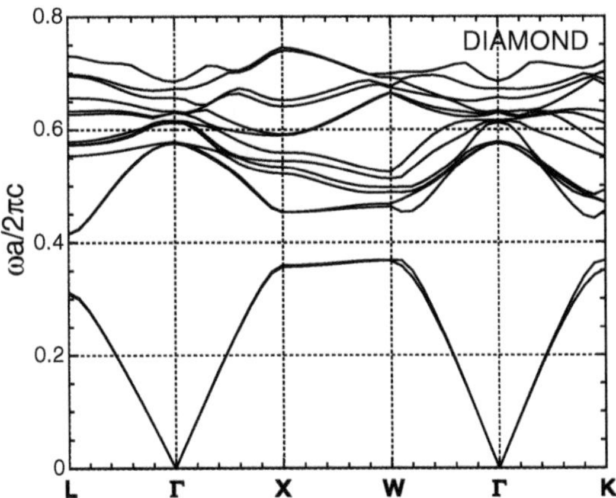

Fig. 13.10. Calculated photonic band structure for the diamond lattice. The dielectric constant of the dielectric rods is 7. The ratio of the rod diameter and the edge of the cubic lattice is 25/70

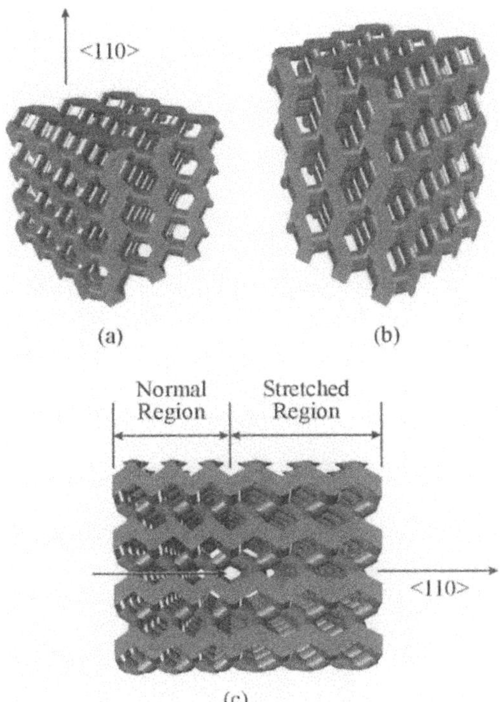

Fig. 13.11. Three-dimensional CAD models of diamond photonic crystals. (**a**) Normal structure, (**b**) modified structure with stretched lattice spacing, (**c**) a direction antenna head composed of normal and stretched diamond structures with graded lattice spacing

0.35–0.42. The inverse diamond lattice structure also exhibits a complete PBG (see Fig. 2.5).

By a rapid prototyping method called stereolithography 3D PCs having diamond-lattice and inverse diamond-lattice structures have been designed using CAD and can be fabricated [23, 24]. The structures were formed with millimeter-order epoxy lattices including titanium oxide based ceramic particles of about 10 μm in average diameter. The value of ε of dielectric rods is 10. A unit cell image of the diamond structures is shown in Fig. 13.11(a). The crystal sample of $50 \times 50 \times 50\ \mathrm{mm}^3$ in size was composed of dielectric rods of 2.88 mm diameter and 4.33 mm length. The lattice spacing for the $\langle 100 \rangle$ direction was 10 mm in length. The volume ratio of dielectrics in the crystal was 33%. Figure 13.11(b) shows the modified diamond structure with the stretched lattice spacing for the $\langle 110 \rangle$ direction at a stretching ratio of 150%. The normal and stretched diamond lattices were joined together to realize the directional transmission of microwaves.

Microwave transmissions through the crystal samples were measured by using two monopole antennas and a network analyzer. A monopole antenna for emission was inserted into the center position of a crystal sample. Another monopole antenna was placed at an interval of 150 mm away from the center to receive the microwaves. The transmission amplitude was measured for every 15° as shown in Fig. 13.12. The normal diamond structure formed a perfect PBG in the frequency range of 13.5–16.5 GHz. Modified diamond structures having a stretched lattice spacing in the $\langle 110 \rangle$ direction shifted the BG toward lower frequency. The location of the BGs agrees with the experimental results of the attenuation of the microwave transmission amplitude through the samples by using a network analyzer and microwave cavities.

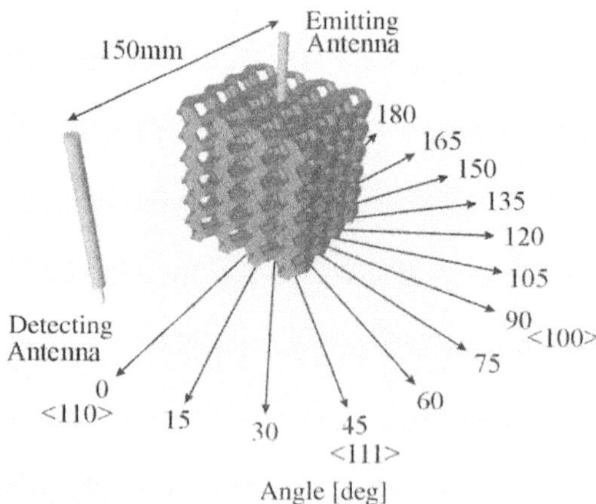

Fig. 13.12. An experimental configuration for measurement of transmission attenuation in the microwave range through a photonic crystal of diamond structure

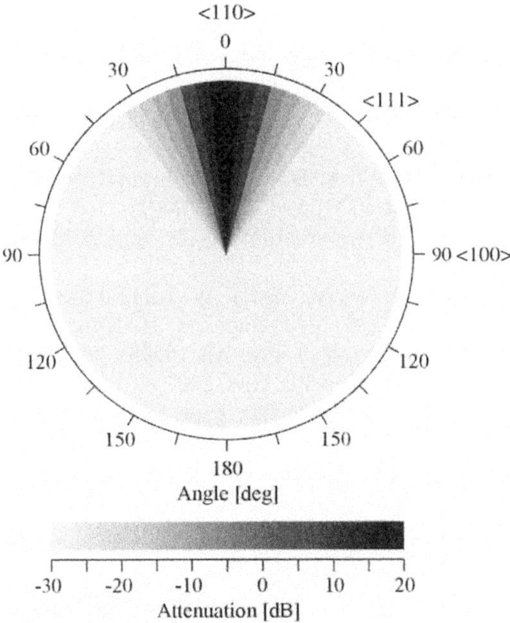

Fig. 13.13. Microwave emission profile through an antenna head composed of the normal and directional diamond structure. Microwaves of 15 GHz were emitted from an inserted monopole antenna

Figure 13.13 shows the emission profile of microwaves of 15 GHz from the directional antenna head to the air space. The transmission efficiency of microwaves was amplified for the limited direction in the range of 0–30°. This result means that the power concentration and direction of electromagnetic wave emission can be controlled by the structural modification of the PC lattice.

References

1. E. Yablonovich, Phys Rev. Lett. **58**, 59 (1987)
2. K. Ohtaka, Phys. Rev. **B19**, 5057 (1979)
3. K. Sakoda, *Optical Properties of Photonic Crystals*, (Springer Verlag, Berlin 2001)
4. E. Ozbay, A. Abeyta, G. Tuttle, M. Tringides, R. Biswas, C. T. Chan, C. M. Soukoulis, and K. M. Ho, Phys. Rev. **B50**, 1945 (1994)
5. E. Ozbay, E. Michel, G. Tuttle, R. Biswas, K.M. Ho, J. Bostak, and D.M. Bloom, Appl. Phys. Lett. **65**, 1617 (1994)
6. Special review issue *Terahertz electromagnetic pulse generation, physics, and applications*. J. Opt. Soc. Am. **B11**, 2454 (1994)
7. M. Tani, S. Matsuura, K. Sakai, and S. Nakashima, Appl. Optics, **36**, 7853 (1997)

8. H. Kitahara, N. Tsumura, H. Kondo, M. W. Takeda, J.W. Haus, Z. Yuan, N. Kawai, K. Sakoda, and K. Inoue, Phys. Rev. **B64**, 045202 (2001)
9. T. Aoki, M. W. Takeda, J. W. Haus, Z. Yuan, M. Tani, K. Sakai, N. Kawai, and K. Inoue, Phys. Rev. **B64**, 045106 (2001)
10. T. Kondo, M. Hangyo, S. Yamaguchi, S. Yano, Y. Segawa, and K. Ohtaka, Phys. Rev. **B60**, 033111 (2002)
11. S. Yano, Y. Segawa, J. S. Bae, K. Mizuno, H. Miyazaki, K. Ohtaka, and S. Yamaguchi, Phys. Rev. **B63**, 153316 (2001)
12. M. Wada, Y. Doi, K. Inoue, J. W. Haus, and Z. Yuan, Appl. Phys. Lett. **70**, 2966 (1997)
13. M. Wada, Y. Doi, K. Inoue, and J. W. Haus, Phys. Rev. **B55**, 10443 (1997)
14. M. Tani, P. Gu, K. Sakai, M. Suenaga, H. Kondo, H. Kitahara, and M. W. Takeda, Technical digest of the 4th Pacific Rim Conference on Laser and Electro-Optics (CLEO/Pacific Rim 2001) (Makuhari-Messe, Japan, 2001)
15. M. Iida, M. Tani, K. Sakai, M. Watanabe, S. Katayama, H. Kondo, and M. W. Takeda, Abstract of International Quantum Electronics Conference (Moscow, Russia, 2002); M. Iida, M. Tani, P. Gu, K. Sakai, M. Watanabe, H. Kitahara, S. Kato, M. Suenaga, H. Kondo, and M. W. Wada, Jpn. J. Appl. Phys. **42**, L1442 (2003)
16. D. R. Hofstadter, Phys. Rev. **B14**, 2239 (1976)
17. M. Kohmoto and L. P. Kada, Phys. Rev. Lett. **50**, 1870 (1983)
18. R. Shimada, T. Koda, T. Ueta, and K. Ohtaka, J. Phys. Soc. Jpn, **67**, 3414 (1998)
19. R. Shimada, T. Koda, T. Ueta, and K. Ohtaka, J. Appl. Phys. **90**, 3905 (2001)
20. K. Ohtaka and Y. Tanabe, J. Phys. Soc. Jpn. **65**, 2265, 2276, and 2670 (1996)
21. H. Kitahara, T. Kawaguchi, J. Miyashita, and M. W. Takeda, J. Phys. Soc. Jpn. **72**, 951 (2003)
22. H. Kitahara, T. Kawaguchi, J. Miyashita, R. Shimada, and M. W. Takeda, J. Phys. Soc. Jpn. **73**, 296 (2004)
23. S. Kirihara, Y. Miyamoto, K. Takenaga, M. W. Takeda, and K. Kajiyama, Solid State Commun. **121**, 435 (2002)
24. S. Kirihara, M. W. Takeda, K. Sakoda, and Y. Miyamoto, Solid State Commun. **124**, 135 (2002)

14 Perspective

S. Noda and K. Ohtaka

The progress of PC research since 1998, when the project started, to the present has been remarkable as discussed in this book. Here we first focus on the technological side of the research.

Nanotechnology to achieve artificial structures with high accuracy has been developed for the realization of 3D PCs with a complete PBG at optical wavelengths. Remarkable progress has also been made in 2D PCs. The PC slab structure has become one of tractions. This structure utilizes the strong light confinement effect. Propagation in a waveguide engraved in a 2D PC has been well understood and we already have ample information on the theoretical condition for lossless propagation in a line defect waveguide. Moreover, photons propagating in a waveguide are successfully trapped by the point defects introduced in the vicinity of the waveguide. These achievements will lead to the realization of very small light add/drop devices of surface-emitting type and in the application of other nonlinear optics.

The concept of the PC has also been applied to optical fibers. The related systems, now called PC fibers, consist of many air holes arryed periodically in the directions perpendicular to the light-forwarding direction. The results obtained in PC fibers seem to show the possibility of ultimate systems which realize an extremely low loss and low nonlinearity propagation of light. Also, even in the field of traditional propagation principles, i.e. under total reflection conditions, a fiber with air holes, known as a holey fiber, has presented new intersting phenomena one after another. This is expected to function as a new device in many ways.

A fabrication technique for uniform 3D PCs using bias sputtering has also been developed. The interesting optical characteristics of the superprism effect observed in a PC fabricated by this method will enable us to apply it as a wavelength filter or as a dispersion compensator.

The coherent operation of a wide-area PC laser has become possible using a 2D PC. The Bragg diffraction involved in lasing is multidirectional due to the two-dimensionality and is physically very interesting because multidirectional Bragg diffraction provides a new feedback mechanism not attainable in a traditional distributed feedback laser. In engineering, too, such control of the laser mode will lead to various important applications such as high-power lasers or surface-emitting lasers with very narrow divergence angles.

The other interesting topics discussed in the literature are (i) "negative refractive index" phenomena, or "left-handed material", (ii) thermal emission control by PCs, (iii) combination of organic material and PCs, (iv) self-assembled fabrication or stamp fabrication of PCs.

Based on these recent developments, we hope that 2D-based devices such as channel add/drop devices, band-edge lasers, dispersion-controlled devices, and so on, will be commercially available in the near future. Monolithic integration of various optical components may be achieved by 2D PCs. It is natural to expect that higher integration would be possible in 3D complete PBG structures. Thermal emission control is also interesting from the viewpoint of the saving of energy for the 21st century.

Although the main purpose of this project was to develop high-quality PCs for potential applications to optical devices, the theoretical study of PCs has also made remarkable progress. In addition to the high expectation for technological application, the recent developments in the study of PCs seems due partly to a precise establishment of the one-particle picture of photons, which allows us to make a compact, sometimes straightforward, treatment of Maxwell's equations. Before the start of the project, however, there were a number of unsolved theoretical problems, such as how to treat the lifetime of leaky PBs in a system of finite thickness. As revealed in this book, many important theoretical issues have been examined and solved, within the framework of Maxwell's equations, since this project started in 1998.

Future development in the field of PCs will depend on how successfully and how rapidly their technological applications proceed in the next few years. For really exciting progress of the field, however, PCs should keep providing us with topics which are not only useful but also brand new and deep, both conceptually and physically. In our view, there are still several interesting problems even within the framework of Maxwell's equation, which remain to be attacked. One example is the physics of optically uncoupled photonic band modes at the Γ point. There are a lot of modes at the Γ point which are confined completely inside a PC, i.e, their leakage out of the PC is completely inhibited for symmetry reasons. Is there an optical means to excite them? Could we use some other methods, charged particle bombardment, for example, if the excitation is indeed impossible optically? If we can somehow send a photon to such a mode of perfect confinement, can we use a PC as a photon container, which precludes leakage due to the perfect confinement effect? If we can accumulate a substantial number of photons in such a mode, can we have a Bose–Einstein condensation of photons? If the answer is yes, what then is the characteristic of the Goldstone boson? Obviously, all these problems are accompanying exciting technological application.

Outside the one-particle properties, there are of course a variety of many-body problems of photons, interacting with other photons, electrons, excitons, and so on. One may quote the photon–polaron problem as an example. The problems so far studied in such topics as nonlinear photon mixing, polariton formation involving excitons and plasmons, and so on, have mostly been treated in the single-photon picture. However, we can hope that they will lead to the future development of many-body processes in PCs. It is reasonable to expect that PCs can serve as arenas for fundamental physical problems of photons, just as quantum wells or arrayed quantum wells of semiconductors have done so in the history of physics of electrons.

January 15, 2004 S. Noda and K. Ohtaka

Appendix A. Reciprocal Lattice Vector and Discretized Wavevector

K. Ohtaka and K. Inoue

1 Reciprocal Lattice Vectors and First Brillouin Zone

Reciprocal lattice vectors of a lattice are defined to be the wavevectors $\boldsymbol{h}$ that satisfy

$$\exp(\mathrm{i}\boldsymbol{h}\cdot\boldsymbol{R}) = 1, \tag{1}$$

for any lattice translation vector $\boldsymbol{R}$ given by

$$\boldsymbol{R} = p_1\boldsymbol{a}_1 + p_2\boldsymbol{a}_2 + p_3\boldsymbol{a}_3. \tag{2}$$

Here p_1, p_2, p_3 are three arbitrary integers and $\boldsymbol{a}_1$, $\boldsymbol{a}_2$, $\boldsymbol{a}_3$ are three primitive translation vectors that define the lattice. For the three special cases of $\boldsymbol{R} = \boldsymbol{a}_1, \boldsymbol{a}_2$ and $\boldsymbol{a}_3$, (1) leads to

$$\begin{aligned}\boldsymbol{h}\cdot\boldsymbol{a}_1 &= 2\pi n_1,\\ \boldsymbol{h}\cdot\boldsymbol{a}_2 &= 2\pi n_2,\\ \boldsymbol{h}\cdot\boldsymbol{a}_3 &= 2\pi n_3,\end{aligned} \tag{3}$$

repectively, using three integers. They are linear coupled equations for $\boldsymbol{h} = (h_x, h_y, h_z)$. Resolving the vectors into the cartesian components, we obtain the solution $\boldsymbol{h}$ from Cramer's rule of linear algebra:

$$\boldsymbol{h} = n_1\boldsymbol{b}_1 + n_2\boldsymbol{b}_2 + n_3\boldsymbol{b}_3, \tag{4}$$

where

$$\begin{aligned}\boldsymbol{b}_1 &= 2\pi\frac{\boldsymbol{a}_2\times\boldsymbol{a}_3}{\boldsymbol{a}_1\cdot(\boldsymbol{a}_2\times\boldsymbol{a}_3)},\\ \boldsymbol{b}_2 &= 2\pi\frac{\boldsymbol{a}_2\times\boldsymbol{a}_3}{\boldsymbol{a}_1\cdot(\boldsymbol{a}_2\times\boldsymbol{a}_3)},\\ \boldsymbol{b}_3 &= 2\pi\frac{\boldsymbol{a}_2\times\boldsymbol{a}_3}{\boldsymbol{a}_1\cdot(\boldsymbol{a}_2\times\boldsymbol{a}_3)}.\end{aligned} \tag{5}$$

The solution (4) obtained as a necessary condition is obviously a sufficient condition for (1) to hold for an arbitrary lattice translation $\boldsymbol{R}$. The three vectors $\boldsymbol{b}_1$, $\boldsymbol{b}_2$ and $\boldsymbol{b}_3$ define the primitive translation vectors of the reciprocal lattice. The lattice points spanned by $\boldsymbol{b}_1$, $\boldsymbol{b}_2$ and $\boldsymbol{b}_3$ are the reciprocal lattice points.

Reciprocal lattice and real lattice have important relations. Using three integers l, m, n, we can define the reciprocal lattice points $\boldsymbol{h}(l, m, n)$ by

$$\boldsymbol{h}(l, m, n) = l\boldsymbol{b}_1 + m\boldsymbol{b}_2 + n\boldsymbol{b}_3.$$

It is shown that the direction of the vector $\boldsymbol{h}(l, m, n)$ is perpendicular to the (real) lattice plane of Miller indices (l, m, n). Also, the length $|\boldsymbol{h}(l, m, n)|$ is inverse of the spacing $d(l, m, n)$ of the (real) lattice planes (l, m, n), i.e.,

$$|\boldsymbol{h}(l, m, n)| = 2\pi/d(l, m, n). \tag{6}$$

Miller indices are defined using three coprime numbers. If l, m, n are not coprime numbers like $\boldsymbol{h}(2, 4, 6)$, then

$$|\boldsymbol{h}(2, 4, 6)| = 2(2\pi/d(1, 2, 3)).$$

The first BZ is the Wigner–Seitz cell around $\boldsymbol{h} = 0$, the origin of the reciprocal space, which is defined to be the region of reciprocal space that is closer to the point $\boldsymbol{h} = 0$ than any other lattice point. The Wigner–Seitz cell contains one lattice point in it and fills all the space when translated through all reciprocal lattice vectors. In other words, the first BZ is the territorial region that belongs to the point $\boldsymbol{h} = 0$ in the reciprocal space and is constructed as the smallest volume entirely closed by a set of planes that are the perpendicular bisectors of various reciprocal lattice vectors drawn from the origin. Notice that such planes are particularly important in the theory of wave propagation in crystals, because a wave with a wavevector drawn from the origin terminating on any of these planes should satisfy without fail the conditions for diffraction. The volume of the Wigner–Seitz cell of the real lattice, i.e., the territorial region of one lattice point, is equal to the unit cell volume v_c, the volume of the parallelepiped formed by $\boldsymbol{a}_1$, $\boldsymbol{a}_2$, $\boldsymbol{a}_3$. The volume of the first BZ, i.e., the volume of the Wigner–Seitz cell in the reciprocal lattice space is equal to the volume of the parallelepiped formed by $\boldsymbol{b}_1$, $\boldsymbol{b}_2$, $\boldsymbol{b}_3$, because obviously both are the region occupied by one reciprocal lattice point. From (5), therefore, the volume of the first BZ is calculated to be $(2\pi)^3/v_c$.

Examples of the first BZ and the names of the special points inside it are given in Fig. 1.

Let us consider, as an example, how to obtain the first BZ of a 2D triangular lattice, depicted in the upper right of Fig. 1. Letting a be the lattice constant, we have $\boldsymbol{a}_1 = a(1, 0)$, and $\boldsymbol{a}_2 = a(1/2, \sqrt{3}/2)$ in the xy plane, and we take $\boldsymbol{a}_3$ such that $\boldsymbol{a}_3 = c\hat{\boldsymbol{z}}$ parallel to the z axis, with an arbitrary constant c. Then, since two vectors $\boldsymbol{b}_1$ and $\boldsymbol{b}_2$ are in the $(\boldsymbol{a}_1, \boldsymbol{a}_2)$ plane, we can construct the 2D reciprocal lattice by using $\boldsymbol{b}_1$ and $\boldsymbol{b}_2$. From (5), we obtain

$$\begin{aligned} \boldsymbol{b}_1 &= \frac{2\pi}{\sqrt{3}a}(\sqrt{3}, -1), \\ \boldsymbol{b}_2 &= \frac{4\pi}{\sqrt{3}a}(0, 1), \end{aligned}$$

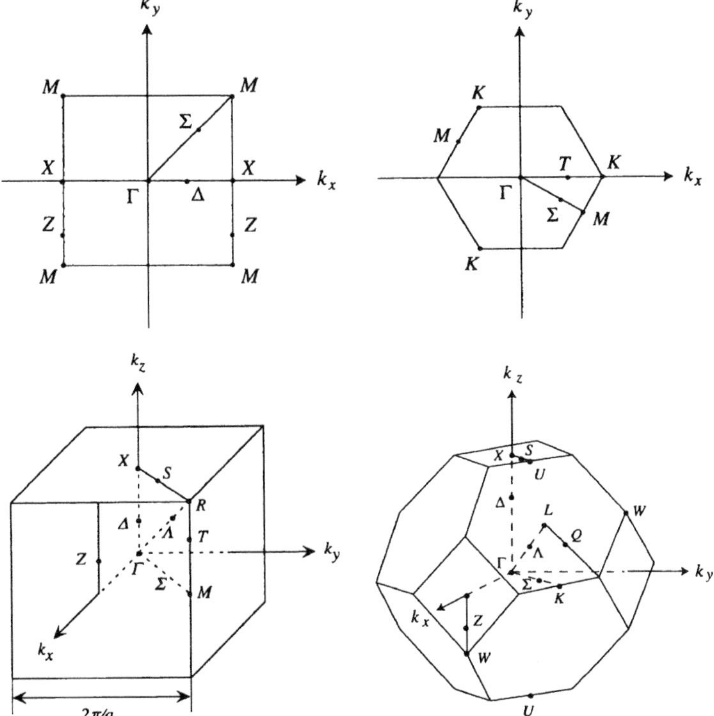

Fig. 1. First Brillouin zone of various lattices: 2D square lattice (upper left), 2D triangular lattice (upper right), simple cubic lattice (lower left) and fcc lattice (lower right). The names of symmetry points are shown

and $|\boldsymbol{b}_1| = |\boldsymbol{b}_2| = 4\pi/(\sqrt{3}a)$. They form a triangular lattice of lattice constant $4\pi/(\sqrt{3}a)$, the direction of the lattice being rotated by $\pi/6$ with respect to the real-space lattice. Bisecting $\boldsymbol{b}_1$ and $\boldsymbol{b}_2$ and their equivalents, we obtain the first BZ. In the first BZ, there are three high-symmetry points, marked as Γ, M and K in the figure, positioned at $(0,0)$, $(\pi/a)(1, 1/\sqrt{3})$, and $(\pi/a)(4/3, 0)$, respectively, and their equivalents. A 2D triangular lattice is an important example of PCs, which will be frequently treated in this book. In this connection, it is remarked that the symbols X and J are used often instead of M and K, respectively. Any 2D or 3D BZ is obained similarly.

2 Density of States

The number of states is calculated only if the values taken by $\boldsymbol{k}$, and hence those taken by the band frequency $\omega_n(\boldsymbol{k})$, are discrete (countable). By imposing the periodic boundary condition using a large integer N_1, given by

$$\begin{aligned} \boldsymbol{E}_{\boldsymbol{k}}(\boldsymbol{r}+N_1\boldsymbol{a}_1) &\equiv \mathrm{e}^{\mathrm{i}\boldsymbol{k}\cdot N_1\boldsymbol{a}_1}\boldsymbol{E}_{\boldsymbol{k}}(\boldsymbol{r}) \\ &= \boldsymbol{E}_{\boldsymbol{k}}(\boldsymbol{r}) \end{aligned} \tag{7}$$

(the first line is the Bloch theorem) and similarly for $\boldsymbol{a}_2$ and $\boldsymbol{a}_3$ with integer N_2 and N_3, respectively, we can discretize the values of the wavevector $\boldsymbol{k}$ of band states. Expressing

$$\boldsymbol{k} = w_1\boldsymbol{b}_1 + w_2\boldsymbol{b}_2 + w_3\boldsymbol{b}_3$$

in the first BZ, we find

$$\boldsymbol{k} = \frac{p_1}{N_1}\boldsymbol{b}_1 + \frac{p_2}{N_2}\boldsymbol{b}_2 + \frac{p_3}{N_3}\boldsymbol{b}_3 \tag{8}$$

with three arbitrary integers p_1, p_2, p_3. The spacings between two allowed values of $\boldsymbol{k}$ in the $\boldsymbol{b}_1, \boldsymbol{b}_2$ and $\boldsymbol{b}_3$ directions are given by

$$\frac{1}{N_1}|\boldsymbol{b}_1|,$$

etc. Therefore the region of the first BZ of volume $(2\pi)^3/v_{\mathrm{c}}$ is divided into $N_1N_2N_3$ small cells, each containing one allowed $\boldsymbol{k}$ point given by (8). Here we call this unit of $\boldsymbol{k}$ space a k cell. The volume of the k cell is thus

$$\frac{(2\pi)^3}{N_1N_2N_3v_{\mathrm{c}}} \equiv \frac{(2\pi)^3}{V_{\mathrm{total}}},$$

V_{total} being the volume of the system on which the periodic boundary condition was imposed. Usually the volume of the system itself is used as V_{total}.

The total number of discretized $\boldsymbol{k}$ points in a given volume $\Delta\boldsymbol{k} = \Delta k_x\Delta k_y\Delta k_z$ is equal to the number of k cells in it and is given by

$$\frac{\Delta\boldsymbol{k}}{(2\pi)^3/V_{\mathrm{total}}} = \frac{V_{\mathrm{total}}}{(2\pi)^3}\Delta\boldsymbol{k}.$$

From this rule, for an arbitrary function $F(k_x,\, k_y,\, k_z)$ it follows that

$$\frac{(2\pi)^3}{V_{\mathrm{total}}}\sum_{\boldsymbol{k}} F(\boldsymbol{k}) \to \int \mathrm{d}\boldsymbol{k}F(\boldsymbol{k}) \tag{9}$$

in the limit $V_{\mathrm{total}} \to \infty$, because the sum of $F(\boldsymbol{k})$ over the discrete $\boldsymbol{k}$ points, each multiplied by the k cell volume, becomes an integral over the $\boldsymbol{k}$ space.

In the 1D system of lattice constant a and system length L, used for the quantization of k, the allowed values of k are

$$k = \frac{2\pi p}{L} = \frac{p}{N}\frac{2\pi}{a}$$

for $p = 0, \pm 1, \pm 2, \ldots$, where N is the number of unit cells in the length L. The total number of allowed points in the first BZ, which is the region

$-\frac{\pi}{a} \le k < \frac{\pi}{a}$, is precisely equal to N. Note that the equality sign to define the first BZ may be added either to $-\frac{\pi}{a}$ or $\frac{\pi}{a}$ but not both, for these two points are equivalent.

In a 2D square lattice spanned by $\boldsymbol{a}_1 = a\hat{\boldsymbol{x}}$ and $\boldsymbol{a}_2 = a\hat{\boldsymbol{y}}$, we obtain

$$k_x = \frac{2\pi p_x}{Na}, \qquad k_y = \frac{2\pi p_y}{Na} \qquad \left(-\frac{N}{2} \le p_x,\, p_y < \frac{N}{2}\right).$$

The number of allowed points of $\boldsymbol{k}$ of the first BZ is N^2, which is equal to the total number of lattice points of the system.

In this way we can show for any lattice that the number of allowed values of $\boldsymbol{k}$ within the first BZ is exactly equal to the total number of lattice points. This is why we can obtain information of any system by considering only the states within the first BZ.

The density of states $\rho(\omega)$ is defined to be the number of states of the system per unit frequency at the frequency ω. By "per unit frequency", we mean that an infinitesimally small interval $[\omega, \omega + \Delta\omega]$ has the number of states $\rho(\omega)\Delta\omega$. To calculate $\rho(\omega)$, we have only to know the volume $\Delta\boldsymbol{k}$ in the first BZ, where all the discrete $\boldsymbol{k}$ points therein give the band energies within the frequency frequency interval. This volume divided by the k cell volume is $\rho(\omega)\Delta\omega$. If we know the $\boldsymbol{k}$ dependence of the band frequency $\omega_n(\boldsymbol{k})$, we can calculate it for each band. In Fig. 2, which depicts a model 1D-dispersion curve, we show that $\rho(\omega)\Delta\omega$ is equal to the number of discrete points of k in the corresponding interval Δk of k space. We find (see also the part following (9) of Appendix B)

$$\rho(\omega) = \frac{L}{2\pi} \frac{1}{\mathrm{d}\omega(k)/\mathrm{d}k}. \tag{10}$$

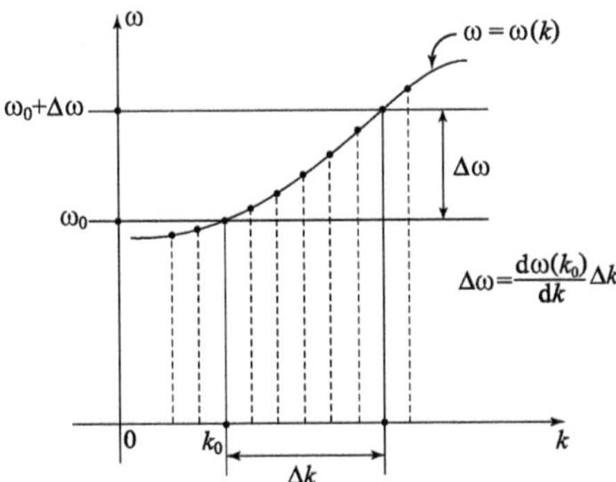

Fig. 2. Density of discretized states in a 1D system. The spacing between neighboring k values is equidistant, while that of the ω values is not. The prescribed allowance $\Delta\omega$ is mapped to Δk on the k axis

Appendix B. Phase Shift of Light and Density of States

K. Ohtaka

In this appendix we give the relationship between the transmission and reflection amplitudes of light by a slab PC and the density of states of PB modes set up in it.

To avoid complexity, we consider the frequency region where no diffracted plane waves appear, i.e., any channel associated with a nonzero 2D reciprocal lattice vector is closed. We consider a slab of a fcc PC, which is bounded by two (001) surfaces in the z direction. We take the origin $z = 0$ at the middle of the slab and assume the mirror symmetry with respect to the plane $z = 0$. We take the x, y axes along the two sides of the 2D square lattice.

Consider *two* plane-wave light waves that are incident on this slab PC simultaneously from above and below. We treat the case where the incident plane of the light is the xz plane, so that the lateral wavevector of the incident light is $\boldsymbol{k}_{\parallel} = (k_x, 0)$. Their amplitudes are are expressed as a (from below) and b (from above) in the scalar notation. The p-polarized incident light, which has even parity with respect to the xz mirror plane, and the s-polarized incident light, which has odd parity, are not mixed because of the parity conservation and can be treated separately. All these simplifying assumptions, including the absence of diffraction channels or sp mixing, can be relaxed, however (see Sect. 4.3.3).

Let the two incident waves have the same polarization, either p or s, and 3D incident wavevectors $(k_x, 0, \pm k_z)$. After the transmission and reflection, the waves come out with *scalar* and *complex* amplitudes t and r, respectively. Far away from the slab, the field is then expressed as

$$\boldsymbol{E}(\boldsymbol{r}) = \begin{cases} \left(b\boldsymbol{e}_- \mathrm{e}^{-\mathrm{i}k_z z} + (ta + rb)\boldsymbol{e}_+ \mathrm{e}^{\mathrm{i}k_z z}\right) \mathrm{e}^{\mathrm{i}k_x x} & \text{for } z \simeq +\infty, \\ \left(a\boldsymbol{e}_+ \mathrm{e}^{\mathrm{i}k_z z} + (ra + tb)\boldsymbol{e}_- \mathrm{e}^{-\mathrm{i}k_z z}\right) \mathrm{e}^{\mathrm{i}k_x x} & \text{for } z \simeq -\infty. \end{cases} \tag{1}$$

Here, $\boldsymbol{e}_+$ and $\boldsymbol{e}_-$ are the unit vectors to specify the polarizations of the upward and downward incident waves, respectively. They are taken so that $\boldsymbol{e}_+$ is a mirror reflection of $\boldsymbol{e}_-$ by the xy mirror, as shown in Fig. 1. In the p-incidence, they lie within the xz plane, while in the s-incidence, they are both in the y direction. The first terms of (1) represent the incident light and the second terms show the waves emerging as the result of the scattering of the incident waves. Denoting the amplitudes of the scattered waves at $z = \pm\infty$ as $S_{\pm}$, we obtain

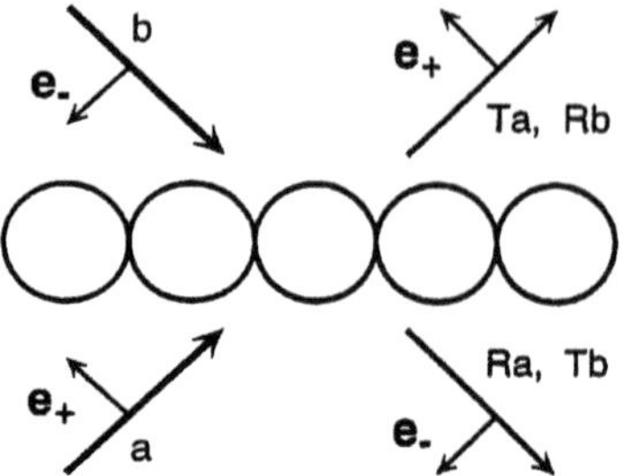

Fig. 1. Formation of a standing wave by two incoming light waves of p-polarization. The vectors $\boldsymbol{e}_{\pm}$ are unit vectors which specify the polarizations of the incident light. The symbol t and r are used in the text for the amplitudes T and R, respctively

$$\begin{pmatrix} S_+ \\ S_- \end{pmatrix} = \begin{pmatrix} t & r \\ r & t \end{pmatrix} \begin{pmatrix} a \\ b \end{pmatrix} . \tag{2}$$

The scattered amplitudes are thus obtained by the linear transformation of the incident amplitudes, using a 2×2 transformation matrix (S matrix). So far, a and b are arbitrary. From now on, we choose a and b so that the column vector $(a, b)^t$ is an eigenvector of the S matrix. Then we have from (2)

$$\begin{pmatrix} S_+ \\ S_- \end{pmatrix} = \lambda \begin{pmatrix} a \\ b \end{pmatrix} , \tag{3}$$

with an eigenvalue λ of the S matrix, which, from the unitarity of the S matrix, is expressed as

$$\lambda = \mathrm{e}^{2\mathrm{i}\delta} . \tag{4}$$

From a given set of t and r, the S matrix yields two eigen phase-shifts. Let us denote them as δ_{e} and δ_{o}, the suffix e (o) referring to even (odd) parity. They depend on the frequency and polarization of the incident light through t and r. From (3) and (4), it follows that

$$\begin{aligned} \mathrm{e}^{2\mathrm{i}\delta_{\mathrm{e}}} &= t + r \quad \text{for the even-parity eigenvector} \quad (a, b)^t_{\mathrm{e}} = \frac{1}{\sqrt{2}}(1, 1)^t , \\ \mathrm{e}^{2\mathrm{i}\delta_{\mathrm{o}}} &= t - r \quad \text{for the odd-parity eigenvector} \quad (a, b)^t_{\mathrm{o}} = \frac{1}{\sqrt{2}}(1, -1)^t . \end{aligned} \tag{5}$$

When $(a, b)^t_{\mathrm{e}}$ and $(a, b)^t_{\mathrm{o}}$ are substituted for $(a, b)^t$ of (1), we obtain two standing waves from (1). The even-parity eigenvector leads to a standing wave of even parity with respect to the mirror reflection $z \to -z$, which is expressed as

$$\boldsymbol{E}_{\mathrm{e}}(\boldsymbol{r}) = \begin{cases} \frac{1}{\sqrt{2}} \mathrm{e}^{\mathrm{i}\delta_{\mathrm{e}}} \mathrm{e}^{\mathrm{i}k_x x} \left(\boldsymbol{e}_- \mathrm{e}^{-\mathrm{i}(k_z z + \delta_{\mathrm{e}})} + \boldsymbol{e}_+ \mathrm{e}^{\mathrm{i}(k_z z + \delta_{\mathrm{e}})} \right) & \text{for } z \simeq +\infty , \\ \frac{1}{\sqrt{2}} \mathrm{e}^{\mathrm{i}\delta_{\mathrm{e}}} \mathrm{e}^{\mathrm{i}k_x x} \left(\boldsymbol{e}_+ \mathrm{e}^{\mathrm{i}(k_z z - \delta_{\mathrm{e}})} + \boldsymbol{e}_- \mathrm{e}^{-\mathrm{i}(k_z z - \delta_{\mathrm{e}})} \right) & \text{for } z \simeq -\infty . \end{cases} \tag{6}$$

Since $(\boldsymbol{e}_+)_x = (\boldsymbol{e}_-)_x$, $(\boldsymbol{e}_+)_y = (\boldsymbol{e}_-)_y$ and $(\boldsymbol{e}_+)_z = -(\boldsymbol{e}_-)_z$, its three cartesian components are written as

$$\boldsymbol{E}_{\mathrm{e}}(\boldsymbol{r}) = \mathrm{e}^{\mathrm{i}\delta_{\mathrm{e}}}\mathrm{e}^{\mathrm{i}k_x x}\begin{pmatrix} (\boldsymbol{e}_+)_x \cos(k_z|z| + \delta_{\mathrm{e}}) \\ (\boldsymbol{e}_+)_y \cos(k_z|z| + \delta_{\mathrm{e}}) \\ \mathrm{i}(\boldsymbol{e}_+)_z (z/|z|) \sin(k_z|z| + \delta_{\mathrm{e}}) \end{pmatrix}. \tag{7}$$

In the same way, the amplitude $(a,b)^t_{\mathrm{o}}$ gives the odd-parity solution $\boldsymbol{E}_{\mathrm{o}}(\boldsymbol{r})$, which is expressed as

$$\boldsymbol{E}_{\mathrm{o}}(\boldsymbol{r}) = \mathrm{e}^{\mathrm{i}\delta_{\mathrm{o}}}\mathrm{e}^{\mathrm{i}k_x x}\begin{pmatrix} (\boldsymbol{e}_+)_x (z/|z|) \sin(k_z|z| + \delta_{\mathrm{o}}) \\ (\boldsymbol{e}_+)_y (z/|z|) \sin(k_z|z| + \delta_{\mathrm{o}}) \\ \mathrm{i}(\boldsymbol{e}_+)_z \cos(k_z|z| + \delta_{\mathrm{o}}) \end{pmatrix}. \tag{8}$$

The above forms apply both to the p- and s-incidences. In the p case, $(\boldsymbol{e}_+)_y$ is zero, while in the s case, only the y component remains. Since the p- and s-polarized standing-wave solutions involve different values of δ_{e} and δ_{o}, we classify them by adding the symbols (p) and (s) as $\delta_{\mathrm{e}}^{(\mathrm{p})}$ and $\delta_{\mathrm{o}}^{(\mathrm{s})}$, etc.; there are accordingly four phase shifts for a given k_x, each associated with one standing wave.

Using these asymptotic forms (7) and (8), we can determine the normal-mode frequency by confining the field to the region bounded by two perfect mirrors, placed parallel to the xy plane at $z = \pm z_m$, z_m being a point at infinity (Fig. 4.7). At the mirror, the field components parallel to the mirror surface must vanish. From (7), we then have

$$k_z = \left(p + \frac{1}{2}\right)\pi/z_{\mathrm{m}} - \delta_{\mathrm{e}}/z_{\mathrm{m}}, \tag{9}$$

with $p = 0, 1, 2, \cdots$ and from (8), we obtain

$$k_z = p\pi/z_{\mathrm{m}} - \delta_{\mathrm{o}}/z_{\mathrm{m}}. \tag{10}$$

Equation (9) determines the discrete eigenvalues of even-parity modes. Regarding k_z and δ_{e} to be $k_z(\omega)$ and $\delta_{\mathrm{e}}(\omega)$, we can find the spacing $\delta\omega$ between the frequencies of the mode p and $p+1$, i.e., the spacing $\delta\omega$ corresponding to $\delta p = 1$:

$$\frac{\partial k_z(\omega)}{\partial \omega}\delta\omega = \frac{\pi}{z_m}\delta p - \frac{1}{z_m}\frac{\partial \delta_{\mathrm{e}}(\omega)}{\partial\omega}\delta\omega,$$

with $\delta p = 1$. Therefore

$$\frac{1}{\delta\omega} = \frac{z_m}{\pi}\frac{\partial k_z(\omega)}{\partial\omega} + \frac{1}{\pi}\frac{\partial\delta_{\mathrm{e}}(\omega)}{\partial\omega}. \tag{11}$$

Noting that the DOS at ω is equal to $1/\delta\omega(\omega)$ and that the first term on the right is the DOS without the slab, the second term gives the increment of DOS $\Delta\rho$. For example, from $\delta_{\mathrm{e}}^{(\mathrm{p})}(\omega)$, we obtain

$$\Delta\rho^{(\mathrm{p})}_{k_x\mathrm{e}}(\omega) = \frac{1}{\pi}\frac{\partial}{\partial\omega}\delta^{(\mathrm{p})}_{k_x\mathrm{e}}(\omega) \tag{12}$$

for the DOS change of the p-polarized even-parity states with $\boldsymbol{k}_\parallel = (k_x, 0)$. The explicit k_x and ω dependences of the DOS increment thus result from those of the phase shifts. The DOS change for the states of other symmetries are similarly obtained.

Finally, we relate the phases of t and r with the phase shift. Since the S matrix in (2) is unitary, the product tr^* is an imaginary quantity, showing that t and r are of the form

$$\begin{aligned} t &= |t|\mathrm{e}^{\mathrm{i}\phi}, \\ r &= \pm\mathrm{i}|r|\mathrm{e}^{\mathrm{i}\phi}, \end{aligned} \tag{13}$$

ϕ being the real phase of the transmitted amplitude t. From (5), it holds that

$$\begin{aligned} \mathrm{e}^{2\mathrm{i}\delta_\mathrm{e}} &= t + r, \\ &= \mathrm{e}^{\mathrm{i}(\pm\phi_0 + \phi)}, \end{aligned} \tag{14}$$

with the phase ϕ_0 introduced through

$$|t| \pm \mathrm{i}|r| = \mathrm{e}^{\pm\mathrm{i}\phi_0}, \tag{15}$$

which holds from the flux conservation $|t|^2 + |r|^2 = 1$. In the same way we find

$$\begin{aligned} \mathrm{e}^{2\mathrm{i}\delta_\mathrm{o}} &= t - r, \\ &= \mathrm{e}^{\mathrm{i}(\mp\phi_0 + \phi)}. \end{aligned} \tag{16}$$

Therefore, we obtain from (14) and (16)

$$\phi = \delta_\mathrm{e} + \delta_\mathrm{o}. \tag{17}$$

This relation holds for both p and s polarizations. The conclusion is that the phase ϕ of the transmission amplitude t of the p light is equal to $\delta^{(\mathrm{p})}_\mathrm{e} + \delta^{(\mathrm{p})}_\mathrm{o}$. The same holds true for the s-light; that is, the frequency derivative of ϕ defined by (5) gives us the *sum* of even- and odd-parity DOS. This completes the derivation. The final result for the increment of the DOS caused by the presence of the PC slab is thus

$$\begin{aligned} \Delta\rho^{(\mathrm{p})}_{\boldsymbol{k}_x}(\omega) &= \frac{1}{\pi}\frac{\partial}{\partial\omega}\phi^{(\mathrm{p})}_{\boldsymbol{k}_x}(\omega) \\ &= \frac{1}{\pi}\frac{\partial}{\partial\omega}\left(\delta^{(\mathrm{p})}_{\boldsymbol{k}_x\mathrm{e}}(\omega) + \delta^{(\mathrm{p})}_{\boldsymbol{k}_x\mathrm{o}}(\omega)\right) \end{aligned} \tag{18}$$

for the p-polarized normal modes (the superscript should be changed in the case of s). We have put the suffix $\boldsymbol{k}_x$ to various quantities. The sum of the p and s contributions gives the total increase. Here we have given the DOS change per unit frequency. The DOS change per unit wavenumber is given replacing $\partial/\partial\omega$ by $\partial/\partial k$. Note that the phase of the transmitted amplitude has by itself information on the sum and fails to distinguish the contributions of the two parities. This feature reflects the fact that once the light comes in, the system no longer has mirror symmetry in the plane $z = 0$. Therefore, the method used above to single out one of the parities using both of t and r is by no means trivial.

Index

Springer Series in
OPTICAL SCIENCES

New editions of volumes prior to volume 70

1 **Solid-State Laser Engineering**
By W. Koechner, 5th revised and updated ed. 1999, 472 figs., 55 tabs., XII, 746 pages
14 **Laser Crystals**
Their Physics and Properties
By A. A. Kaminskii, 2nd ed. 1990, 89 figs., 56 tabs., XVI, 456 pages
15 **X-Ray Spectroscopy**
An Introduction
By B. K. Agarwal, 2nd ed. 1991, 239 figs., XV, 419 pages
36 **Transmission Electron Microscopy**
Physics of Image Formation and Microanalysis
By L. Reimer, 4th ed. 1997, 273 figs. XVI, 584 pages
45 **Scanning Electron Microscopy**
Physics of Image Formation and Microanalysis
By L. Reimer, 2nd completely revised and updated ed. 1998, 260 figs., XIV, 527 pages

Published titles since volume 70

70 **Electron Holography**
By A. Tonomura, 2nd, enlarged ed. 1999, 127 figs., XII, 162 pages
71 **Energy-Filtering Transmission Electron Microscopy**
By L. Reimer (Ed.), 1995, 199 figs., XIV, 424 pages
72 **Nonlinear Optical Effects and Materials**
By P. Günter (Ed.), 2000, 174 figs., 43 tabs., XIV, 540 pages
73 **Evanescent Waves**
From Newtonian Optics to Atomic Optics
By F. de Fornel, 2001, 277 figs., XVIII, 268 pages
74 **International Trends in Optics and Photonics**
ICO IV
By T. Asakura (Ed.), 1999, 190 figs., 14 tabs., XX, 426 pages
75 **Advanced Optical Imaging Theory**
By M. Gu, 2000, 93 figs., XII, 214 pages
76 **Holographic Data Storage**
By H.J. Coufal, D. Psaltis, G.T. Sincerbox (Eds.), 2000
228 figs., 64 in color, 12 tabs., XXVI, 486 pages
77 **Solid-State Lasers for Materials Processing**
Fundamental Relations and Technical Realizations
By R. Iffländer, 2001, 230 figs., 73 tabs., XVIII, 350 pages
78 **Holography**
The First 50 Years
By J.-M. Fournier (Ed.), 2001, 266 figs., XII, 460 pages
79 **Mathematical Methods of Quantum Optics**
By R.R. Puri, 2001, 13 figs., XIV, 285 pages
80 **Optical Properties of Photonic Crystals**
By K. Sakoda, 2001, 95 figs., 28 tabs., XII, 223 pages
81 **Photonic Analog-to-Digital Conversion**
By B.L. Shoop, 2001, 259 figs., 11 tabs., XIV, 330 pages
82 **Spatial Solitons**
By S. Trillo, W.E. Torruellas (Eds), 2001, 194 figs., 7 tabs., XX, 454 pages
83 **Nonimaging Fresnel Lenses**
Design and Performance of Solar Concentrators
By R. Leutz, A. Suzuki, 2001, 139 figs., 44 tabs., XII, 272 pages
84 **Nano-Optics**
By S. Kawata, M. Ohtsu, M. Irie (Eds.), 2002, 258 figs., 2 tabs., XVI, 321 pages
85 **Sensing with Terahertz Radiation**
By D. Mittleman (Ed.), 2003, 207 figs., 14 tabs., XVI, 337 pages

Springer Series in
OPTICAL SCIENCES

---

86 **Progress in Nano-Electro-Optics I**
Basics and Theory of Near-Field Optics
By M. Ohtsu (Ed.), 2003, 118 figs., XIV, 161 pages

87 **Optical Imaging and Microscopy**
Techniques and Advanced Systems
By P. Török, F.-J. Kao (Eds.), 2003, 260 figs., XVII, 395 pages

88 **Optical Interference Coatings**
By N. Kaiser, H.K. Pulker (Eds.), 2003, 203 figs., 50 tabs., XVI, 504 pages

89 **Progress in Nano-Electro-Optics II**
Novel Devices and Atom Manipulation
By M. Ohtsu (Ed.), 2003, 115 figs., XIII, 188 pages

90/1 **Raman Amplifiers for Telecommunications 1**
Physical Principles
By M.N. Islam (Ed.), 2004, 488 figs., XXVIII, 328 pages

90/2 **Raman Amplifiers for Telecommunications 2**
Sub-Systems and Systems
By M.N. Islam (Ed.), 2004, 278 figs., XXVIII, 420 pages

91 **Optical Super Resolution**
By Z. Zalevsky, D. Mendlovic, 2004, 164 figs., XVIII, 232 pages

92 **UV-Visible Reflection Spectroscopy of Liquids**
By J.A. Räty, K.-E. Peiponen, T. Asakura, 2004, 131 figs., XII, 219 pages

93 **Fundamentals of Semiconductor Lasers**
By T. Numai, 2004, 166 figs., XII, 264 pages

94 **Photonic Crystals**
Physics, Fabrication and Applications
By K. Inoue, K. Ohtaka (Eds.), 2004, 209 figs., XV, 320 pages

95 **Ultrafast Optics IV**
Selected Contributions to the 4th International Conference
on Ultrafast Optics, Vienna, Austria
By F. Krausz, G. Korn, P. Corkum, I.A. Walmsley (Eds.), 2004, 281 figs., XIV, 506 pages

96 **Progress in Nano-Electro Optics III**
Industrial Applications and Dynamics of the Nano-Optical System
By M. Ohtsu (Ed.), 2004, 155 figs., XIV, 226 pages

97 **Microoptics**
From Technology to Applications
By J. Jahns, K.-H. Brenner, 2004, 303 figs., XI, 335 pages

98 **X-Ray Optics**
High-Energy-Resolution Applications
By Y. Shvyd'ko, 2004, 181 figs., XIV, 404 pages

99 **Few-Cycle Photonics and Optical Scanning Tunneling Microscopy**
Route to Femtosecond Angstrom Technology
By M. Yamashita, H. Shigekawa, R. Morita (Eds.) 2004, 241 figs., XX, 393 pages

GPSR Compliance
The European Union's (EU) General Product Safety Regulation (GPSR) is a set of rules that requires consumer products to be safe and our obligations to ensure this.

If you have any concerns about our products, you can contact us on

ProductSafety@springernature.com

In case Publisher is established outside the EU, the EU authorized representative is:

Springer Nature Customer Service Center GmbH
Europaplatz 3
69115 Heidelberg, Germany

www.ingramcontent.com/pod-product-compliance
Ingram Content Group UK Ltd.
Pitfield, Milton Keynes, MK11 3LW, UK
UKHW021835190726
13853UKWH00003B/1300

* 9 7 8 3 6 4 2 5 3 5 7 0 3 *